Solid–Liquid Thermal Energy Storage

Solid–Liquid Thermal Energy Storage

Modeling and Applications

Edited by

Moghtada Mobedi, Kamel Hooman, and Wen-Quan Tao

CRC Press
Taylor & Francis Group
Boca Raton London New York

CRC Press is an imprint of the
Taylor & Francis Group, an **informa** business

First edition published 2022
by CRC Press
6000 Broken Sound Parkway NW, Suite 300, Boca Raton, FL 33487-2742

and by CRC Press
2 Park Square, Milton Park, Abingdon, Oxon, OX14 4RN

CRC Press is an imprint of Taylor & Francis Group, LLC

ISBN: 978-1-032-10018-0 (hbk)
ISBN: 978-1-032-10026-5 (pbk)
ISBN: 978-1-003-21326-0 (ebk)

DOI: 10.1201/9781003213260

Typeset in Times
by codeMantra

Contents

Preface

Providing an introduction to both solid–liquid phase change and advanced aspects of the many problems, this book is prepared keeping both industry and academia in mind. Fundamental aspects of the phase change problem as well as practical implementation of the technology developed based on it are analyzed. The book offers a combination of theoretical, numerical, and experimental techniques that are implemented to understand the underlying physics and also to facilitate the use of solid–liquid thermal energy systems in lab-, prototype-, and full-scale applications. Theoretical modeling of these systems, the governing equations, assumptions, and solutions by improved numerical methods are explained in different chapters of this book. Different important topics in this area from classification of phase change material (PCM) heat exchangers to enhancement of heat transfer by using fins, high conductive porous media, encapsulated PCMs, and nanoparticles are discussed. Recent advances in PCM-based thermal energy storage systems, due to the rapid development of the technology and the increasing number of studies in this area, are surveyed here. Preparing the chapters, the authors' intention was to provide a guideline, where possible, for policy-makers, practicing engineers, and researchers to improve their designs and help with decision-making processes. The book consists of three sections: Introduction, Fundamental Studies, and Applications.

INTRODUCTION

Solid–liquid thermal storage has been the subject of many studies, but we preferred to start the book with an introduction chapter briefly explaining the importance of thermal storage, their classification, as well as difficulties faced in their design and deployment. Many studies have been carried out on PCMs to improve their latent heat and thermophysical properties and also to manufacture new PCMs for different working temperatures. Chapter 2 discusses recent studies on PCMs focusing on materials. The chapter identifies eight material-related challenges that currently hinder robust operation of PCMs and in developing novel ones. Chapter 3 focuses on the experimental methods for this kind of thermal storage system. It is subdivided into two main sections: the first one goes over fundamental studies trying to demonstrate different experimental techniques, while the second one presents different applications focusing on the most important performance parameters to be evaluated experimentally.

FUNDAMENTAL STUDIES

Chapter 4 aims at providing design guidelines for optimal performance of storage tanks containing solid–liquid PCMs. The authors reviewed and listed some of the approximate solutions for melting and solidification in storage tanks filled with PCMs under the effect of thermal conductivity enhancers or thermal spreaders. Being highly flexible and low cost to implement, numerical modeling plays an indispensable role in investigating thermo-hydraulics of latent heat thermal energy storage systems. Chapter 5 highlights the multiscale numerical methods and their coupling schemes for solid–liquid phase change conjugate heat transfer including molecular dynamics simulation, lattice Boltzmann method, and finite volume method as well as the applications of multiscale modeling technique. Chapter 6 discusses a number of silicon alloys studied with respect to their melting temperatures (850°C–1400°C) and latent heat of fusion to provide insight on the use of silicon alloys for thermal energy storage. Chapter 7 briefly introduces the different heat transfer enhancement techniques employed in solid–liquid thermal energy storage systems, particularly fins and heat pipes. Methods for enhancement of heat transfer continue with Chapter 8 which is on a fin-metal foam hybrid structure. Thermal energy charging/discharging performance for the proposed hybrid structure is experimentally evaluated by closely monitoring the full melting time, temperature response, and melting front evolution. Finally, combination of carrier fluids and PCMs has been studied in Chapter 9 where the current state of research on encapsulated phase change slurries has been touched on.

APPLICATIONS

Increasing the TRL (Technology Readiness Level) for PCM heat exchangers has been a goal sought by this book. Hence, an entire section has been devised to focus on industrial applications by listing the challenges, opportunities, and success stories. Chapter 10 reviews 150 studies on PCM heat exchangers and classifies them so that their advantages and disadvantages can be known. Chapter 11 summarizes different types of water/ice thermal energy storage systems and provides an overview of alternative PCMs to be used in cool thermal energy storage systems. The evolution of melt front in a horizontal shell and tube latent heat storage system for concentrated solar power plants is investigated in Chapter 12. The possibility of having a melt trap around a tube in horizontal orientation of a shell and tube thermal energy storage system is investigated by using thermal stress analysis. The application of honeycomb systems is discussed in Chapter 13. The performed analysis simulates heat and fluid flow in the honeycomb system as a porous medium both for the sensible and latent heat thermal energy storage systems. A novel compact cooling system for thermal management of lithium-ion battery packs is proposed in Chapter 14. The suggested cooling system is a hybrid consisting of PCM and heat pipe. Experiments were conducted to verify the effectiveness of the hybrid thermal management system. Finally, Chapter 15 provides application examples of PCM-based cold energy storage devices through integration within a cold chain, including warehouses and transportation.

Editors

Moghtada Mobedi is a professor of heat transfer and works in Mechanical Engineering Department, Shizuoka University in Japan. He received his Ph.D. from Middle East Technical University, Turkey in 1994. After working in an HVAC company as a project manager, he worked in the Mechanical Engineering Department of Izmir Institute of Technology in Turkey between 2003 and 2015. Since 2015, he has been working at Shizuoka University and continues his research in Japan. He has taught many bachelor's, master's, and Ph.D. courses such as heat transfer, computational fluid dynamics, convective heat transfer, and numerical methods in Turkey, Japan, and European countries. His research interests include heat transfer enhancement in solid–liquid phase change, heat and mass transfer in porous media, adsorption heat pump, and computational fluid dynamics. He published more than 70 papers in international journals as well as 100 papers in national and international conferences and 3 book chapters on the various applications of heat transfer. He has supervised many master's and Ph.D. students both in Turkey and Japan. He received fellowships from the Japan Society for the Promotion of Science, European Union, and Cracow University of Technology to visit laboratories of different universities in Japan, Poland, Italy, Sweden, and Austria. He has led many projects funded by "State Planning Department of Turkey", "Scientific and Technological Research Council of Turkey", "Japan Socicty for the Promotion of Science", and "Suzuki Foundation" to study on discovering innovative methods for heat transfer enhancement for single convection heat transfer, adsorbent beds as well as for solid–liquid phase change thermal storage.

Kamel Hooman is a professor of heat transformation technology at the Delft University of Technology. He received his Ph.D. from The University of Queensland in 2009 where he has worked for almost two decades. He is working closely with the industry in the field of thermo-fluids engineering. He was named Australia's Research Field Leader in Thermal Sciences in 2019. His book "Convective Heat Transfer in Porous Media" has been published in 2019 (CRC Press) to help both undergraduate and postgraduate students who work on porous media flows. An author of over 150 archival journal articles, 8 book chapters, and over 50 conference papers, he has given numerous national and international invited lectures, keynote addresses, and presentations. He has been awarded fellowships from Emerald, Australian Research Council, National Science Foundation China, Australian Academy of Sciences, and Chinese Academy of Sciences with visiting professor/researcher positions at the University of Padova, La Sapienza University of Rome, Krakow Institute of Technology, Ecole Centrale Paris, University of Malaya, Karlsruhe Institute of Technology, Xi'an Jiaotong University, Harbin Institute of Technology, North Western Polytechnical University, Tianjin University, and Shandong University.

He is the associate editor for the *International Journal of Heat and Mass Transfer*, *Heat Transfer Engineering*, and *Journal of Porous Media* while serving on the editorial/advisory board of some international journals and conferences in the field of energy storage, conversion, and management. As an editor for Heat Exchanger Design Handbook (Begell House); he relies on his practical experience to ensure the latest development in the field of heat exchangers is kept up to date and shared with the practicing engineers. He has been the organizer and chair of the International Conference on Cooling Tower and Heat Exchanger sponsored by IAHR (International Association for Hydro-Environment Engineering and Research). He has supervised 11 doctoral students and has directed over 10 post-docs. He has an h-index of 46 with an i10-index of 154. He has carried out various sponsored research projects through companies, governmental funding agencies, and national labs. He has also consulted for various companies and governments in Australia and overseas.

Wen-Quan Tao is a professor at Key Laboratory of Thermo-Fluids Science & Engineering of MOE, and Int. Joint Research Laboratory of Thermal Science & Engineering, Xi'an Jiaotong University, China. He graduated from Xi'an Jiaotong University in 1962 and received his graduate Diploma in 1966 under the supervision of Professor S. M. Yang. From 1980 to 1982, he worked with Professor E. M. Sparrow as a visiting scholar at the Heat Transfer Laboratory of the University of Minnesota. He was selected as a member of the Chinese Academy of Science in 2005. He has published more than 300 technical papers in international journals. He has published eight books in heat transfer and numerical heat transfer, among which the book titled *Numerical Heat Transfer* has been cited more than 15,000 times at home and abroad. He has supervised more than 140 graduate students. His recent research interests include multiscale simulations of fluid flow and heat transfer problems, thermal management of fuel cell, cooling technique of data center, thermal energy storage and saving, and enhancement of heat transfer.

Contributors

Bakytzhan Akhmetov
School of Mechanical and Aerospace Engineering
Nanyang Technological University
Singapore, Singapore

Sandra K. S. Boetcher
Department of Mechanical Engineering
Embry-Riddle Aeronautical University
Daytona Beach, Florida

Serge Bondarenko
Climate Change Technologies Pty Ltd
Lonsdale, Australia

Frank Bruno
Future Industries Institute
University of South Australia
Mawson Lakes, Australia

B. Buonomo
Department of Mechanical Engineering
Università degli Studi della Campania "Luigi Vanvitelli"
Aversa, Italy

M. Calati
Department of Management and Enginnering
University of Padua
Padova, Italy

Wenjiong Cao
Department of New Energy Technologies
Chinese Academy of Sciences
Beijing, China

Li Chen
Key Laboratory of Thermo-Fluid Science and Engineering of MOE, School of Energy and Power Engineering
Xi'an Jiaotong University
Xi'an, China

Lin Cong
Birmingham Centre for Energy Storage & School of Chemical Engineering
University of Birmingham
Birmingham, United Kingdom

Siyuan Dai
Birmingham Centre for Energy Storage & School of Chemical Engineering
University of Birmingham
Birmingham, United Kingdom

L. Darvishvand
Department of Mechanical Engineering, Yadegar-e-Imam Khomeini (RAH) Shahre-Rey Branch
Islamic Azad University
Tehran, Iran

Yulong Ding
Birmingham Centre for Energy Storage & School of Chemical Engineering
University of Birmingham
Birmingham, United Kingdom
and
School of Energy and Environmental Engineering
University of Science & Technology Beijing
Beijing, China

Michael Evans
UniSA STEM
University of South Australia
Mawson Lakes, Australia

L. J. Fischer
Competence Center Thermal Energy Storage
Lucerne University of Applied Sciences and Arts
Luzern, Switzerland

D. Guarda
Department of Management and Enginnering
University of Padua
Padova, Italy

Rhys Jacob
Future Industries Institute
University of South Australia
Mawson Lakes, Australia

Fangming Jiang
Department of Energy Technologies and Engineering
Chinese Academy of Sciences
Beijing, China

Kamel Hooman
School of Mechanical and Mining Engineering
The University of Queensland
Brisbane, Australia

B. Kamkari
Department of Mechanical Engineering, Yadegar-e-Imam Khomeini (RAH) Shahre-Rey Branch
Islamic Azad University
Tehran, Iran

Ming-Jia Li
Department of Engineering Thermophysics
Xi'an Jiaotong University
Xi'an, China

Yongliang Li
Birmingham Centre for Energy Storage & School of Chemical Engineering
University of Birmingham
Birmingham, United Kingdom

Ming Liu
Future Industries Institute
University of South Australia
Mawson Lakes, Australia

A. Mahmoudi
Faculty of Engineering Technology, Department of Thermal and Fluid Engineering (TFE)
University of Twente
Enschede, The Netherlands

Peter Majewski
Future Industries Institute
University of South Australia
Mawson Lakes, Australia

Yelaman Maksum
Birmingham Centre for Energy Storage & School of Chemical Engineering
University of Birmingham
Birmingham, United Kingdom

O. Manca
Dipartimento di Ingegneria
Università degli Studi della Campania "Luigi Vanvitelli"
Aversa, Italy

S. Mancin
Department of Management and Enginnering
University of Padua
Padova, Italy

S. Maranda
Latent Storage Group
Lucerne University of Applied Sciences and Arts
Luzern, Switzerland

M. Mehrali
Faculty of Engineering Technology, Department of Thermal and Fluid Engineering (TFE)
University of Twente
Enschede, The Netherlands

Moghtada Mobedi
Mechanical Engineering Department
Shizuoka University
Hamamatsu, Japan

S. Nardini
Dipartimento di Ingegneria
Università degli Studi della Campania "Luigi Vanvitelli"
Aversa, Italy

Binjian Nie
Birmingham Centre for Energy Storage & School of Chemical Engineering
University of Birmingham
Birmingham, United Kingdom

A. di Pasqua
Dipartimento di Ingegneria
Università degli Studi della Campania "Luigi Vanvitelli"
Aversa, Italy

Peng Peng
Laboratory of Advanced Energy Systems, CAS Key Laboratory of Renewable Energy, Guangdong Key Laboratory of New and Renewable Energy Research and Development, Guangzhou Institute of Energy Conversion
Chinese Academy of Sciences
Beijing, China

Qinlong Ren
Key Laboratory of Thermo-Fluid Science and Engineering of MOE, School of Energy and Power Engineering
Xi'an Jiaotong University
Xi'an, China

Soheila Riahi
Future Industries Institute
University of South Australia
Mawson Lakes, Australia

S. Sabet
Dipartimento di Ingegneria
Università degli Studi della Campania "Luigi Vanvitelli"
Aversa, Italy

M. Shahi
Faculty of Engineering Technology, Department of Thermal and Fluid Engineering (TFE)
University of Twente
Enschede, The Netherlands

Shane Sheoran
Future Industries Institute
University of South Australia
Mawson Lakes, Australia

A. Stamatiou
Latent Storage Group
Lucerne University of Applied Sciences and Arts
Luzern, Switzerland

Nikki Stanford
Future Industries Institute
University of South Australia
Mawson Lakes, Australia

Wen-Quan Tao
Key Laboratory of Thermo-Fluid Science and Engineering of MOE, School of Energy and Power Engineering
Xi'an Jiaotong University
Xi'an, China

Yu-Bing Tao
Key Laboratory of Thermo-Fluid Science and Engineering of MOE, School of Energy and Power Engineering
Xi'an Jiaotong University
Xi'an, China

Lige Tong
School of Energy and Environmental Engineering
University of Science & Technology Beijing
Beijing, China

Chunyang Wang
Institute of Engineering Thermophysics
Chinese Academy of Sciences
Beijing, China

Yiwei Wang
Laboratory of Advanced Energy Systems, CAS Key Laboratory of Renewable Energy, Guangdong Key Laboratory of New and Renewable Energy Research and Development, Guangzhou Institute of Energy Conversion
Chinese Academy of Sciences
Beijing, China
and
University of Chinese Academy of Sciences
Beijing, China

Li Wang
School of Energy and Environmental Engineering
University of Science & Technology Beijing
Beijing, China

J. Worlitschek
Competence Center Thermal Energy Storage
Lucerne University of Applied Sciences and Arts
Luzern, Switzerland

Xiaohu Yang
Department of Architecture and Built Environment
Xi'an Jiaotong University
Xi'an, China

Chunrong Zhao
School of Mechanical and Mining Engineering
The University of Queensland
Brisbane, Australia

Yanqi Zhao
Birmingham Centre for Energy Storage & School of Chemical Engineering
University of Birmingham
Birmingham, United Kingdom
and
School of Mechanical Engineering
Jiangsu University
Zhenjiang, China
and
Jiangsu Collaborative Innovation Centre of Photovoltaic Science and Engineering
Changzhou University
Changzhou, China

Boyang Zou
Birmingham Centre for Energy Storage & School of Chemical Engineering
University of Birmingham
Birmingham, United Kingdom

1 An Introduction to Solid–Liquid Thermal Energy Storage Systems

Moghtada Mobedi
Shizuoka University

Kamel Hooman
The University of Queensland

Wen-Quan Tao
Xi'an Jiaotong University

CONTENTS

1.1 INTRODUCTION

With global population growth and our modern lifestyle, our energy demand keeps increasing. Our ever-increasing demand for energy should be met by renewable energy for a sustainable future. Both sustainability and environmental concern contributed to the growth of renewable energy at the international level, while most of the countries set national goals and made significant progress. In Australia, for example, 27.7% of the generated electricity was from renewable sources in 2020 with 50% annual increase in solar generation [1]. Tasmania, the Australian island state with just over 10 TWh annual energy demand, is currently running on 100% renewable energy. Heavily relying on wind, renewables contribute to 90% of the electricity supply in Scotland. Over the first quarter of 2020, clean energy sources contributed to almost half of national electricity generation in the United Kingdom. Similarly, over the first three quarters of the same year, renewables contributed to almost half of Germany's electricity

DOI: 10.1201/9781003213260-1

consumption. Currently, across all the key sectors, i.e. power, heating, industry and transport, demand for renewables is on the rise. The power sector, in particular, is in the lead with its demand for renewables increasing by 8% to reach 8300 TWh; the highest annual growth which was recorded in 2020, according to the IEA (International Energy Agency) [2]. Two-thirds of this amazing renewables growth is attributed to wind and solar PV (photovoltaic); expectedly both sectors are expected to expand in 2021. For instance, over 900 TWh is expected to be generated from these two sources in China this year. These numbers are 589 and 550 TWh for the European Union and the United States, respectively. On global scale, solar PV and wind grew by 12% and 23%, respectively, in 2020. As anticipated, the levelized cost of electricity keeps dropping thanks to technology and market development. However, the capacity factor, for either of solar PV or wind, is too low to allow for baseload energy production with no storage. Battery storage remains essential yet prohibitively expensive. Hence, alternative forms of energy storage including thermal energy storage are called for.

Ultimately, with our move toward energy generation from renewable sources, energy storage remains the key issue. This is mainly because most of the renewable energy sources are intermittent in nature. It can be claimed that a fully renewable energy future is impossible without inexpensive and reliable storage of energy. While chemical, electrical, mechanical and potential energy storage options have been investigated before, the focus of this book is on thermal energy storage in phase change materials (PCMs). Within thermal energy storage techniques, we limit the scope of this book to those relying on phase change. In particular, systems with phase transition from a solid to a liquid state are to be investigated. Applications for such system encompass a wide range including food safety and transport, building energy management, electronic cooling, waste heat recovery, health and well-being, refrigeration, solar thermal power generation, automobile industry, as well as thermal management of satellites. There are certain challenges to be addressed before these PCM-based technologies can be commercialized and we will cover some of those later on. On top of practical questions, there are fundamental research questions to be answered. Some of these questions center on the tools we use to analyze the thermo-chemo-mechanical behavior of thermal storage systems. This book will offer a combination of theoretical, numerical and experimental techniques to address these questions in forthcoming chapters. Moreover, while development of a technology is important, its deployment and market success will depend on a number of parameters among which cost can be investigated and minimized as will be shown in this book.

1.2 CLASSIFICATION OF THERMAL ENERGY STORAGE

Thermal energy storage (TES) technologies may be classified according to following four aspects: working temperature, duration of storage, circulation of heat transfer medium and energy storage mechanism [3–5] (Figure 1.1).

- *Temperature*: When the stored energy temperature is below 100°C, the system may be referred to as a low temperature energy storage one. When the stored energy temperatures are in 100°C–200°C or beyond 250°C, the terms medium and high temperature energy storage are used, respectively.

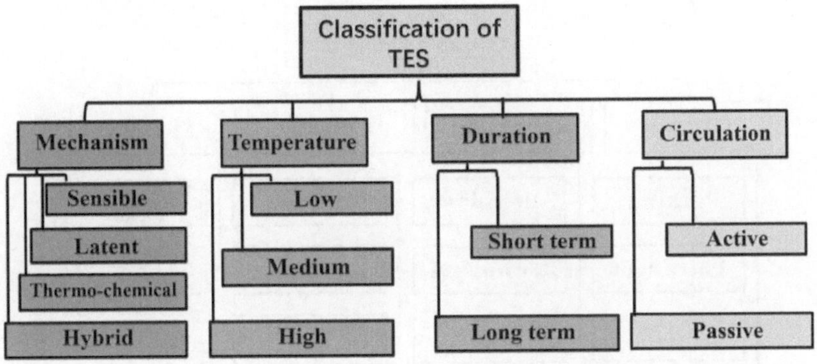

FIGURE 1.1 Classifications of thermal energy storage technologies.

- *Duration*: If the energy storage system is charged or discharged daily or weekly, it is called short-term storage. While in a long-term storage system energy is stored for months or seasons [6].
- *Circulation*: Since the TES system usually requires heat transfer fluid (HTF) during the charging and discharging process, an obvious question would be about the driving force for moving HTF in the loop. If the HTF is circulated by buoyancy effects without external force (such as a fan or a pump), the storage system is called passive, while in an active system, the HTF is driven by an external force.
- *Mechanism*: The TES mechanism is, arguably, the most important aspect of the classification [7–10]. Here, we use the following four indicators for our classifications, and the technology used for energy storage can be further subdivided into four different types (Figure 1.2).

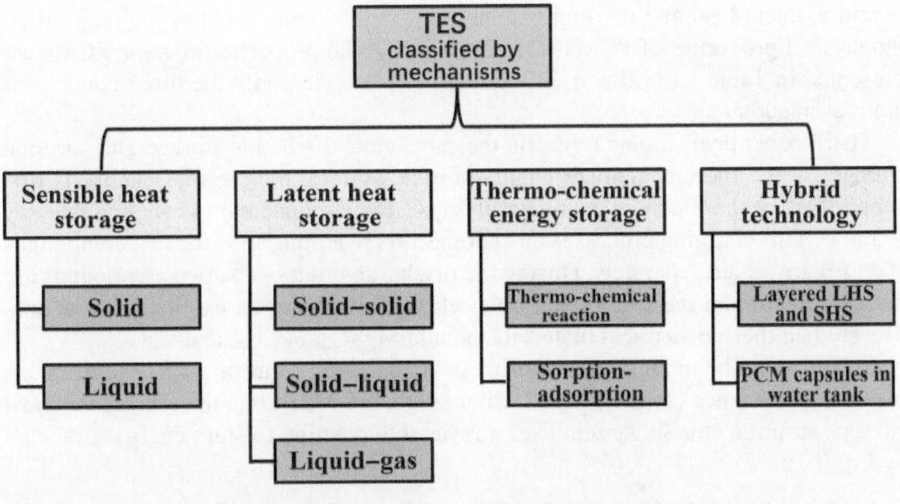

FIGURE 1.2 Further classification of TES by mechanism.

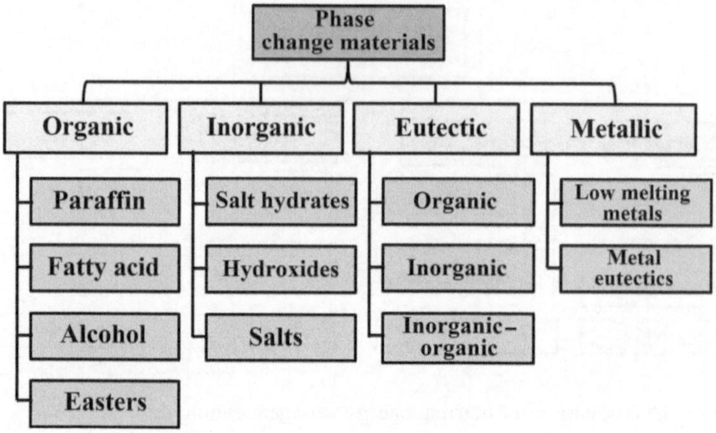

FIGURE 1.3 Classification of PCMs.

In the sensible heat storage (SHS), heat may be stored in solids (rocks, concrete, bricks, sand, or metals). Heat can also be stored in liquids: most often liquids water, molten salts and mineral oils. In the latent heat storage (LHS), energy is stored, in PCMs, when material changes its state from solid to solid [11], solid to liquid or from liquid to gas while the solid-liquid phase change is the most widely adopted technology, hence the exclusive focus of this book. The frequently used PCMs are organic, inorganic and eutectic materials, as well as metallic liquids which have been studied at laboratory scale [12] (Figure 1.3).

Relying on high latent heat for PCMs is essential. Ideally, a low mass of the PCM should hold as much as possible of latent heat of fusion. The phase change process initiates at a given melting point, and the released heat must be conducted through the solid and liquid phases. In view of the above, these three, melting temperature, latent heat and thermal conductivity, can be considered as the major thermophysical properties of PCMs [13]. Thermophysical properties of some PCMs are presented in Table 1.1 [12] and, in the interest of brevity, only the three major ones are presented here.

Thermochemical storage relies on thermochemical reaction storage and sorption storage. In the thermochemical energy storage system, the thermal energy is first supplied to the thermochemical material, say C, to dissociate it into two products, say A and B. This charging process is an endothermic reaction. Then the two compounds A and B are stored separately. During the discharge process, the two compounds are reunited again and the stored energy is released thanks to an exothermic reaction. The studied thermochemical materials include $MgSO_4$, $FeCO_3$, and $CaCO_3$.

Sorption can be thought of as a process of capturing a gas or a vapor (called sorbate) by a substance (solid or liquid, called sorbent). Much like the thermochemical storage, sorption storage system uses a reversible reaction to store energy as shown by Eq. (1.1):

$$AB + \text{Heat} \Leftrightarrow A + B \tag{1.1}$$

TABLE 1.1
Thermophysical Properties of Some PCMs

Phase Change Materials	Classification	Temperature, °C	Latent Heat of Fusion, kJ/kg	Conductivity, W/(m·K)
Paraffin wax (C_{13}–C_{18})	Organic	32	251	0.214
Polyglycol E600	Organic	22	127.2	0.189
Vinyl stearate	Organic	29	122	0.25
Butyl stearate	Organic	19	140	0.21
1-Dodecanol	Organic	26	200	0.169
Octadecane	Organic	28	243.5	0.148
Palmitic acid	Organic	57.8	185.4	0.162
Capric acid	Organic	32	152.7	0.153
Caprylic acid	Organic	16	148	0.149
Propyl palmitate	Organic	10	186	-
$KNO_3/NaNO_3$	Inorganic	220	100.7	0.56
$CaCl_2{\cdot}6H_2O$	Inorganic hydrates	29	187	0.53
$LiNO_3/KNO_3/NaNO_3$	Inorganic eutectic mixture	121	310	0.52
(a) $KNO_3/LiNO_3$	Inorganic eutectic mixture	124	155	0.58
(b) $KNO_3/NaNO_3/LiNO_3$	Inorganic eutectic mixture	130	276	0.56
$LiNO_3/CaNO_3$ with $KNO_3/NaNO_3$ (30%/70%)	Inorganic noneutectic mixture	260	305	0.54
Capric acid and lauric acid	Organic eutectic mixture	19.67	126	0.21
Polymethyl methacrylate/capric-stearic acid mixture	Organic eutectic mixture	21.37	116.25	0.15

where A and B are called working pair. When heat is supplied (charging process), the compound AB is dissociated into components A and B (desorption) which are stored separately in different tanks. When heat is needed, A and B are brought together and the chemical potential is unleashed through generating heat (discharging process, or absorption/adsorption process) [3,14].

SHS and LHS have their own advantages and disadvantages. Hence, a hybrid TES system, a combination of sensible and LHS technologies, can be a practical solution. Such systems possess the advantages of both SHS and LHS. One typical example is consisted by layered PCM (upper) and rocks (layered with air as the HTF [15]). During the charging process, hot air enters the system, transfers heat to PCM and rocks in sequence and then leaves the system bottom. During the discharge process, cold air is admitted to collect the stored heat. Such an arrangement of LHS and SHS can stabilize the temperature of the heated air around the PCM's melting temperature [15].

Here, we briefly discuss the advantages and disadvantages of different TES technology. Among the TES technologies discussed, the simplest one (both design and maintenance) is the SHS which is a low-cost mature technology. The drawbacks, however, are the very low energy storage capacity, non-isothermal behavior during the discharging process and high heat losses associated with SHS. On the other hand, LHS possesses high energy storage capacity and can offer a more uniform and constant temperature for the HTF during the discharge process. Its major disadvantage is the low thermal conductivity of the PCMs, usually less than 1.0 W/(m·K) as shown in Table 1.1. Accordingly, a great deal of information is available in the literature on efforts to increase the thermal conductivity of the compound. This has been discussed in detail in subsequent chapters. Compared with sensible and LHS, the advantages of the thermochemical storage include higher volumetric energy storage capacity, compact size, less energy losses while their high operating temperature (beyond 600°C) can be a challenge for implementing this technology for low-temperature applications.

1.3 DIFFICULTIES ASSOCIATED WITH SOLID–LIQUID THERMAL STORAGE

Despite their attractive features, deploying the solid–liquid phase change systems face the following two challenges: (1) phase change material challenges and (2) system design difficulties. In this section, these difficulties are acknowledged, and some suggestions are put forward in literature and the subsequent chapters of this book are touched on.

1.3.1 CHALLENGES WITH PCM

PCMs and the required improvements have been analyzed in Chapter 2 of this book. The desired thermal, physical, chemical and kinetic properties for a solid–liquid PCM are elaborated on in Chapter 2. In some cases, it is difficult to offer quantitative measures, for instance, manufacturing of a PCM with higher latent heat as well as higher thermal conductivity is always aimed at while there is no upper bound for

TABLE 1.2
A Summery on the Required Features of a PCM [Chapter 2, 16–18]

Thermophysical properties	• High latent heat • High thermal conductivity • High specific heat • Appropriate melting/freezing temperature
Chemical properties	• Non-inflammable, non-explosive and non-poisonous • Chemically stable • Compatibility with material containers • No chemical decomposition
Physical properties	• High density • During freezing, there must be little or no supercooling • Low vapor pressure
Economical parameters	• Low cost • Available in large quantities • Commercially viable

the numerical value of these parameters. Therefore, Table 1.2 can be interpreted as a starting point offering challenges and endless opportunities for researchers who study PCMs.

In what follows, the major challenges faced to improve the material properties are touched on.

- *Low thermal conductivity*:

  As Table 1.1 demonstrates, the numerical value of the thermal conductivity of most of the PCMs is limited to be very close to that of water; in fact, ranging from 0.148 to 0.58 W/(m·K). Thus, many studies were reported in literature aiming at increasing the thermal conductivity of the PCM thereby augmenting the heat transfer and shortening the charge/discharge process by applying one (or a combination) of the following techniques:

  a. Extended surfaces:

     There are many studies on the extended surfaces (or fins) to enhance heat transfer to or from a PCM. The enhancement of heat transfers by using extended surface is explained by Kamkari et al. in Chapter 7 of this book. They discussed the cylindrical, rectangular and spherical fins for enhancement of heat transfer for different applications. With significant buoyancy effects, asymmetrical fins extended in the direction of gravity, are proposed in [19] for further enhancement of heat transfer. The fins enhance conduction heat transfer while limiting the space available for free convection of the PCMs. As expected, with fins, phase change is accelerated in the regions close to the fin surfaces.

  b. Metal foams:

     Metal foams can be imagined as interconnected fins. They can enhance heat transfer in the entire domain, almost uniformly, albeit at the expense of weakening the natural convection heat transfer of the PCM. With the use of thermal spreader or thermal conductivity enhancers, a part of the PCM is replaced by a different material that does not undergo phase change along with the PCM. Hence, the storage tank size has be to bigger to contain the same mass of PCM (with no conductivity enhancer). Therefore, it is a common practice to use high porosity metal foam to avoid a significantly bigger tank to contain the PCM. The use of metal foam as well as combination of fin and metal foams are discussed in Chapter 8 by Yang et al. They performed an experiment study and compared the use of fin, metal foam, as well as a combination of both and compared the thermal performances of these three designs with the benchmark case being the pure PCM in a storage tank.

c. Highly conductive nanoparticles:

Highly conductive additives such as metallic nanoparticles can also be used to increase the thermal conductivity of the PCM. The effect of the nanoparticles on the enhancement of heat transfer of PCM is discussed in Chapters 2 and 7 of this book. In Chapter 2, Stamatiou et al. stated that there are large discrepancies between the thermal conductivity values reported by different researchers for enhancement of heat transfer by nanoparticles due to sample instabilities (segregation of conductive additives with time), sample in-homogeneities (effect of sampling location and method), thermal conductivity measurement device/methodology and exact properties of additives (e.g., porosity of expanded graphite). Kamkari et al., in Chapter 7, also went over difficulties associated with the use of metal nanoparticles for instance sedimentation and aggregation under high particle loadings as well as viscosity augmentation of nano enhanced PCMs.

- *Density change between liquid and solid phase of PCM*:

In many solid–liquid thermal storage systems, PCM exists in a container while undergoing cyclic heating and cooling leading to successive phase changes. This leads to significant density variation deviating from the (already different) solid- and liquid-phase values. Thus, expansion and contraction due to both temperatures change and phase change have to be accounted for otherwise mechanical deformation in the container, leakage or infiltration may occur. Two solutions are offered [20]; either to have an elastic wall or an air gap to compensate for density change during phase change.

a. Super cooling:

When a liquid is frozen below its normal freezing point, supercooling occurs. For a PCM, this can cause a delay in starting of the solidification process. Methods such as adding nucleating agent or metal additives to the PCM are some suggested remedies for supercooling [21–22]. In Chapter 2 of this book, the mechanism of supercooling as well a suggested method to prevent supercooling of PCMs are discussed.

b. Phase segregation:

Generally, segregation appears in multicomponent PCMs. When the density of multicomponent in a PCM is different, separation of the components is observed due to buoyancy effect [21]. It is reported that major problems with most of the salt hydrates are supercooling, phase segregation, corrosion and thermal instability. Addition of thickening agents, preventing density induced flow, to the aqueous solution helps reducing this irreversible segregation.

1.3.2 System Design Challenges

Difficulties associated with the design of solid–liquid thermal storage systems, and the degree of freedom in the designer's choice, can be classified as follows:

a. *Gravity direction*:

Design of passive systems where HTF is circulated in the system by only relying on free convection would certainly need to consider the orientation of the piping and the tank. For the PCM itself, the direction along which gravity pools can significantly affect some designs. A mobile system, where the PCM tank keeps tilting during the journey (such as airplanes or PCM-assisted batteries for Electric vehicles (EV) should then be designed carefully to ensure a seamless performance.

b. *Round trip efficiency*:

Melting and solidification are significantly different. The former starts from a solid and during the process, the liquid fraction is increased. The latter, however, can suffer from skin effects where a thin layer of solid can be formed around the heat sink. Therefore, the heat transfer from the liquid to the heat sink must experience an additional resistance posed by the solid layer which thickens with time. Consequently, the designer has to compromise as the optimal design for a storage system under charging condition may be suboptimal during the discharge process.

c. *Acceleration*:

The design of solid–liquid thermal storage systems in high-speed vehicles (such as airplane, space shuttles, rockets, high-speed trains) has to take into account the effect of additional acceleration or deceleration of the device which will affect the heat transfer process, in particular the buoyancy-induced flow.

d. *Design method*:

Unfortunately, there is no accurate widely accepted design method to design of solid–liquid thermal storage systems for various applications. The designer must rely on existing codes and standards, for example density difference can get the storage tank to classify as a pressure vessel. Currently, designers heavily rely on computational techniques in the absence of more conventional design tools and techniques such as charts or tables. Till now, the number of studies on establishing a general design method (similar to those of single-phase heat exchangers) is limited. Hence, further studies need to be conducted and documented to facilitate the design of PCM heat exchangers.

e. *Long computational time*:

As mentioned above, the design of PCM heat exchangers is done computationally due to lack of a systematical designing method. Since most of practical solid–liquid thermal storage devices have to be modeled as three-dimensional objects, the design of a PCM thermal storage device for freezing and melting period requires long computational time. This can be particularly difficult when an optimization process is to be conducted. Finally, combining a three-dimensional simulation of the PCM tank with HTF pipes flowing through it is currently impractical.

1.4 SOLID–LIQUID THERMAL ENERGY STORAGE APPLICATIONS

Solid–liquid TES systems offer unique benefits, as described above, so that they can be applied to solve a number of practical issues faced by the industry. In what follows, a number of industrial applications which are lending themselves well to the deployment of PCMs are listed and briefly discussed.

a. *Solar applications*

While solar energy is abundant, the intermittent nature of the energy calls for inexpensive storage of the heat for use at night or when overcast. Many studies on the application of solid–liquid phase change on solar water heating systems, solar air heating systems, solar cookers as well as solar greenhouses can be found in the literature [23]. Development of PVT (photovoltaic thermal) systems leads to new opportunities. Cooling the PV panels increases their conversion efficiency and allows for the use of heat which would have otherwise been wasted. Current PVT systems use water (or a mixture) for this purpose, but direct use of PCMs or charging a storage tank using the collected hot water can be a very interesting option.

Recent advances in CSP (Concentrated Solar Panels) led to renewed interest in high temperature storage mainly to increase the capacity factor. Lie et al. propose the use of silicone metal alloys and their applications for high temperature storage systems in Chapter 6. It is demonstrated that while metal and metal alloy PCMs can overcome the common disadvantage of low thermal conductivity of salt-based PCMs, they also offer high energy densities, good cycling stability and low volumetric expansion. In parallel, for high-temperature CSP systems, Riahi et al. investigate the thermal stress–strain distribution in a heat exchanger with PCM that has a melting point of 705.8°C.

b. *Food industry*

The use of PCM for food conservation makes perfect engineering sense. PCM can be used to maintain a set temperature for medicine, food and drinks (vaccine, blood, organs, meats, wines, vegetable, soft drinks) over a prolonged period. In Chapter 15 of this book, Maksum et al. discusses the application of PCM for thermal management of a warehouse either by applying the PCM in conjunction with the refrigeration system or by embedding it in the building material. Design of warehouses must be done while minimizing the risk of overheating while slowing down the rate of microbial growth.

c. *Transportation*

Meals, medicine and frozen products should be transported and delivered ideally with the product remaining at the same temperature as the initial one or below a certain temperature. PCM lends itself well to an application of this nature as Chapter 15 of this book demonstrates. Commercial refrigerated vehicles rely on combustion engine powering mechanical refrigeration technology to provide a temperature regulation to the compartment. However, the use of PCM-based cold energy storage technology offers a cost-effective, zero-carbon and zero-emission alternative.

d. *Micro- and nano-encapsulated PCMs*

Encapsulated PCMs are hollow particles filled with PCMs. Encapsulated PCMs can be stagnant or flow with a fluid to assist with thermal management. Chapter 9 investigates micro- and nano-encapsulated PCM and compares their performance with that of PCM as a single product. Another advantage of encapsulated PCMs is the ability to overcome material incompatibility (corrosion for instance) issues. Here, the shell, which contains the PCM, separates it from the surrounding, thereby limiting the potential risk to each shell of the encapsulated PCM while the piping and pieces of equipment in the line can be saved.

e. *Air-conditioning systems*

Solid–liquid TES systems are suitable devices for increasing performance of air-conditioning systems through higher COP (Coefficient of performance) while reducing their capital and maintenance cost. The PCM can either be integrated to the air-conditioning system (allowing for thermal inertia, lower or higher temperature for heat sink or source) or be used in conjunction with inlet air pre-cooling or pre-heating devices. They can store temperature at a given temperature during the off-peak period and utilize it during peak hours. This will be particularly interesting for heat-driven systems and heat pumps. Chapter 10 reviews the literature on PCM heat exchangers and offers a classification method for these devices that will facilitate their design and deployment. Chapter 11 investigates the use of ice thermal storage for chiller and air-conditioning systems.

f. *Other applications*

The applications of solid–liquid thermal storage systems are not limited to the aforementioned areas and they are used and suggested for many different areas such as using in the building walls for insulation, space applications to store heat during sun light period and use it during the eclipse, electric vehicle to cool batteries, or even using in textile providing constant temperature for the body. In Chapter 14 of this book, Wang et al. propose a novel compact cooling system for thermal management of cylindrical lithium-ion battery packs which is a hybrid of PCM and heat pipe cooling systems. They state that the lithium-ion battery modules/packs equipped with the present heat pipe and PCM hybrid BTMS can be used to power vehicles such as motorbikes or scooters.

1.5 CONCLUSION

The need for the use of TES in this century is discussed in this chapter. Shifting of the used energy sources from fossil sources into the renewable energy sources forces researchers to study on TES systems more and more since most of renewable energy sources are intermittent. There are there kinds of thermal energy systems as sensible, latent and chemical energy storage systems. Among them the solid–liquid TES systems have many advantages such as high heat thermal storage capacity due to the use of latent heat, simple operation and allowing the use of environment friendly materials. However, it has difficulties such as low thermal conductivity, volume change during the phase change and sub-cooling.

All of difficulties of solid–liquid phase changes and their suggested remedies as well as their significant applications are discussed in different chapters of this book.

ACKNOWLEDGMENT

The third author acknowledges his colleague Mr. Zheng for his help in the preparation of the manuscript.

REFERENCES

1. C. Zhao, M. Opolot, M. Liu, F. Bruno, S. Mancin, K. Hooman, Phase change behaviour study of PCM tanks partially filled with graphite foam, *Applied Thermal Engineering* 196 (2021) 117313. doi:10.1016/j.applthermaleng.2021.117313.
2. IEA, World Energy Outlook 2020, https://www.iea.org/reports/world-energy-outlook-2020.
3. C. Suresh, R. P. Saini, Review on solar thermal energy storage technologies and their geometrical configurations, *Energy Research* 44 (2020) 4163–4195.
4. G. Alva, Y. Lin, G. Fang, An overview of thermal energy storage system, *Energy* 144 (2018) 341–378.
5. H. Zhang, J. Baeyens, G. Caceres, J. Degreve, Y. Lv, Thermal energy storage: recent development and practical aspects, *Progress in Energy and Combustion Science* 53 (2016) 1–40.
6. C. Bott, I. Dressel, P. Bayer, State-of-technology review of water-based closed seasonal thermal energy storage systems, *Renewable and Sustainable Energy Reviews* 113 (2019) 109241.
7. J. Lizana, R. Chacartegui, A. Barrios-Padura, J. M. Valverde, Advances in thermal energy storage materials and their applications towards zero energy buildings: a critical review, *Applied Energy* 203 (2017) 219–239.
8. L. F. Cabeza, A. de Gracia, G. Zsembinszki, E. Borri, Prospective on thermal energy storage research, *Energy* 231 (2021) 120943.
9. Z. Ding, W. Wu, M. Leung, Advanced/hybrid thermal energy storage technology: material, cycle, system and perspective, *Renewable and Sustainable Energy Reviews* 145 (2021) 111088.
10. Z. Li, Y. Lu, R. Huang, J. Chang, X. Yu, R. Jiang, X. Yu, A. P. Roskilly, Applications and technological challenges for heat recovery, storage and utilization with latent thermal energy storage, *Applied Energy* 283 (2021) 116277.
11. A. Fallahi, G. Guldentops, M. Tao, S. Granados-Focil, S. V. Dessel, Review on solid-solid phase change materials for thermal energy storage: molecular structure and thermal properties, *Applied Thermal Engineering* 127 (2017) 1427–1441.

12. M. K, Rathod, J. Banerjee, Thermal stability of phase change materials used in latent heat energy storage systems: a review, *Renewable and Sustainable Energy Reviews* 18 (2013) 246–258.
13. A Dinker, M. Agarwal, G. D. Agarawal, Heat storage materials, geometry and applications: a review, *Journal of the Energy Institute* 90 (2017) 1–11.
14. K. E. N'Tsoukpoe, H. Liu, N. Le Pierres, L. Luo, A review on long-term sorption solar energy storage, *Renewable and Sustainable Energy Review* 13 (2009) 2385–2396.
15. G. Zanganeh, M. Commerford, A. Haselbacher, A. Pedretti, Stabilization of the outflow temperature of a packed-bed thermal energy storage by combining rocks with phase change materials, *Applied Thermal Engineering* 70 (2014) 316–320.
16. N. I. Ibrahim, F. A. Al-Sulaiman, S. Rahman, B. S. Yilbas, A. Z. Sahin, Heat transfer enhancement of phase change materials for thermal energy storage applications: a critical review, *Renewable and Sustainable Energy Reviews* 74 (2017) 26–50.
17. E. Osterman, V. V. Tyagi, V. Butala, N. A. Rahim, U. Stritiha, Review of PCM based cooling technologies for buildings, *Energy and Buildings* 49 (2012) 37–49.
18. M. Liu, W. Saman, F. Bruno, Review on storage materials and thermal performance enhancement techniques for high temperature phase change thermal storage systems, *Renewable and Sustainable Energy Reviews* 16(4) (2012) 2118–2132.
19. M. Gürtürk, B. Kok, A new approach in the design of heat transfer fin for melting and solidification of PCM, *International Journal of Heat and Mass Transfer* 153 (2020) 119671.
20. J. Dallaire, L. Gosselin, Various ways to take into account density change in solid-liquid phase change models: formulation and consequences, *International Journal of Heat and Mass Transfer* 103 (2016) 672–683.
21. A. Bland, M. Khzouz, T. Statheros, E. I. Gkanas, PCMs for residential building applications: a short review focused on disadvantages and proposals for future development, *Buildings* 87 (2015) 654–662.
22. V. Kapsalis, D. Karamanis, Solar thermal energy storage and heat pumps with phase change materials, *Applied Thermal Engineering* 99(25) (2016) 1212–1224.
23. A. Sharma, V. V. Tyagi, C. R. Chen, D. Buddhi, Review on thermal energy storage with phase change materials and applications, *Renewable and Sustainable Energy Reviews* 13 (2009) 318–345.

2 Solid–Liquid Phase Change Materials for Energy Storage

Opportunities and Challenges

A. Stamatiou, S. Maranda, L. J. Fischer, and J. Worlitschek
Lucerne University of Applied Sciences and Arts

CONTENTS

2.1 INTRODUCTION: BACKGROUND AND MOTIVATION

Phase change materials (PCMs) surround us. PCM for energy applications (not to be confused with PCM for data storage [1]) can be in principle all materials that undergo a phase transition while absorbing/releasing a significant amount of heat. They can thus serve as both heat sources and heat sinks in two major categories of applications: (1) thermal energy storage applications and (2) temperature stabilization applications. The former topic is becoming increasingly important in the last decades due to the urgency of developing energy storage solutions to address the intermittency

DOI: 10.1201/9781003213260-2

of renewable energies. The latter is mostly independent of sustainability questions and aims at providing solutions for industrial and other types of applications where temperature precision is of crucial importance mostly for quality management. Some examples include transportation of sensitive goods such as pharmaceuticals, efficient cooling of high-precision machinery, thermal management of batteries or electronic devices and gastronomical applications.

PCM research is intrinsically interdisciplinary as it requires a deep knowledge of material science, thermodynamics, heat transfer and crystallization kinetics. While the heat transfer challenges and engineering aspects are well addressed in literature, the material and chemistry related questions and challenges often lack the deeper understanding that would enable the next breakthroughs in the field. This book chapter aims to address some of these critical, material-related challenges but also presents opportunities that will enable them to be tackled thus driving the topic of latent heat storage forward.

2.2 WORKING PRINCIPLE

Most PCM applications employ solid–liquid PCM (as opposed to liquid–gas or solid–solid) as storage materials. Here, heat absorbed or released during melting or solidification of a material, i.e., the phase change enthalpy of this solid–liquid phase transition, is used. However, solid–solid and liquid–gas phase transitions can also be of interest in specific cases. For solid–solid systems, a change in the crystalline structure of the materials takes place, which is associated with an absorption or release of heat. As the PCMs remain in solid state at all times, they present no leakage issues and can be thus easier to handle but typically they have lower phase transition enthalpies compared to solid–liquid PCM. Gas–liquid systems are interesting as they can possess much higher gravimetric phase change enthalpies (in the order of 5–10 times) in comparison to the solid–liquid phase change but the great volume expansion during evaporation makes these materials often hard to implement.

This book chapter concentrates on solid–liquid PCMs which are currently, the most widespread PCMs because they offer a good compromise between ease of handling and high phase change enthalpy. The chosen material can either be a pure component, a mixture, or a mixture behaving in its phase transition like a pure component (eutectic mixture). Therefore, the fundamentals relevant in this context are on the one hand solid–liquid phase equilibria and on the other hand the kinetics of the melting and solidification processes. While the phase equilibria of pure components and eutectics without impurities show a solid–liquid phase transition at one temperature at a given pressure, phase diagrams of mixtures can be rather complex and show temperature ranges of the phase change.

The kinetics of the melting process are typically rather fast as they are dominated by diffusion and therefore measured melting temperatures are normally close to the equilibrium phase change temperature. In case of crystallization, the kinetics of solidification is dominated by nucleation, i.e., the formation of a critical first solid cluster, and the further growth of the crystal nuclei. This leads to hysteresis effects of solidification being observed at temperatures significantly lower than the phase equilibrium temperature due to the chemical potential required to drive the process.

In the field of PCM research and applications this observed hysteresis is typically referred to and quantified as supercooling or metastable zone width.

Additionally, the fundamentals of heat and mass transfer are relevant in the description of melting and solidification. Heat transfer phenomena are typically dominant with respect to crystallization once crystals are present and mass transfer plays a role regarding convection in the liquid phase or in segregation phenomena. The typically observed melting and solidification process of a solid–liquid PCM for an idealized, pure component is shown in Figure 2.1.

At the beginning of the melting, the PCM is in a solid state. During heating up the temperature of the PCM increases, storing heat in a sensible form, until it reaches the temperature where it starts to melt. This temperature is called melting temperature (T_m). If more energy is added, the storage medium starts to melt at a constant temperature, storing heat in a latent form. Once the substance is entirely melted, any further addition of heat will result in a temperature increase and further storage of heat in a sensible form. During cooling down, a very similar process occurs. The material is now in liquid form and its temperature is decreasing as it is releasing heat in a sensible form. When it reaches the solidification temperature, T_s, crystallization begins with a simultaneous release of the latent heat at almost isothermal conditions. When the entire material is solidified, any further cooling down leads to sensible cooling (and further temperature decrease) of the PCM. The phase change enthalpy, Δh_{PC}, describing the difference between the enthalpy of the substance in liquid and

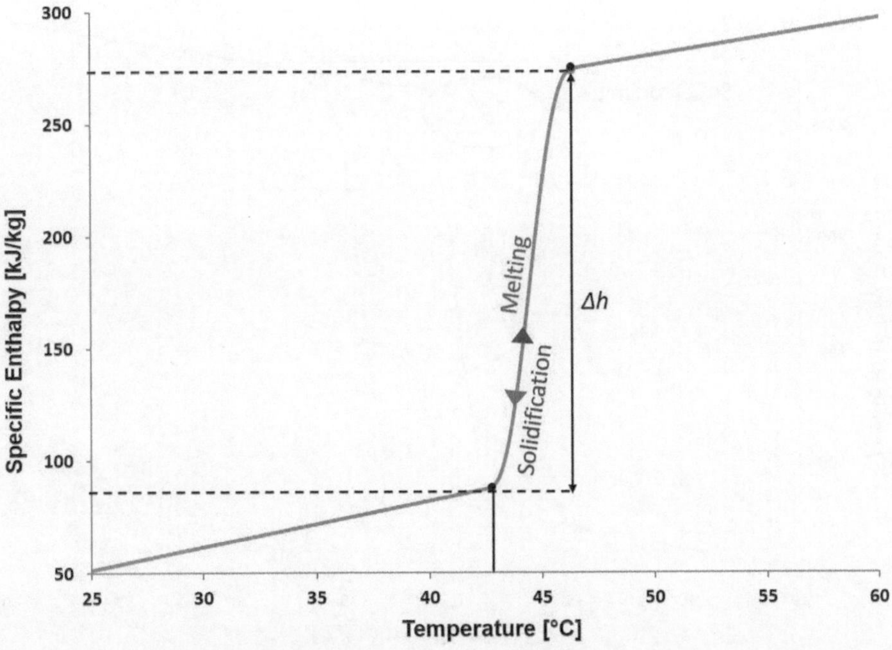

FIGURE 2.1 Specific enthalpy as a function of temperature during idealized melting and solidification process of a fictional PCM.

solid form, is released or absorbed during both the phase change from liquid to solid and the reverse process.

In theory, the phase change occurs exactly at the phase change temperature T_m, at which the solid and liquid phases of the PCM are in thermodynamic equilibrium. In technical applications, the phase change during melting does not take place at exactly the phase change temperature, but extends over a melting range. This is typically due to impurities and the finite speed of the temperature increase. During cooling, most PCMs do not solidify close below the melting temperature, but start crystallizing after a temperature well below the melting temperature. This effect of supercooling can be observed in Figure 2.2. When crystallization starts, the temperature of the originally supercooled liquid increases due to the release of the phase change enthalpy up to a maximum of the equilibrium phase change temperature T_m. In most cases, supercooling is not desired in a latent thermal energy storage (TES) (with some exceptions, e.g., for seasonal storage applications [2]). The presence of supercooling makes it necessary to reduce the temperature of the PCM well below the phase change temperature to start crystallization and to release the latent heat stored in the material. This results in energy and exergy efficiency losses of the storage system. In extreme cases, the solidification of the PCM might not occur at all, making the extraction of the stored latent heat impossible. Supercooling is one of the major challenges of PCM research and will be explained in more detail in Section 2.4.2.

In many cases, the observed temperature hysteresis between the melting and solidification process does not exceed 5 K. Additionally to the kinetics of crystallization,

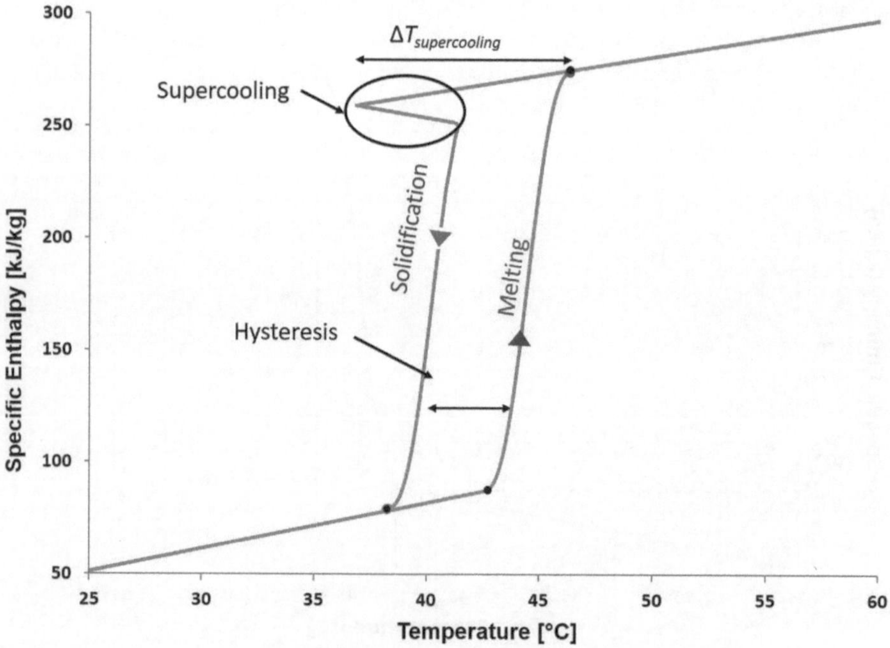

FIGURE 2.2 Specific enthalpy as a function of temperature during a realistic melting and solidification process where hysteresis and supercooling take place.

the hysteresis observed in a technical process can be influenced by heat transfer from the heat source/sink to the PCM (apparent hysteresis).

2.3 PHASE CHANGE MATERIALS AND CLASSIFICATIONS

When selecting a phase change material for an application careful consideration of the requirements list is necessary. Of course, the requirements vary from application to application. Table 2.1 attempts to summarize the most commonly desired thermal, physical, kinetic and chemical properties for latent TES applications using solid–liquid PCM. Along with an appropriate phase change temperature, a high enthalpy of fusion is possibly the most important property for PCMs. Other important properties are high physical and chemical stability over the lifetime of the material, high thermal conductivity and specific heat capacity, high crystal nucleation and growth rates as well as compatibility with surrounding materials (e.g., metals and polymers used for encapsulation, heat transfer enhancement). In terms of kinetic requirements, increased awareness has been raised recently on the adverse effects of polymorphism, especially on organic materials [3]. Specific applications that are particularly temperature sensitive might require additionally narrow phase change temperature (e.g., transportation of sensitive goods) or might have strict safety requirements that influence the acceptable toxicity and flammability of the materials (e.g., building applications). An increasing interest in sustainable materials is recently also promoting bio-origin and biodegradability as important requirements for specific applications. Additionally, to the properties listed in the table other practical considerations must also be taken into account such as low cost and high availability of the PCM in the market.

In reality, all of these criteria are almost never met fully by any PCM. It is therefore important that the criteria are weighted based on the priorities of the specific

TABLE 2.1
Overview of Commonly Desired Thermal, Physical, Chemical and Kinetic Properties for Solid–Liquid PCM

Thermal Properties	Physical Properties
• Phase change temperature in the desired range for the application • High phase change enthalpy per unit volume and/or mass • High specific heat capacity • High thermal conductivity	• A small volume change during phase change • A low vapor pressure at the operating temperature • No separation in case of mixtures • A high density
Kinetic Properties	**Chemical Properties**
• Low supercooling effect • A high enough crystallization rate • No polymorphism	• No chemical degradation • Compatibility with surrounding materials • Non-toxic, non-flammable and non-explosive to ensure safety, • Biobased nature

application so that the PCM that constitutes the best solution can be selected in each case. It is also worth mentioning, that many of these properties can be modified on a material level (e.g., by using nucleating agents, thickeners, highly conductive substances) or can be compensated for by modification in the storage design (e.g., heat transfer enhancement structures). These topics will be discussed further in more detail in the following sections of this chapter.

Figure 2.3 shows an overview of the most popular classes of materials that have been investigated to date and are potential PCMs for technical applications. There are several ways of classifying PCMs. In this case PCMs are divided in organic vs inorganic materials and pure substances vs mixtures.

The most popular pure-substance PCMs that have been investigated and are used to date are paraffins, fatty acids, esters, alcohols sugar alcohols and salts. A very popular class of mixture PCMs, salt hydrates, are ionic compounds containing one or more types of salt and water at distinct molar ratios. They are often reported as pure PCM because upon crystallization the water and salt molecules are enclosed in a single crystal lattice. In this chapter they are classified as mixtures because they are in fact combinations of water and salt that often behave as inhomogeneous mixtures. A further distinction is made between eutectic and non-eutectic mixtures. Eutectic mixtures are generally considered more suitable as PCM because they have a narrow phase change temperature, which is generally desired in PCM applications.

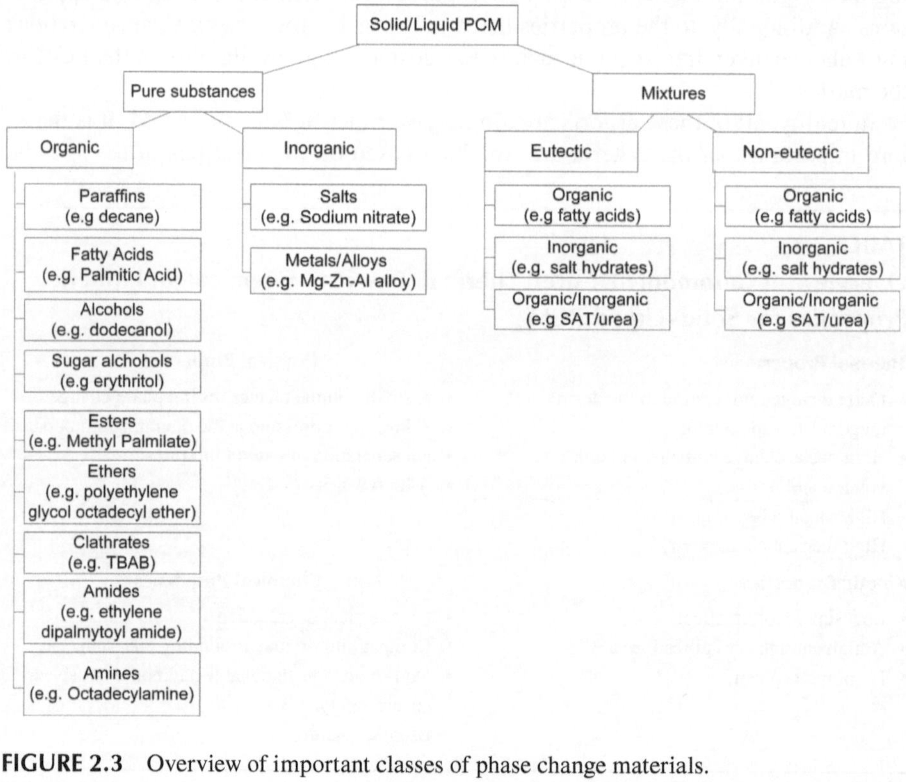

FIGURE 2.3 Overview of important classes of phase change materials.

Non-eutectic mixtures are often used if the phase change temperature of the pure substances or eutectics does not exactly match the desired application temperature. For example, intermediate values for T_m can be achieved by mixing different types of paraffins or fatty acids. However, these mixtures have a phase change temperature range rather than a sharp phase change temperature and a lower phase change enthalpy than the individual pure substances or eutectic mixtures. It should be mentioned that phase change dispersions are not included in this classification because they comprise a continuous phase that does not undergo solidification and are considered mixtures of a PCM and an inert phase.

An overview of phase change enthalpy as a function of phase change temperature for the most important classes of phase change materials is shown in Figure 2.4.

The most commonly used organic PCMs are paraffin waxes which are characterized by high flammability and higher prices in a relatively limited melting point range. However, they present a high cyclability, low corrosiveness and low supercooling [5]. Paraffins, despite their moderate phase change enthalpies, were the first PCM to be widely commercialized due to their ease in application.

Salt hydrates consist of inorganic salts (AB; A: Anion, B: Cation) and water (H_2O). They form a crystalline solid phase with a general formula of $AB{\cdot}nH_2O$, where n has values of, e.g., 3 (trihydrate), 4 (tetrahydrate), and 10 (decahydrate). Phase change in salt hydrates is a process of dehydration and dissolving of the salt in the released

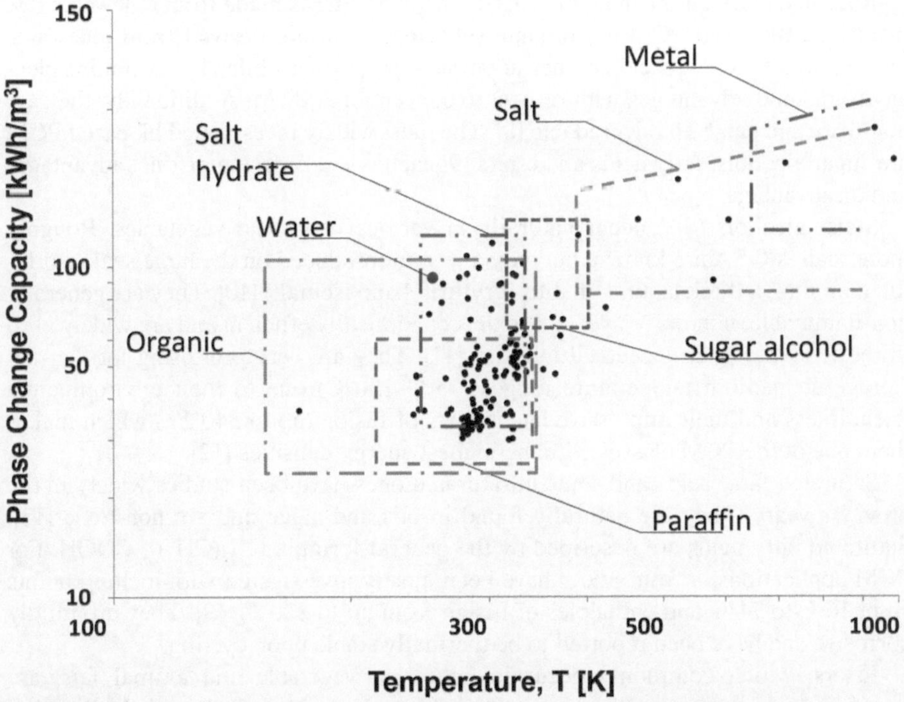

FIGURE 2.4 Overview of phase change enthalpy as a function of phase change temperature for the most important classes of phase change materials. (Adapted from Ref. [4].)

water in case of melting and hydration in case of freezing; both processes with strong interaction of heat and mass transfer. Salt hydrates can be grouped according to their phase equilibrium behavior into congruent melting, semi-congruent-melting and incongruent-melting materials. In a congruent melting system, the solid and the liquid phases at the melting point consist of the same compositions. A semi-congruent material typically produces lower hydrated salts along with the hydrated salt with a composition corresponding to that of the original salt solution. Incongruent-melting materials are similar to semi-congruent materials but instead of the presence of lower hydrates in the solution anhydrous solid particles are produced while melting.

Salt hydrates have been extensively studied for latent heat storage (LHS) applications on accounts of their high volumetric latent heat of fusion (up to 350 J/mL), typically higher thermal conductivities than organic PCM and lower costs (<1 Euros/kg). Furthermore, there are no flammability issues with salt hydrates due to their low vapor pressures. For LHS, many salt hydrates have been identified for the temperature range 7°C–100°C, making it possible to select PCM for a specific heating or cooling application [6]. The potential of salt hydrates as PCM has been showcased in many application-based studies (e.g., $MgCl_2 \cdot 6H_2O$; $CaCl_2 \cdot 6H_2O$; sodium acetate trihydrate (SAT)). SAT is a particularly well-studied salt hydrate PCM, with a melting point of 58.1°C, and continues to gain research interest owing to its relatively high latent heat of fusion (up to 260 J/g), small phase change expansion coefficient, chemical stability, non-toxicity and low cost [7].

Recently, there has been an increased interest in PCMs made from renewable raw materials. Biobased PCM are organic substances that are derived from renewable resources and present excellent thermophysical properties while also exhibiting clear environmental advantages with respect to conventional PCM. Additionally, they are non-toxic and can be biodegradable [8]. The most widely investigated biobased PCM are sugar alcohols, fatty acids and esters [9] each one possessing distinct advantages and disadvantages.

Sugar alcohols (SA) occur naturally in various fruits and vegetables. Roughly more than 900 SA are known, but only a few are produced on the large scale (sorbitol, mannitol, xylitol, lactitol, malitol, erythritol and isomalt) [10]. They are generally non-flammable, non-corrosive, non-toxic, economically efficient and are widely used in the food and pharmaceutical industry [11]. They are very promising for thermal storage at medium temperature ranges (75°C–150°C) due to their environmental friendliness and their impressive latent heats of fusion (up to 340 J/g) which makes them one of the PCM classes with the highest energy densities [12].

Saturated fatty acids and some unsaturated ones have been studies widely in the previous years. They are naturally found in oils and algae and are non-toxic [13]. Saturated fatty acids are described by the general formula $CH_3(CH_2)_{2n}$ COOH. For PCM applications, n from 3 to 9 have been mostly investigated with melting points from 16°C to 74°C and enthalpies of fusion from up to 220 J/g [9]. They are mildly corrosive and have been reported to be thermally stable upon cycling.

Esters, which commonly occur in nature in vegetable and animal fats, are compounds formed from the union of a carboxylic acid and alcohol. Millions of esters exist, each with their own specific thermophysical property making them an impressively large, mostly unexplored class of possible PCM. Esters show little to no supercooling, high chemical and thermal stability and no corrosiveness and are considered only moderately flammable [14]. Recently, several classes of novel ester based materials were investigated revealing melting temperature ranges in the order of −50°C to 150°C and enthalpies of fusion up to 230 J/g [15–18] such as fatty acid esters, diesters and triglycerides. Particularly, triglycerides present exceptionally high enthalpies of fusion for organic PCMs (up to 220 J/g) and for a wide range of temperatures (−25°C to 90°C).

2.4 CHALLENGES

2.4.1 Segregation of Salt Hydrates

Salt hydrates are considered particularly promising in storage applications which require small temperature ranges, such as heat pump systems, offering high storage capacity at low volume and cost. Despite their clear advantages, the applicability of salt hydrates is limited so far mainly due to their bias toward segregation. So far, for most hydrated salt/water systems, no solutions for overcoming segregation for more than a few cycles exist.

An ideal crystallization process requires a composition of salt and water on molecular level that corresponds exactly to the composition of the desired hydrate. If different hydrates or even anhydrates of the salt are formed during the phase change, segregation normally occurs due to the different densities of the hydrated salts. Therefore, any deviation in concentration due to inaccurate mixing, addition or loss of moisture or effects occurring during cycling leads to segregation. Different mechanisms promote or prevent segregation which are influenced by a series of material properties. The mechanisms and properties are interconnected and are not fully understood yet. However, due to segregation, after multiple melting and solidification events, the salt hydrates separate in different phases; typically in a salt-rich phase at the bottom and a water-rich phase on top. Figure 2.5 illustrates four segregated samples of sodium sulfate decahydrate with different additives. One can clearly indicate a salt-rich phase at the bottom and a water-rich phase at the top.

This segregation of salt hydrates reduces the storage capacity significantly and may lead to the formation of different phases, whose phase transition temperatures are out of the range of the desired application and therefore limits the applicability of salt hydrates as thermal storage material substantially. Semi-congruent and incongruent-melting salt hydrates have a stronger tendency to segregate since lower hydrated solid salts, which normally have a higher density than the solution, are present in the melt.

In literature, physico-chemical strategies are suggested to avoid segregation in salt hydrates. These techniques often rely on gelling/thickening agents or by developing eutectic mixtures of salt hydrates by adding or replacing extra water, which experience congruent melting and, thereby, do display less segregation (more cycles).

For selected phase change materials, the addition of thickening agents, preventing density induced flow, to the aqueous solution helped to reduce irreversible segregation. These thickening agents are for example xanthan gum or carboxymethyl cellulose (CMC). However, adding thickening agents does not always solve the issue

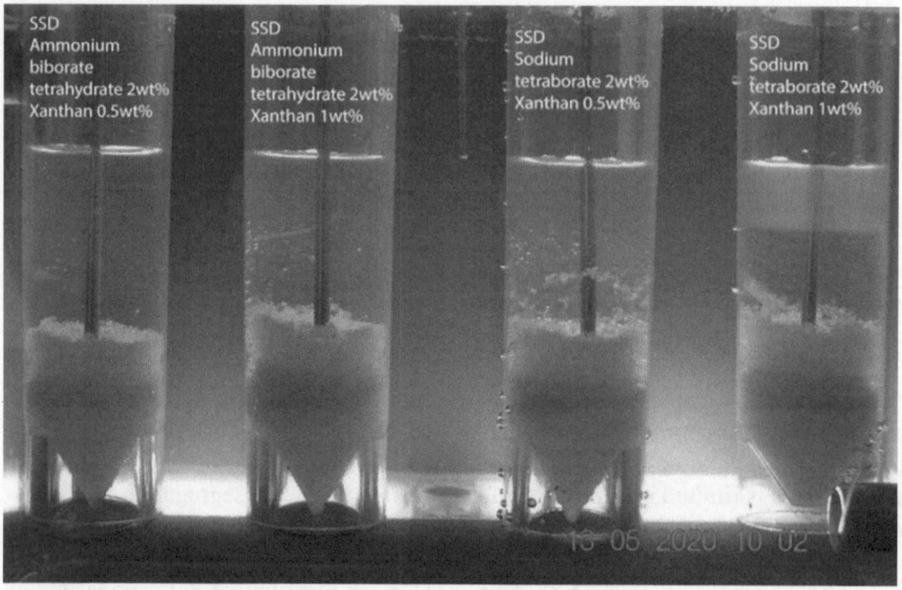

FIGURE 2.5 Photograph of thermally cycled and segregated salt hydrate samples of sodium sulfate decahydrate with different concentrations of thickening and nucleating agents taken in the analytical lab at the Lucerne University of Applied Sciences and Arts.

of segregation. It must be noted that the modification of segregating hydrated salt mixtures with thickeners has only been investigated with purely empirical methodologies based on trial and error. The exact mechanisms that allow for successful thickening in some materials but don't work in others haven't been elucidated in the framework of PCM research. Employment of multi-physics modeling as well as novel experimental techniques such as X-ray computed tomography which allows the *in situ*, non-invasive determination of the formation of new phases during thermal cycling, are expected to help in understanding the mechanisms of segregation and open new paths for the understanding of the key influencing factors and engineering of novel methodologies for the prevention of segregation.

2.4.2 Supercooling

Supercooling is a phenomenon that occurs in all types of materials and is defined as the temperature difference between the onset melting point of a material and the temperature of the material at which it starts to crystalize (nucleate) (see Figure 2.2). In most applications where PCM is used, a large degree of supercooling is undesirable. Supercooling of more than a few degrees will interfere with proper heat extraction from the storage unit or can even prevent it entirely. The degree of supercooling strongly depends on the material, the environment and the purity. Salt hydrates and SA in general exhibit a larger degree of supercooling than other classes of PCM.

The degree of subcooling depends on a number of factors, for example, volume of material in sample or application (e.g., volume of capsule in case of encapsulated

PCM); purity of the material, surface structure of surrounding material (e.g., presence of coarseness or irregularities).

A general lack of understanding on the mechanisms of preventing supercooling and initiating nucleation remains and supercooling continues being a limitation in the development of some types of PCM (e.g., salt hydrates). Thus, finding appropriate nucleating agents, which can withstand different operational conditions and are effective for the PCM in question is essential to glean information on the mechanisms behind heterogeneous nucleation and allowing for their implementation into LHS.

Some known methods that are employed to reduce supercooling are: (1) induction of mechanical stress, shock wave; (2) seeding (adding crystals of the PCM material); (3) addition of a nucleating agents to cause heterogeneous nucleation; (4) creation of a local cold spots to induce homogeneous nucleation at temperatures below the freezing point; and (5) electrical activation. Despite the plethora of existing methods to prevent supercooling, no global solution exists that works for all PCM and can be applied to all relevant applications. Similarly to other challenges encountered in this chapter, these solutions are based on trial-and-error approaches often without a deeper knowledge on the underlying mechanisms. Novel methods such as molecular modeling are expected to increase the understanding of the governing processes connected to nucleation control and to lead to the rational engineering of novel solutions.

2.4.3 Measurement of Thermophysical Properties

Determining the relevant thermophysical properties of a PCM to enable its suitability evaluation and system design might appear trivial at a first glance but is proven to be quite complicated in reality. A good example is the determination of the phase transition temperatures and enthalpies where standard, commercial measurement equipment exists (differential scanning calorimeter, DSC). Despite the misconception that such measurements are straightforward, round robin tests comprising DSC measurements of the same paraffin-based, substance from different institutions revealed very large discrepancies between the measured melting and crystallization curves [19,20]. These discrepancies can be traced back to a variety of influencing factors, such as device, heating/cooling rate, calibration method, sample size and type of crucible. The results were significantly improved by ensuring that all participants used a unified measurement methodology rather than the one they typically used in their laboratory but they still remained far from ideal. The efforts to further improve the DSC measurement methodologies and develop new standards that will be used by the entire PCM community is still ongoing [21].

Even more discrepancies are expected in measurements with substances that have a high tendency to supercool (e.g., salt hydrates, SA) and with mixtures rather than single components (multi-component PCM, mixtures containing additives). In these cases, an additional source of error is introduced by the very small sample size required by the DSC (in the order of some mg) as on the one hand supercooling increases with decreasing sample volume and on the other hand it is very challenging to acquire a representative DSC sample from non-homogeneous bulk sample. The latter issue becomes even more pronounced in the cases of incongruently or

FIGURE 2.6 From left to right: pictures of DSC, drop calorimeter and three-layer calorimeter available at Lucerne Universities of Applied Sciences and Arts for measurement of phase change temperature and enthalpy at different sample sizes (5–20 mg, 50 mL and 100 mL, respectively).

semi-congruently melting samples (most prominently salt hydrates) where it becomes very challenging to determine the properties of the pure material before it separates. In such cases often thickeners and gelling agents are added to the PCM before sampling for the DSC measurement which of course alter the properties of the materials but are often a necessary compromise to achieve an approximation of the phase change temperature and enthalpy of a segregating material.

For these reasons, several calorimetric techniques are being developed that can handle higher sample volumes (in the order of tens to hundreds of milliliters; see Figure 2.6). An example commonly used is three-layer calorimetry which is commercially available and has different types of customized setups based on the T-history and drop calorimetry method [22]. These methods face their own collection of challenges as the fact that most of them are developed in-house makes them susceptible to errors in design, setup development, operation, sample preparation, data collection and post-processing. It is apparent that there is a long way ahead toward developing accurate, precise and practically implementable calorimetric methods. The PCM community is continuously working on tackling these issues by improving and standardizing the development of the calorimetric setups and their data processing and developing new methods to determine calorimetric data on PCM. New commercial equipment is also becoming available to meet these challenges which is expected to also facilitate the measurement of calorimetric data in the future.

These challenges on measurement of thermophysical properties is not encountered solely on measurement of phase transition enthalpies and temperatures. Similar issues are encountered in the measurement of other important properties like thermal conductivity and viscosity [19].

2.4.4 Long-Term Stability and Material Compatibility Determination

The determination of the long-term stability of PCM is a topic of extreme importance as most PCM developers aspire for material lifetimes higher than 20 years. Rigorous testing of the long-term PCM performance stability is particularly crucial in materials that either show tendencies for chemical/thermal degradations (e.g., SA and other organic substances) or for physical separation (e.g., salt hydrates, other types of mixtures). The vast majority of PCM groups attempting to determine long-term stability of materials are choosing thermal cycling as the testing method. In this case the PCM, enclosed in a sample container, is placed in thermal cycling device that heats up and cools down the sample at temperatures resembling the ones expected in the foreseen application. Both the setups and the followed methodologies vary wildly among the different research groups. Some efforts for standardization have been made by RAL institute [23] albeit without describing the acceptable design of the thermal cycling setups or the methods to determine the material properties before, during and after cycling. An effort to summarize and compare different methodologies that are used by different research institutions was performed in the IEA ES Annex 33/SHC Task 58 "Material and Component Development for Compact Thermal Energy Storage" of the International Energy Agency (IEA) [24,25]. The work revealed that there is large deviation in the thermal cycling tests conducted by the different institutions in terms of temperature ranges, heating/cooling rates, sample sizes, choice of gaseous atmosphere surrounding the PCM, material container, and the type of temperature measurement performed. The same holds for the experimental techniques used to assess and quantify the degree of the possible degradation of the PCM. Some of the methods often used for this evaluation are the temperature curves from the cycling setup itself, DSC measurements on extracted samples, visual inspection, X-ray diffraction, etc. All these different methodologies and the lack of any standardization make the classification of materials in terms of long-term stability extremely challenging and comparison of results from different laboratories very complicated. Also, efforts from researchers to translate performed cycles to equivalent years of operation are of course disputable since the degradation mechanisms are typically not identified and no models are used that would allow such extrapolations [26]. PCM research would profit significantly from the development of standardized setups and procedures to conduct thermal cycling stability tests, taking into account of course that different types of materials and applications might have very different requirements.

A much more challenging but potentially very valuable alternative to thermal cycling is the design of accelerated life testing based on lifetime relationship models [26]. In such tests, the degradation factors are identified and their effect on the material stability are quantified and utilized to create lifetime relationships models. Tests are subsequently performed where the degradation factors are stressed, and the developed models are used to translate the results of the accelerated testing to the equivalent years of operation. One interesting example comes from the field of phase change dispersions (PCDs) where the phase change material is dispersed in a continuous phase, (e.g., organic PCM droplet dispersed in water) typically with the help of surfactants. In this case, a LUMiSizer device has been used for the accelerated lifetime tests. This device comprises a centrifuge able to expose samples to up to 2300 times the gravitational acceleration g (9.81 m/s^2) [27,28]. As gravity is considered the main performance degradation factor for PCDs, this allows for stress tests, simulating the effects of various years of shelf-life which is calculated by multiplying the measuring time with the applied gravitational acceleration within the centrifuge.

FIGURE 2.7 (a) LUMiSizer rotor chamber with rectangular sample cells (from Ref. [30]). (b) Segregated SAT sample (with thickener) after being subjected to 525 g for 6 hours.

During the measurement, transmission profiles are recorded which trace any alterations in concentration profiles within the PCD, indicating occurring instabilities and separation processes. With this method, the PCD stability after several months or years can be estimated with testing times of a few minutes or hours.

A similar centrifuge-based approach for accelerated lifetime tests could be used for other materials undergoing separation-related degradation. For example, a gravitational acceleration-based lifetime test for salt hydrate segregation could be conceivable to accelerate segregation processes. The rotor chamber of the LUMiSizer setup and the segregation resulting from subjecting a thickened SAT sample to an accelerated gravitational test employing 525 g for 6 hours is shown in Figure 2.7. Model-based accelerated tests would also be suitable for materials where the main degradation driver is chemical reaction in which case a kinetic model would be necessary to predict PCM properties after long periods of time [29].

Both methods to determine the lifetime of PCM have clear advantages and disadvantages. Thermal cycling is based on an application oriented, black-box approach which despite the lack of theoretical understanding has often shown to be a good indicator of material stability. Nevertheless, the lack of understanding of the basic mechanisms, lack of protocols and standardized tests and setups as well as the uncertainty that is associated with different researchers trying to interpret and replicate "applications conditions" can cause quite some nervousness to developers that are attempting commercialization of a product. There are in fact quite a lot of anecdotes about commercial PCMs that were proven to be instable during operation, presumably because the actual application conditions were quite different than the ones assumed by the producers [31]. Model-based approaches on the other hand, are much more time intensive and complex. Developing accurate models to describe a degradation mechanism can become extremely complicated, especially when chemical reactions are involved [29]. One wonders in which cases such an extensive effort is justified. Is it necessary to know the underlying mechanisms to be able to approximate lifetime stabilities in an acceptable accuracy or are carefully designed thermal cycling tests enough? As it is often the case, the truth probably lies somewhere in between but we expect to see several developments both in the standardization of thermal cycling tests and in the development of new model-based accelerating tests in the upcoming years.

A similar challenge is present in material compatibility determination where PCMs are tested with respect to their suitability to be used in combination with component materials they could come in contact within a PCM application (e.g., heat exchanger, encapsulation and/or tank material). A limited number of investigations can be found in literature exploring the compatibility of PCMs with various metals and plastics. There is no standardized methodology for material compatibility tests in the PCM community and most researchers develop their own testing procedure and definitions of the acceptable values with respect to corrosion rates (when it comes to metals) and swelling or extraction rates (when it comes to plastics) [32]. It is therefore of the essence, that global standards are developed and established that will allow for a uniform determination of the long-term compatibility of PCMs with metals and plastics over the entire envisioned lifetimes.

2.4.5 Polymorphism

Polymorphism describes the ability of a compound to develop different crystal conformations, of identical chemical compositions but different molecular arrangements. The most encountered natural process that is affected by polymorphism is the freezing of water. Nineteen ice polymorphs have been experimentally confirmed at present [33]. While the significance of polymorphism, is established for many scientific fields and classes of molecules, there are other classes that are deeply affected by polymorphism, where its influence remains completely unexplored.

Polymorphism is a challenge of both kinetic and thermodynamic nature encountered in esters, SA and paraffins alike [34–37]. In theory, any chemical can undergo polymorphism, but some compounds are more prone to polymorphism than others. It is known to influence the thermophysical properties of materials including melting point and enthalpy of fusion and to cause the appearance of several endothermic and exothermic events during phase transformations from one polymorph to another. An example of such events can be seen in a DSC measurement with triglyceride shown in Figure 2.8.

One characteristic example which was described in a recent review study is the polymorphism of triglycerides [38]. This phenomenon is relatively well documented because of their utilization in the food industry. Triglycerides are known to crystallize in three crystal phases, namely α, β' and β-forms.

A summary of the effect of the crystal formation on the enthalpy of fusion and peak melting point of saturated symmetrical triglycerides (SSTs) with different chain lengths is shown in Figure 2.9. It can be observed that the crystal phase affects the melting points leading up to a 50% reduction in the melting enthalpy in cases where the less desired transformation occurs. Generally, for SSTs it can be said that the formation of the β-form is the most desired one, as it leads to the highest enthalpies.

All these effects make the controlling of polymorphism highly relevant for the development of efficient LHS systems [37]. Despite this, it has only recently been recognized as an issue from the LHS community and has been the subject of only

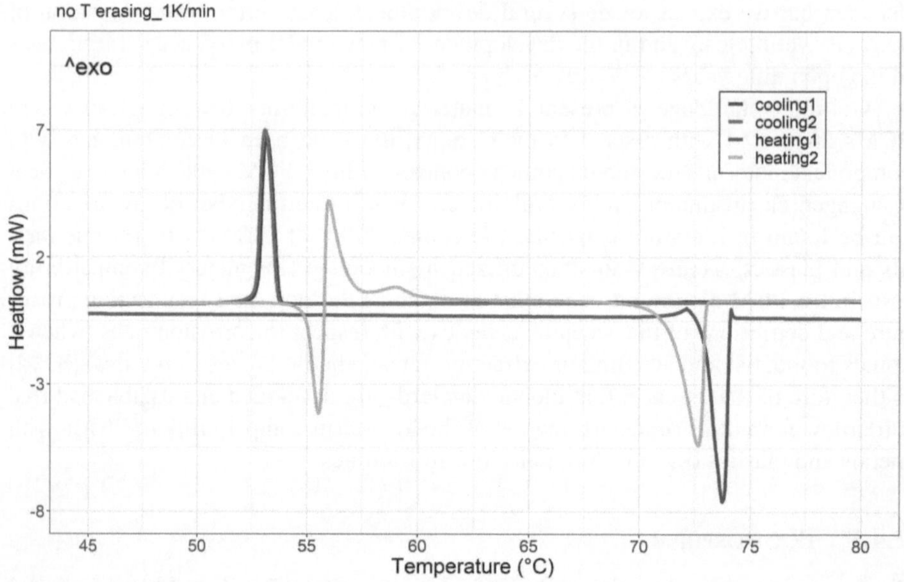

FIGURE 2.8 Heating and cooling curves from DSC measurements of tristearin at heating rates of 1 K/min.

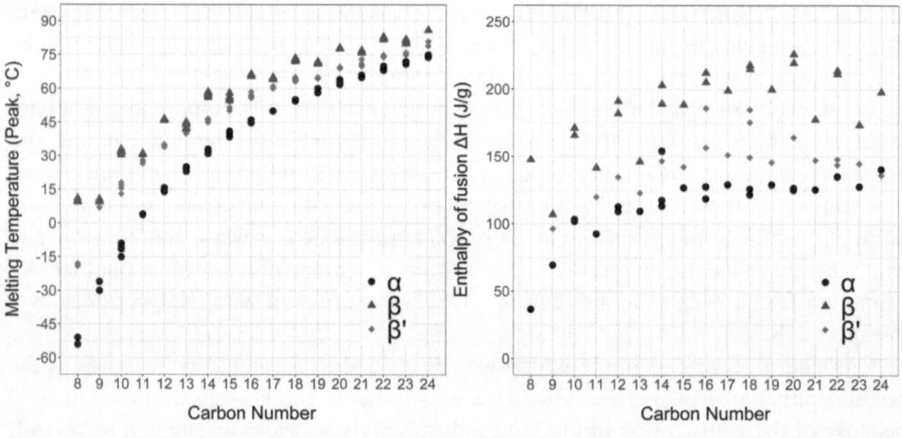

FIGURE 2.9 Summary of effect of carbon number and crystal form on peak melting temperature and enthalpy of fusion on SSTs [38].

a few isolated PCM-related investigations. Control of polymorphism presents a big challenge yet to be addressed and the understanding of polymorphic transformations from a kinetic point of view, is a necessary step to achieve this. Polymorphism, similarly, to other kinetic effects, is influenced by several factors such as thermal history of the sample, impurities, heating and cooling rates, absolute temperatures that the material is subjected to. Thus, controlling polymorphism would require understanding the influence of these factors and careful control of the PCM pre-conditioning and treatment during operation. Designing of additives to promote the desired crystal transformation is a promising approach to achieve reproducible formation of the appropriate crystal phase and therefore to ensure that the PCM exhibits the properties required by the application in a stable manner.

2.4.6 Determination of Crystallization Kinetics and Kinetic Modeling

As mentioned in the previous sections, PCMs are subject to several kinetic limitations that have adverse effects on their performance (e.g., supercooling, polymorphism, slow crystallization kinetics). There is an urgent need for understanding these phenomena for the development of novel, high-performance PCM which can be only achieved by better understanding of the governing mechanisms. This gap in knowledge needs to be addressed by establishing novel methodologies to study the fundamental mechanisms of melt crystallization and developing models that can describe the nucleation and growth processes.

Very few detailed kinetic investigations have been dedicated to characterizing and modeling the crystallization processes of PCM. Consequently, most of the heat and mass transfer models employed to describe latent storage systems consider solidification as a heat transfer-limited process and kinetic effects are neglected. While this is a fair assumption for many PCM systems with fast crystallization rates, it can lead to large errors in simulations of LHS containing PCM that exhibit slow nucleation [39] or growth rates [34] and even more importantly, polymorphism. Attempts have been made to model crystallization rates using macroscopic models such as the apparent specific heat capacity formulation. These are black-box models that can describe the crystallization rate in isolated experiments, but are not connected to the crystallization mechanisms and therefore they are not expected to respond well to any change in the experimental conditions [40]. Waser et al. [39] attempted to describe the solidification behavior of an LHS prototype by combining a crystallization kinetic model with a heat and mass transfer model. The authors implemented a probability function, to describe the heterogeneous nucleation rate as a function of the supercooling degree and proximity to an existing crystal. The approach was based on the classical nucleation theory formulation and was implemented into a computational fluid dynamics (CFD) model of a finned tube LHS to account for the effect of the PCM supercooling to the LHS performance. Despite the kinetic model being based on macroscopic observations with several fitted parameters, the approach was considered promising because it was able to describe the effect of supercooling on the evolution of the outlet temperature of the heat transfer fluid (see Figure 2.10).

2.4.7 Thermal Conductivity Enhancement

It is known that low to medium temperature PCMs suffer from a very low heat conductivity: in the order of 0.1–0.4 W/(m·K) for paraffins, fatty acids and esters (solid state); 0.4–0.8 for SA (solid state) and 0.7–1.4 for salt/water mixtures (solid state). This of course has a strong influence on the overall achievable heat transfer rates during charging and discharging of the storage and can make utilization of PCMs for application that require high power input/output (less than 1 hour charging and

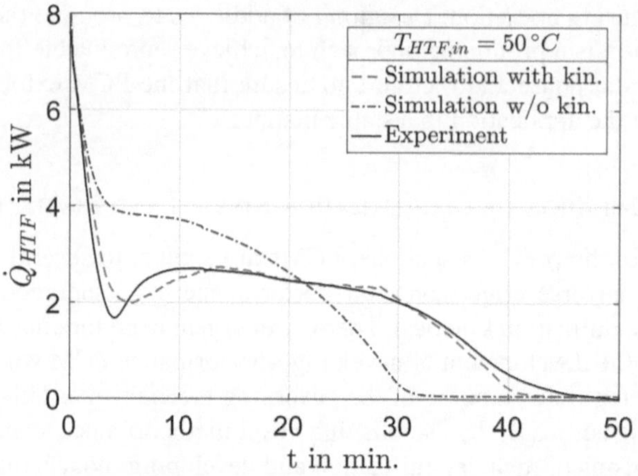

FIGURE 2.10 Experimentally measured and numerically simulated heat flow rate during discharging of a latent heat storage kinetic effects (dash) or neglecting them (dot-dash) [39].

discharging time) quite challenging. There are two main categories of approaches to circumvent this problem. On the one hand, components with high surface areas and made from highly conductive materials (e.g., metallic finned tube heat exchangers and foams) can be incorporated in the LHS to increase the effective thermal conductivity of the system. On the other hand, highly conductive additives can be employed (e.g., expanded graphite, metallic nanoparticles) to increase the effective thermal conductivity of the material level. While the former is a more mature method that has proven to work in a stable manner, the latter if proven to be effective could potentially be a more cost-effective solution involving a less complicated manufacturing process. The modification of the effective thermal conductivity of PCMs using additives has been extensively studied in literature [41], unfortunately with very large discrepancies between the thermal conductivity values reported by different researchers. The root causes of the discrepancies can be various: sample instabilities (segregation of conductive additives with time), sample inhomogeneities (effect of sampling location and method), thermal conductivity measurement device/methodology, exact properties of additives (e.g., porosity of expanded graphite), and utilization or not of additional additives (e.g., thickeners). Despite of the very promising results reported by some of the studies (e.g., ninefold increase in the thermal conductivity with 5.5% addition of expanded graphite [42]), the large inconsistencies between studies have led to a dispute in the scientific community about the effectiveness and stability of the method. However, we can expect that the efforts of the scientific community to shed light to this topic will continue in the next years with more standardized methodologies and conclusive results.

2.4.8 Identification of Novel PCM

Despite the large number of PCMs already investigated, there is a vast number of possible PCMs, the properties of which remain unknown. This concerns classes of materials that haven't been considered as PCM yet (e.g., ionic liquids), classes that are too vast and with little commercial availability (esters) as well as mixtures such as blends of different salts and water. While a purely experimental screening of these classes is associated with immense effort in synthesis/mixing of substances and intensive characterization, developments of theoretical and numerical nature are expected to enable a very efficient pre-screening of novel substances and mixtures that will facilitate the efficient identification of new PCMs. Such methods included the application of modified BET models to predict properties of mixtures based on hydrated salts [43] and the application of machine learning methods to screen through esters which are not commercially available. Machine learning methodologies have not yet been applied to latent storage applications but promising results demonstrated in other fields opening new possibilities also in the PCM topic [44].

2.5 CONCLUSIONS AND OUTLOOK

Despite the long history of PCM research, there are still many challenges on a fundamental and practical level and many open questions regarding the effects governing the performance of PCM and also the methods to determine their properties. While all of these challenges pose a problem in the development process of PCM, their identification also presents an opportunity to enable a more rational and faster development of high-performance PCM by a better coordination/communication within the PCM community and utilization of novel methods/tools to advance the scientific knowledge. Despite the challenges, in the last years the scientific community has managed to develop a number of high-performance PCM and a lot of companies have emerged commercializing PCM or systems comprising them. In the authors' opinion, the path forward should combine one the one hand fundamental research to enable the deeper understanding of the root causes for the identified challenges and on the other hand applied research and commercialization to guarantee that LHS can play its role in the energy turnaround.

REFERENCES

1. M. Wuttig, N. Yamada, Phase-change materials for rewriteable data storage, *Nat. Mater.* 6 (2007) 824–832.
2. E.P. Del Barrio, R. Cadoret, J. Daranlot, F. Achchaq, New sugar alcohols mixtures for long-term thermal energy storage applications at temperatures between 70°C and 100°C, *Sol. Energy Mater. Sol. Cells.* 155 (2016) 454–468.
3. A. Genovese, G. Amarasinghe, M. Glewis, D. Mainwaring, R.A. Shanks, Crystallisation, melting, recrystallisation and polymorphism of n-eicosane for application as a phase change material, *Thermochim. Acta.* 443 (2006) 235–244.
4. L.J. Fischer, N7 Phasenwechselmaterialien (PCM) für Latent-Wärmespeicher, in: *VDI-Wärmeatlas: Fachlicher Träger VDI-Gesellschaft Verfahrenstechnik und Chemieingenieurwesen*, Springer, Berlin, Heidelberg, 2019: pp. 1989–2008.
5. K. Pielichowska, K. Pielichowski, Phase change materials for thermal energy storage, *Prog. Mater. Sci.* 65 (2014) 67–123.
6. C. Voelker, O. Kornadt, M. Ostry, Temperature reduction due to the application of phase change materials, *Energy Build.* 40 (2008) 937–944.

7. D.E. Oliver, A.J. Bissell, X. Liu, C.C. Tang, C.R. Pulham, Crystallisation studies of sodium acetate trihydrate – suppression of incongruent melting and sub-cooling to produce a reliable, high-performance phase-change material, *CrystEngComm.* 23 (2020) 700–706.
8. L. Liu, X. Fan, Y. Zhang, S. Zhang, W. Wang, X. Jin, B. Tang, Novel bio-based phase change materials with high enthalpy for thermal energy storage, *Appl. Energy.* 268 (2020) 114979.
9. J. Lizana, R. Chacartegui, A. Barrios-Padura, J.M. Valverde, C. Ortiz, Identification of best available thermal energy storage compounds for low-to-moderate temperature storage applications in buildings, *Mater. Construcción.* 68 (2018) 160.
10. H. Schiweck, A. Bär, R. Vogel, E. Schwarz, M. Kunz, C. Dusautois, A. Clement, C. Lefranc, B. Lüssem, M. Moser, S. Peters, R.O. Vogel, E. Merck, E.U.S. Chwarz, E. Merck, Sugar alcohols, in: *Ullmann's Encyclopedia of Industrial Chemistry*, Wiley, 2012: pp. 2–32.
11. V. Pethurajan, S. Sivan, A.J. Konatt, A.S. Reddy, Facile approach to improve solar thermal energy storage efficiency using encapsulated sugar alcohol based phase change material, *Sol. Energy Mater. Sol. Cells.* 185 (2018) 524–535.
12. E.P. del Barrio, A. Godin, M. Duquesne, J. Daranlot, J. Jolly, W. Alshaer, T. Kouadio, A. Sommier, Characterization of different sugar alcohols as phase change materials for thermal energy storage applications, *Sol. Energy Mater. Sol. Cells.* 159 (2017) 560–569.
13. J.A. Noël, P.M. Allred, M.A. White, Life cycle assessment of two biologically produced phase change materials and their related products, *Int. J. Life Cycle Assess.* 20 (2015) 367–376.
14. R. Burke, *Hazardous Materials Chemistry for Emergency Responders*, CRC Press, Boca Raton, FL, 2013.
15. R. Ravotti, O. Fellmann, N. Lardon, L.J. Fischer, A. Stamatiou, J. Worlitschek, Investigation of lactones as innovative bio-sourced phase change materials for latent heat storage, *Molecules.* 24 (2019) 1300.
16. R. Ravotti, O. Fellmann, N. Lardon, L.J. Fischer, A. Stamatiou, J. Worlitschek, Synthesis and investigation of thermal properties of highly pure carboxylic fatty esters to be used as PCM, *Appl. Sci.* 8 (2018) 1069–1087.
17. R. Ravotti, N. Lardon, A. Stamatiou, J. Worlitschek, L. Fischer, O. Fellmann, Analysis of bio-based fatty esters PCM's thermal properties and investigation of trends in relation to chemical structures, *Appl. Sci.* 9 (2019) 225.
18. A. Stamatiou, M. Obermeyer, L.J. Fischer, P. Schuetz, J.J. Worlitschek, Investigation of unbranched, saturated, carboxylic esters as phase change materials, *Renew. Energy.* 108 (2017) 401–409.
19. S. Gschwander, A. Lazaro, L.F. Cabeza, E. Günther, M. Fois, J. Chui, *Development of a Test-Standard for PCM and TCM Characterization Part 1: Characterization of Phase Change Materials*, International Energy Agency (IEA), Paris, 2011.
20. S. Gschwander, T. Haussmann, G. Hagelstein, A. Solé, L.F. Cabeza, G. Diarce, W. Hohenauer, D. Lager, A. Ristic, C. Rathgeber, P. Hennemann, H. Mehling, C. Peñalosa, A. Lázaro, Standardization of PCM characterization via DSC, in: *13th International Conference on Energy Storage*, Greenstock, 2015.
21. L. Müller, G. Rubio-Pérez, A. Bach, N. Muñoz-Rujas, F. Aguilar, J. Worlitschek, Consistent DSC and TGA methodology as basis for the measurement and comparison of thermophysical properties of phase change materials, *Materials (Basel).* 13 (2020) 4486.
22. H. Schmit, C. Rathgeber, P. Hoock, S. Hiebler, Critical review on measured phase transition enthalpies of salt hydrates in the context of solid-liquid phase change materials, *Thermochim. Acta.* 683 (2020) 178477.

23. RAL Institute, Quality & Testing Specifications for Phase Change Materials, RAL Deutsches Institut für Gütesicherung und Kennzeichnung e.V., 2018. http://www.pcm-ral.org/pcm/en/quality-testing-specifications-pcm/.
24. IEA ES Annex 33/SHC Task 58, Experimental devices to investigate degradation of PCM, 2020. https://task58.iea-shc.org/.
25. C. Rathgeber, S. Hiebler, R. Bayón, L.F. Cabeza, G. Zsembinszki, G. Englmair, M. Dannemand, G. Diarce, O. Fellmann, R. Ravotti, D. Groulx, A.C. Kheirabadi, S. Gschwander, S. Höhlein, A. König-Haagen, N. Beaupere, L. Zalewski, Experimental devices to investigate the long-term stability of phase change materials under application conditions, *Appl. Sci.* 10 (2020) 1–30.
26. R. Bayón, E. Rojas, Development of a new methodology for validating thermal storage media: Application to phase change materials, *Int. J. Energy Res.* 43 (2019) 6521–6541.
27. L.J. Fischer, S. Von Arx, U. Wechsler, S. Züst, J. Worlitschek, Phase change dispersion properties, modeling the specific heat capacity, *Int. J. Refrig.* 74 (2017) 240–253.
28. T. Detloff, T. Sobisch, D. Lerche, Instability index, *Dispers. Lett. Tech.* T4 (2013) 1–4.
29. R. Bayón, A. Bonanos, E. Rojas, Assessing the long-term stability of fatty acids for latent heat storage by studying their thermal degradation kinetics, in: *EuroSun 2020 Proceedings*, 2020.
30. LUMiSizer, 2020. https://www.lum-gmbh.com/lumisizer_en.html (accessed May 22, 2020).
31. P. Tan, P. Lindberg, K. Eichler, P. Löveryd, P. Johansson, A.S. Kalagasidis, Effect of phase separation and supercooling on the storage capacity in a commercial latent heat thermal energy storage: Experimental cycling of a salt hydrate PCM, *J. Energy Storage.* 29 (2020) 101266.
32. A. Stamatiou, L. Müller, R. Zimmermann, J. Hillis, D. Oliver, K. Fisher, M. Zaglio, J. Worlitschek, Experimental characterization of phase change materials for refrigeration processes, *Energies.* 14 (2021) 3033.
33. T. Loerting, V. Fuentes-Landete, C.M. Tonauer, T.M. Gasser, Open questions on the structures of crystalline water ices, *Commun. Chem.* 3 (2020) 1–4.
34. M. Duquesne, A. Godin, E. Palomo Del Barrio, F. Achchaq, Crystal growth kinetics of sugar alcohols as phase change materials for thermal energy storage, *Energy Procedia.* 139 (2017) 315–321.
35. A. Godin, M. Duquesne, E. Palomo Del Barrio, J. Morikawa, Analysis of crystal growth kinetics in undercooled melts by infrared thermography, *Quant. Infrared Thermogr. J.* 12 (2015) 237–251.
36. M. Duquesne, A. Godin, E. Palomo del Barrio, J. Daranlot, Experimental analysis of heterogeneous nucleation in undercooled melts by infrared thermography, *Quant. Infrared Thermogr. J.* 12 (2015) 112–126.
37. C. Barreneche, A. Gil, F. Sheth, A. Ine, L.F. Cabeza, Effect of D -mannitol polymorphism in its thermal energy storage capacity when it is used as PCM, *Sol. Energy.* 94 (2013) 344–351.
38. A. R. Ravotti, Worlitschek, J., Pulham, C., Stamatiou, Triglycerides as novel phase-change materials: Thermal properties, *Molecules.* 25 (2020) 5572.
39. R. Waser, S. Maranda, A. Stamatiou, M. Zaglio, J. Worlitschek, Modeling of solidification including supercooling effects in a fin-tube heat exchanger based latent heat storage, *Sol. Energy.* 200 (2018) 1–12.
40. T. Davin, B. Lefez, A. Guillet, Supercooling of phase change: A new modeling formulation using apparent specific heat capacity, *Int. J. Therm. Sci.* 147 (2020) 106121.
41. Y. Lin, Y. Jia, G. Alva, G. Fang, Review on thermal conductivity enhancement, thermal properties and applications of phase change materials in thermal energy storage, *Renew. Sustain. Energy Rev.* 82 (2018) 2730–2742.

42. X. Gu, S. Qin, X. Wu, Y. Li, Y. Liu, Preparation and thermal characterization of sodium acetate trihydrate/expanded graphite composite phase change material, *J. Therm. Anal. Calorim.* 125 (2016) 831–838.
43. C. Rathgeber, H. Schmit, S. Hiebler, W. Voigt, Application of the modified BET model to concentrated salt solutions with relatively high water activities: Predicting solubility phase diagrams of $NaCl+H_2O$, $NaCl+LiCl+H_2O$, and $NaCl+CaCl_2+H_2O$, *Calphad.* 66 (2019) 101633.
44. S. Chibani, F.-X. Coudert, Machine learning approaches for the prediction of materials properties, *APL Mater.* 8 (2020) 080701.

3 Experimental Techniques and Challenges in Evaluating the Performance of PCMs

S. Mancin, M. Calati, and D. Guarda
University of Padua

CONTENTS

3.1 INTRODUCTION

This chapter focuses on the experimental techniques used by the different researchers to evaluate the performance of the phase change materials (PCMs). When looking to the literature, it is possible to subdivide the works into two main categories: fundamental studies and applied ones.

The first category refers to those works that experimentally characterize the behavior of the PCMs when embedded in different kinds of heat exchangers (e.g. shell and tubes, macrocapsules, plate heat exchangers) or enhanced surfaces (e.g. fins, porous media) in terms of: temperature distribution, liquid fraction evolution, stored energy, instant power, charging and discharging times, etc. The term fundamental refers to the fact that the results collected for the specific PCM can be, in general, extended to others but do not refer to any specific applications. These kinds of experiments can involve small amount (tens of grams) or large amount (tens to hundreds of kilograms) of PCMs adopting many different experimental methodologies.

The second category gathers the studies in which the PCM is coupled to a heat transfer solution to satisfy the specific requirements of a particular application:

DOI: 10.1201/9781003213260-3

thermal management of electronics or electrical batteries, building cooling of heating, refrigeration, solar or other exotic applications. In these cases, the experimental methods aim at evaluating the performance of the system that commonly is compared against a reference case which operates without PCM. Each application presents its own specificities and requirements, and thus the experimental techniques vary accordingly.

This chapter is subdivided into two main sections. The first discusses the fundamental studies trying to show the different experimental methodologies used, highlighting advantages and disadvantages of those methods. Specific experimental tips are also included to guide the reader into the application of the most accurate experimental techniques. The second section presents the different applications focusing on the most important performance parameters to be evaluated and the experimental methods implemented by the different authors trying to highlight pros and cons of those approaches.

3.2 FUNDAMENTAL STUDIES

As already stated, the fundamental studies are meant to understand the underlining heat and mass transfer mechanisms at the basis of the solid–liquid phase change process. It is well known that the thermal conductivity of most of the PCMs is rather low and it does not permit an efficient heat transfer during the charging and discharging processes. Thus, many of the fundamental studies aim at investigating the improvement achievable when enhanced surfaces (e.g. fins, 3D periodic structures, porous media, and macro-/micro-encapsulations) are used to increase the heat transfer capabilities of the PCMs [1–16].

These fundamental studies implement different experimental techniques to study the phase change process even if the performance of the PCMs is usually evaluated as a function the following parameters: the temperature distribution inside the PCM, the heat transfer rate exchanged, the liquid fraction evolution, the melting/solidification (i.e. charging/discharging) time, and the stored energy. One of the main challenges of these kinds of experiments is represented by the intrusive nature of the most used techniques, which sometimes may affect the phase change process, especially in solidification. Considering the temperature distribution inside the PCM, this measurement can be used to have a general overview of the phase change process or to detect the liquid fraction. In fact, when the melting front passes through the temperature probe, the value of the temperature suddenly increases and, thus, it is possible to identify the position of the solid–liquid interface at a certain time. This methodology can only be used if the temperature probes are uniformly distributed in the PCM volume.

Seddegh et al. [9] used this technique to determine the liquid fraction in single pass shell and tube heat exchanger. Four cylinders of 0.5 m height and 0.0512 m shell radius with different inside tubes were studied. The PCM was poured into the annulus and eight T-type thermocouples were used to measure its temperature. Four TCs were located 20 mm far from the outer surface of the inner tube at different heights while the other four were positioned only 5 mm far from the outer surface of the inner tube at the same heights.

Hot and cold water was used for charging and discharging phases, respectively; the water temperature at the inlet and outlet of the tested tubes as well as the water flow rate were measured. From the measurements of the temperature made by the T-type thermocouples, the authors were able to estimate the liquid fraction by implementing a computational procedure, called "weighting method".

A similar work was carried out by Kamkari and Shokouhmand [4] who coupled the direct visualization of the phase change using a video camera with the temperature field reconstruction using thin thermocouples. In fact, the authors positioned 32, 30, and 26 inside the PCM of three samples: unfinned, 1-fin, and 3-fin, respectively. Figure 3.1a shows the temperature measuring grids created by the authors, while Figure 3.1b reports the images and the temperature fields recorded during the melting process at different imposed wall temperatures.

The easiest way to follow qualitatively the melting front is the actual visualization of the phase change process as done by several authors [1,5,8,12,14,15,17–22].

For example, Mancin et al. [15] experimentally studied the performance of different copper foams to improve the heat transfer capabilities of paraffin waxes for efficient thermal energy storages. The authors used a video camera to visualize the melting process and four T-type thermocouples to monitor the temperature field inside the composite material made by the PCM embedded in the copper foams. An example of the collected results is reported in Figure 3.2.

The tested samples were heated from the bottom by means of an electrical cartridge and the heat losses were estimated to be less than 5%.

One of the most critical issues of the heat transfer experiments is how to account for the energy losses during the experiments; in fact, in the case of steady state experimental tests, it is always possible to measure the heat transfer rate in two different ways (e.g. cold and hot fluid or electrical power and heat transfer fluid) and then compare the two measurements through a thermal balance. In the case of solid–liquid phase change, the tests are transient, and the heat supplied to the PCM is stored as latent heat, thus it might be difficult to estimate the heat losses but this value is of fundamental importance since it might affect the estimation of the whole heat stored by the PCM, which is one of the most interesting parameters when comparing different storage solutions.

Righetti et al. [14] conducted a series of experiments to study the effect of 3D periodic structures on the solid–liquid phase change of different PCMs. The authors designed a simple setup which consists of a heater which is located beneath the sample. Several ultra-thin T-type thermocouples were installed inside the PCM box to measure the temperature distribution inside the PCM. The authors also developed an original methodology to estimate the heat losses to the surroundings during the test with a dedicated experimental test, which used an aluminum block with the same dimensions as the investigated samples: 42 mm × 42 mm × 60 mm. A T-type thermocouple was inserted in the middle of a lateral face in a 20 mm deep hole, drilled 30 mm below the upper face (i.e. in the sample center). Another T-type thermocouple was inserted in the center of the top surface a 1 mm deep hole. The top surface was then insulated with 25 mm thick rock wool layer in order to mimic the same operating condition of the PCM samples. Four additional T-type thermocouples were located in as many 0.5 mm deep holes drilled in the center of the lateral faces.

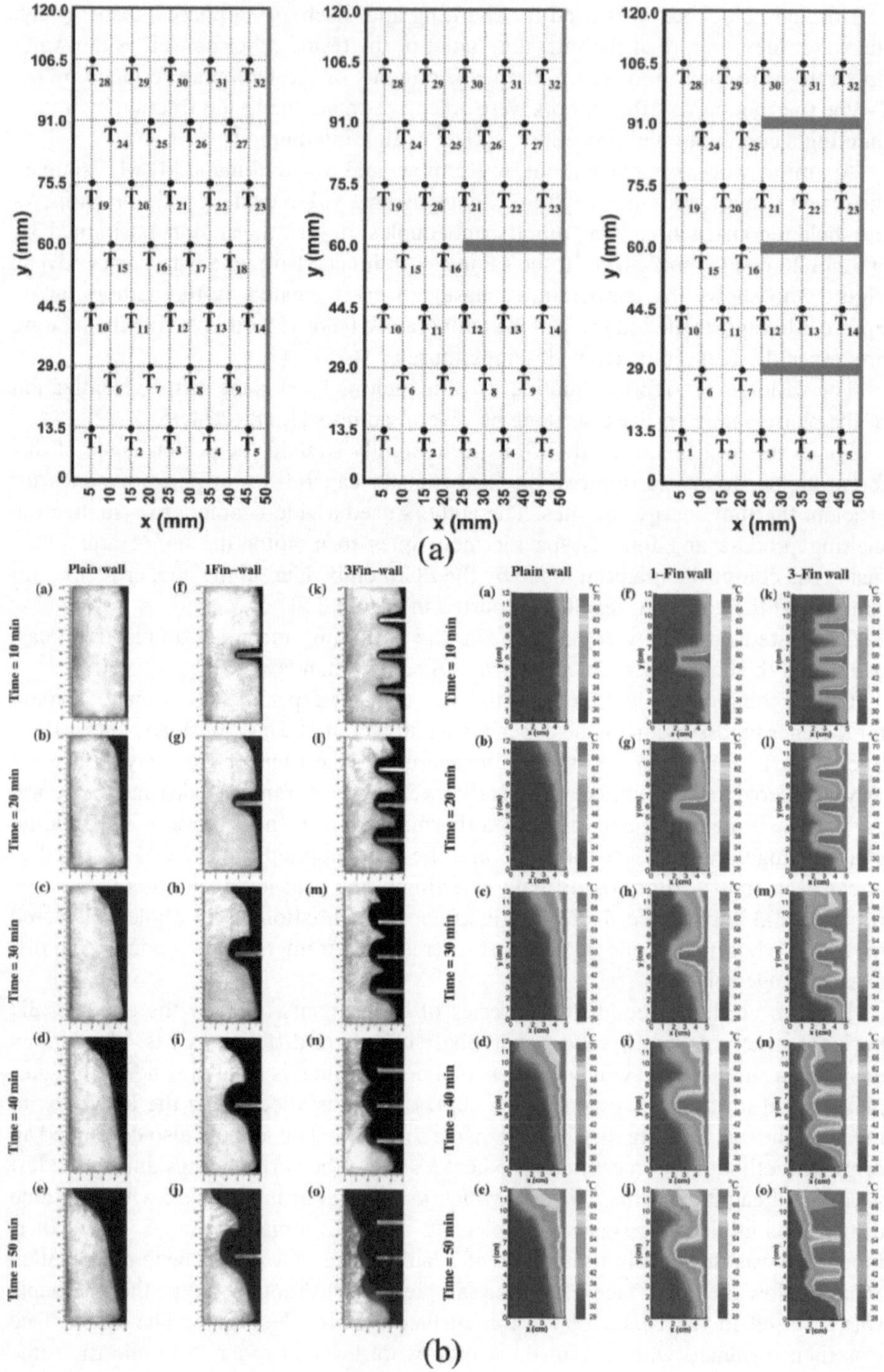

FIGURE 3.1 (a) Temperature measuring grids. (b) Visualization of the melting front and temperature contours. (Adapted from Kamkari and Shokouhmand [4]).

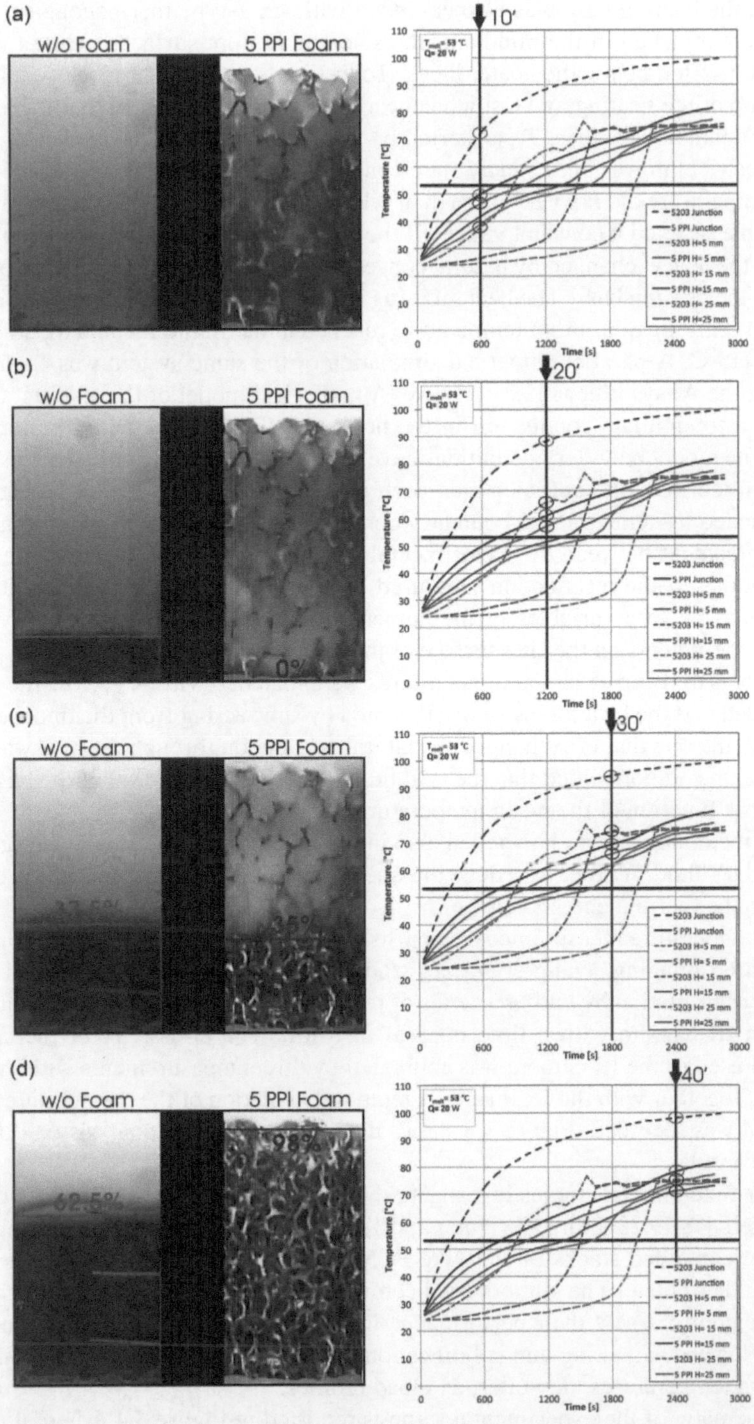

FIGURE 3.2 Visualization of the melting front and temperature profiles. (Adapted from Mancin et al. [15]).

Besides, the heating base was also equipped with six T-type thermocouples, five of those were attached in the middle of the side and bottom surfaces, while the sixth one was inserted below the heater block. To test it, the aluminum block was located on the top of the heating and a silicone heat transfer paste was used to minimize the thermal contact resistance. Tests started by switching on the cartridge heater with a fixed electrical power until steady state conditions were achieved (i.e. all the monitored temperature did not vary more than 0.1 K over 5 min). When the temperature of the sample reached a constant value, all the supplied electric power was required to balance the heat exchanged by natural convection to the surrounding plus the one lost through the heating base. The heat flux was increased, and the testing procedure was repeated until the aluminum temperature reached a maximum mean temperature of around 115°C. A parallel numerical simulation of the same system was carried out by using the Ansys Fluent 18.2 software. A numerical model of the test section was generated to simulate conduction in the heater block assembly and the aluminum block. The model boundary conditions have a constant heat flux applied at the heater block and convection coefficients on each external surface. A formal procedure is implemented to iterate on the boundary conditions in the model in order to achieve a fair agreement between the experimental and numerical values of temperature at the locations in the test section measured by thermocouples. For each calibration data point, the free variables in the numerical simulation are the convective heat transfer coefficients on the side surface of the aluminum block and insulation. When all the measured experimental temperatures were matched within ±0.1 K, the procedure ended and the heat losses were calculated by subtracting from the imposed heat flow rate, the one that is exchanged by natural convection through the side walls. As expected, the authors found that the heat flux lost into the insulation of the basement varied as a function of the mean temperature.

This methodology can be applied also in the case of non-electrical heating as for a secondary fluid in order to reduce the uncertainty on the estimation of the energy stored in the thermal energy storage.

The temperature measurements can also be done using a non-intrusive technique: the infrared imaging. Wang et al. [26] studied the melting process inside different macro-encapsulation by taking a series of infrared (IR) images of the PCM capsules using an IR camera with a time interval of 5 min. The emissivity of the capsule material used in the IR camera was calibrated by direct measurements with thermocouples. Together with the IR images, a temporal evolution of the temperature inside the PCM was recorded. Figure 3.3 shows the results of the IR analysis on different type of capsules.

Other fundamental studies investigate the performance of different types of heat exchangers [18,19,23–36]. D'Avignon and Kummert [36] studied a horizontal PCM tank that contained stacks of slab-like PCM capsules between which heat transfer fluid could circulate. The authors used commercially available capsules which were instrumented to assess their performance in addition to that of the PCM tank as a whole. Numerous melting and solidification cycles were carried out by varying the inlet fluid temperatures, flowrates, and load profiles.

A schematic of the experimental setup is reported in Figure 3.4, where it can be seen that the auxiliary loops were used in series to service the sole inlet and outlet

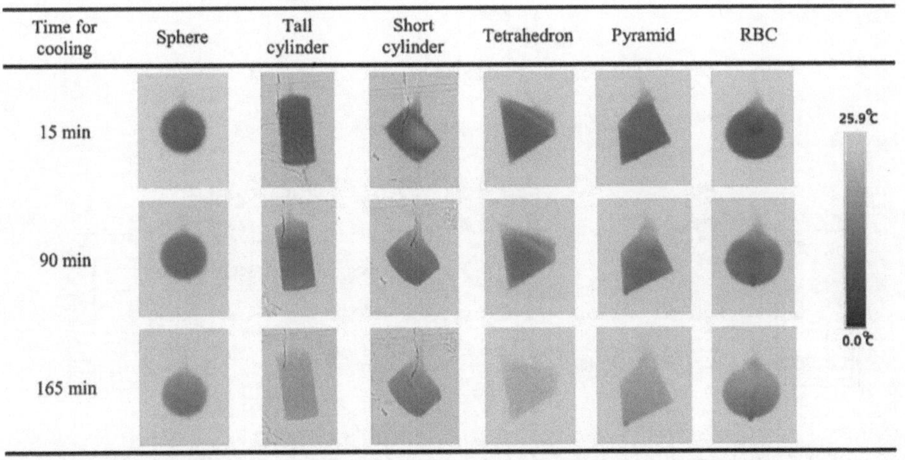

FIGURE 3.3 Infrared camera images of capsules during melting in natural convection with air. (Adapted from Wang et al. [26]).

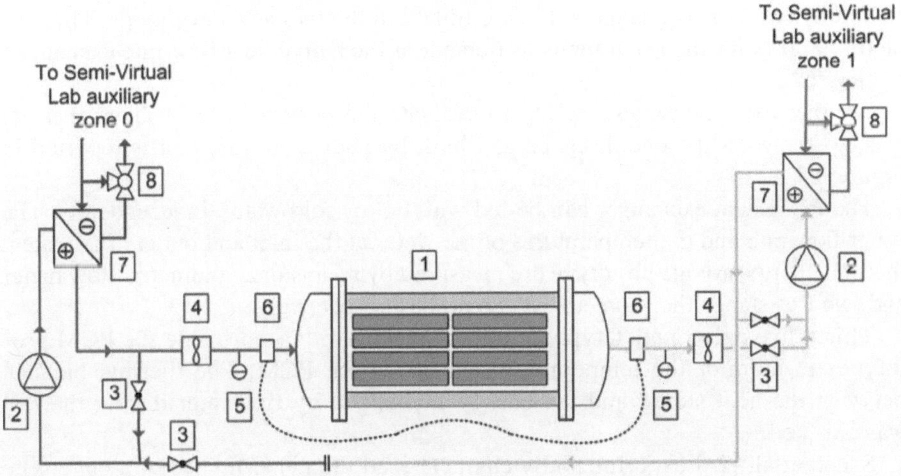

FIGURE 3.4 Schematic of the experimental setup. (Adapted from D'Avignon and Kummert [36]).

to the PCM tank. Each test loop is equipped with its own variable speed pump permitting flowrates of up to 7.6 L/s. In this case, balancing valves were used, so part of the flow would bypass the tested equipment in order to reduce the flowrate imposed to the PMC tank without affecting the pump's performance. Turbine flow meters with magnetic pick-up measured the flowrate inside the PCM tank with an uncertainty of ±1% of the reading. A set of calibrated PT-100 platinum resistance sensors and a custom-made thermopile were installed at the tank's inlet and outlet in order to accurately measure the fluid temperature imposed on the equipment and

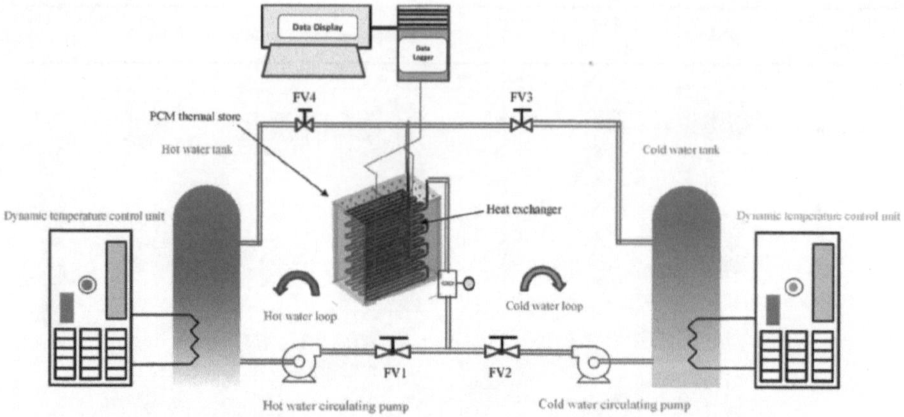

FIGURE 3.5 Schematic of the experimental setup. (Adapted from Fadl et al. [31]).

its response. The uncertainties on their measurements are ±0.165 K for the platinum sensors and ±0.04 K for the thermopile. The temperature of each test loop is controlled by two heat exchangers that are linked to the two auxiliary loops. This test facility allows for the accurate measurement of the actual heat flow rate exchanged by the PCM tank.

Another example of test facility to evaluate the performance of latent thermal energy storage is the one designed and built by Fadl et al. [31] that is reported in Figure 3.5.

The PCM heat exchanger can be fed with hot or cold water, independently. The water flow rate and the temperatures of the water at the inlet and outlet of the latent thermal energy storage prototype are measured by means of a volumetric flow meter and two Resistance thermometer (RTD) probes, respectively.

Thirty-five calibrated T-type thermocouples were placed inside the PCM prototypes to monitor the temperature field inside the PCM. The thermal balance between the heat stored and the energy exchanged by the water during the test was checked.

Saeed et al. [24] experimentally characterized the performance of a plate type heat exchanger which used roll bond heat exchangers inserted inside the PCM. Also in this case, two independent water loops can feed the energy storage, which consists of 20 plate heat exchangers connected in parallel. The water temperatures at the inlet and outlet of the storage are measured as well as the water flow rate in both loops. Figure 3.6 shows two drawings of the plate type heat exchangers. The authors performed a detailed analysis of the performance comparing the energy stored inside the PCM with the energy exchanged by the water including also the heat losses.

Finally, another interesting experimental work is the one published by Li et al. [23], who conducted an experimental investigation of the discharging process of direct contact thermal energy storage. In direct contact thermal energy storages, the PCM is not separated from the working fluid and the heat transfer occurs in direct contact.

FIGURE 3.6 Plate type heat exchanger for latent thermal energy storage. (Adapted from Saeed et al. [24]).

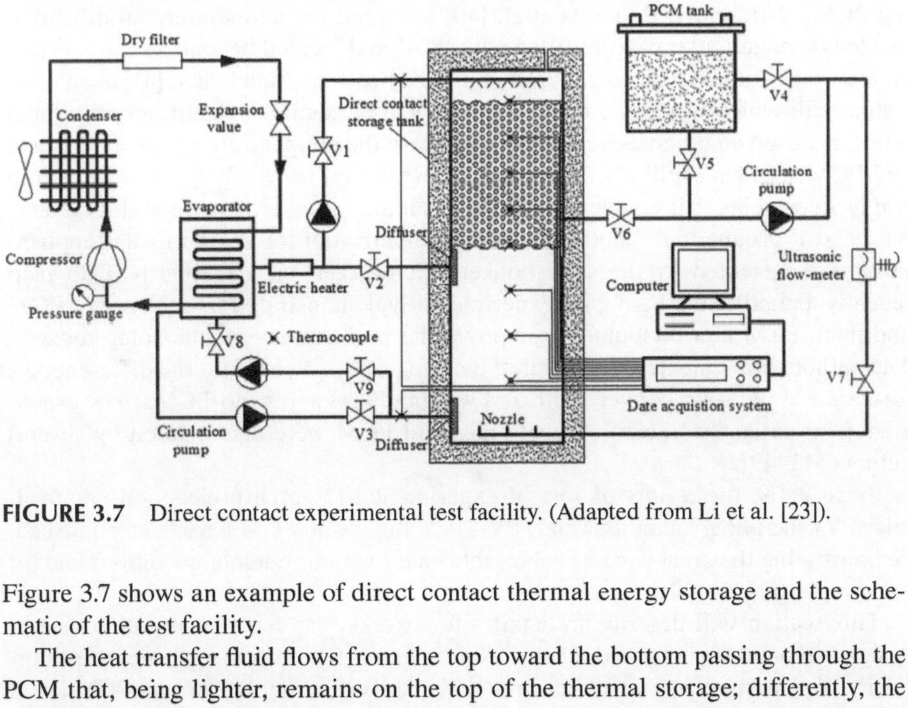

FIGURE 3.7 Direct contact experimental test facility. (Adapted from Li et al. [23]).

Figure 3.7 shows an example of direct contact thermal energy storage and the schematic of the test facility.

The heat transfer fluid flows from the top toward the bottom passing through the PCM that, being lighter, remains on the top of the thermal storage; differently, the PCM enters in the thermal storage through a set of nozzles from the bottom and moves upward.

Two auxiliary loops are needed: one for the heat transfer fluid and the other for the PCM which needs to be maintained in liquid phase to be pumped back to the thermal storage. Eight T-type thermocouples were installed inside the tank while other two were positioned at the inlet and outlet of the tank. The PCM mass flow rate was measured by an ultrasonic flowmeter.

3.3 APPLIED STUDIES

This section explores the different methodologies implemented to evaluate the performance of the PCMs embedded in latent thermal energy storages (LTESs) for specific applications. The section presents the most interesting works on: solar applications, refrigeration, building heating, cooling and domestic hot water, building envelope (i.e. walls), thermal management of electrical batteries, and thermal management of electronic devices.

3.3.1 Solar Applications

There is a great number of works related to the use of PCMs in solar applications. Among those: Sajawal et al. [37] investigated the thermal performance of double pass air heater implementing PCM contained in metallic finned tubes; Kumar et al. [38] experimentally investigated a forced convection solar air heater using packed bed absorber plates with PCMs while Saxena et al. [39] proposed to improve the efficiency of solar air heaters by integrating helical tubes carrying PCM. Differently, Bouadila et al. [40] analyzed the temperature stratification inside two rectangular cavities filled with PCM and located beyond a solar thermal collector for domestic hot water production. Kabeel and Abdelgaied [41] developed a solar still which was powered by a cylindrical parabolic concentrator with focal pipes that used an oil heat exchanger to increase the temperature of the water basin and PCM storage. Similarly, Yousef and Hassan [42] tested the performance of a single slope solar still couple with PCMs as a thermal storage unit and its effects on the still production. Another interesting application of PCM in solar applications is represented by the solar cookers, Schwarzer and da Silva [43]. In fact, recently, Palanikumar et al. [44] experimentally demonstrated that the use of PCM and nano-PCM can undoubtedly improve the performance of the solar cookers. The authors implemented the infrared thermography to compare the different performance of the solar cookers with and without PCM and nano-PCMs. The evacuated solar collectors including heat pipes and PCM were also studied by several authors [45–49].

Since 2009, the results of several experimental research projects on photovoltaic (PV) and photovoltaic-thermal (PV-T) cooling using PCM have been published, demonstrating the great benefits achievable under various outdoor conditions and for various periods [50–57].

This section will describe in detail a few works that can give useful advice to design and build accurate and reliable experimental apparatus to evaluate the performance of solar modules, which integrate PV and PV-T technologies with PCM latent thermal energy storages.

Hasan et al. [53] reported an interesting work on the annual evaluation of the energy performance of a photovoltaic-phase change material system in comparison with a PV standard unit. These measurements must take into account both the PV-PCM system and the weather conditions, including the solar radiation. The author developed an original experimental apparatus consisting of a pyranometer and a weather station to measure the solar radiation intensity and the ambient temperature and wind speed, respectively. They added a set of T-type thermocouples to measure the temperature of the front and back surfaces of the PVs. Particular attention was devoted to cover the thermocouples located on the front surface with an opaque insulation tape to shield them from the direct solar radiation. The experimental campaign was conducted on two PV panels: one equipped with the PCM and the other took as reference. The experimental setup developed by Hasan et al. [53] can be considered a reference for this kind of measurements.

Another interesting recent work has been proposed by Li et al. [54], who conducted an experimental campaign to study the performance of a solar PV panel integrated with phase change material. The authors tested a PV-only (i.e. reference), a PV-PCM module, and a PV-T PCM module which also implemented a thermal solar collector. Li et al. [54] included several interesting information about possible experimental issues that can occur during these experiments. One of those is the high thermal contact resistance between the PV panel and the PCM container that needs also to withstand to a non-negligible stress due to the pressure generated by the liquid PCM under the influence of gravity. The thermal contact resistance can alter the experimental performance of the PV-PCM system because it introduces an additional temperature drop increasing the PV temperature and, thus, reducing the benefits of the addition of the thermal storage. The authors proposed to avoid the use of thermal adhesive and prefer the use of epoxy. Moreover, the modules are commonly installed tilted or even vertical and, thus, the pressure of the liquid can be strong enough to lead to PCM leaking. Considering the experimental setup, Li et al. [54] used PT-100 sensors to monitor the temperature of the different modules. It has to be pointed out that, in the case of solar PV characterization, the choice of the temperature sensor does not present any critical issue: both thermocouples and PT-100 are suitable.

Simón-Allué et al. [56] studied the effects of PCM on different models of PV-T collectors. The authors developed a setup according to the ISO 9806:2013 Solar Energy–Solar thermal collectors–Test methods [58] in order to collect useful and reliable data for comparison.

More recently, Carmona et al. [57] experimentally compared the performance of a hybrid PV-T solar energy collector with integrated PCM against that of a traditional PV module by analyzing the data collected over 5 days. Considering the PV-T construction, it has to be pointed out that a high conductive thermal paste was used to reduce the thermal resistance between the PV module and the base of the absorber plate. Considering the PV-T modules, two subsystems were designed: thermal one and PV one. The thermal one was designed and built according to ISO 9806:2013 [58], as reported in Figure 3.8, while the PV subsystem included an MPPT (Maximum Power Point Tracking) charge controller that ensured the PV module to operate close to its maximum efficiency all the time.

As shown in Figure 3.9, this subsystem is completed by a battery tank and a pure sine wave inverter to supply power to alternating current loads. The MPPT was used to monitor in real time all the electrical performance variables which then were recorded every minute for 4 months.

The temperature of the PV module, PV-T PCM module, and thermal subsystem was measured by 24 K-type thermocouples connected directly to the data acquisition

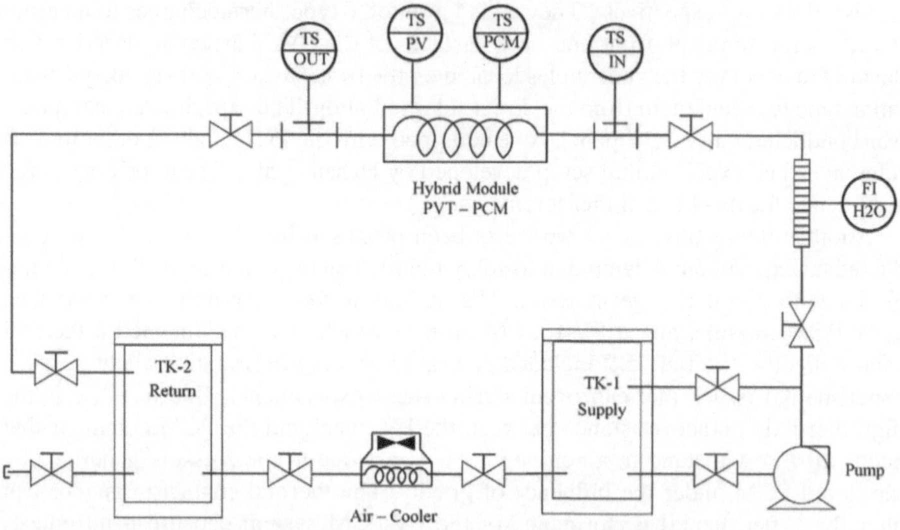

FIGURE 3.8 Thermal subsystem. (Adapted from Carmona et al. [57]).

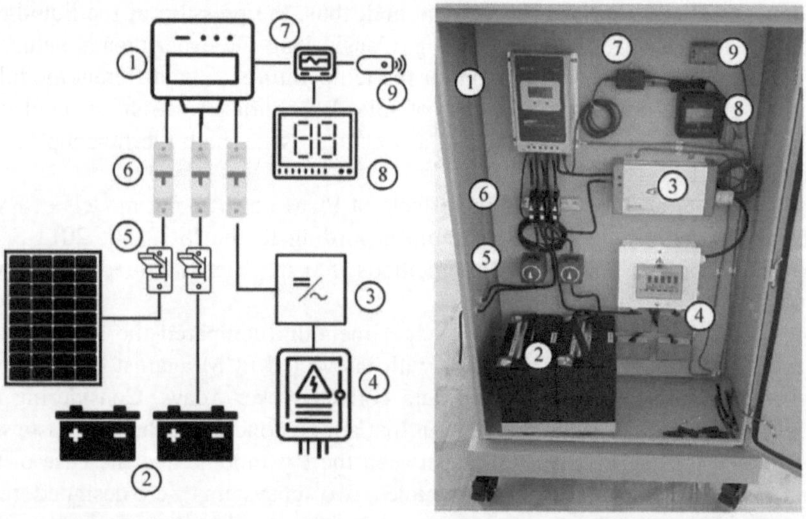

1. MPPT Charge Controller; 2. DC / AC pure Sine Wave Inverter; 3. Batteries; 4. Maintenance Switches; 5. 20 A DC Circuit Breaker; 6. 40A DC Circuit Breakers; 7. PV Subsystem Data Logger; 8. HMI Display; 9. WiFi Router; 10. AC Circuits Board

FIGURE 3.9 PV subsystem. (Adapted from Carmona et al. [57]).

unit. The apparatus was completed by a pyranometer and weather station to monitor the solar radiation and the ambient conditions according to ISO 9806 [58].

From the collected data, Carmona et al. [57] were able to evaluate several interesting performance parameters: daily electrical energy generated, daily electrical efficiency, daily thermal energy storage, daily thermal efficiency, and daily hybrid efficiency.

The results showed that the PV-T PCM hybrid module exhibited a daily thermal efficiency remarkably higher (around 31%) than that of the reference PV module (around 13%).

An interesting work was also proposed by Ramm Dheep and Sreekumar [59] to verify the thermal reliability and corrosion characteristics of an organic PCM to be implemented inside a solar air heater. Thermal cycling reliability and the material compatibility are two of the most important characteristics to assess the practical possible implementation and performance of PCMs but, in many cases, they are neglected or ignored. The authors performed 2000 accelerated thermal cycles using an experimental setup consisting of an electrical hot plate that could control the temperature between 25°C and 250°C and a PCM capsule. The latter was made of glass and contained the PCM. During the accelerated thermal cycling test, the hot plate was maintained at 90°C, around 10–15 K above the melting temperature range of the phenyl acetic acid (75°C–80°C) till the PCM was completely melted; then, the sample was cooled down in still ambient air by natural convection. The cycle was repeated 2000 times. Every 100 cycles, 0.5 g of PCM was extracted from the capsule to measure the phase transition temperature and latent heat of fusion using the differential scanning calorimeter technique.

The compatibility tests investigated the interaction between the PCM and three metals: aluminum, copper, and stainless steel. The corrosion tests followed the same cycling procedure of the thermal stability ones, but, in this case, a small piece of each metal was inserted inside a dedicated test capsule. The gravimetric analysis was conducted to investigate the mass loss and corrosion rate; moreover, the surface of the metals after 2000 cycles were inspected by scanning electron microscopy to determine the type of corrosion (e.g. pitting, cracks). Furthermore, Raam Dheep and Sreekumar [59] run also an experimental campaign to verify the performance of the solar air heater with built-in PCM-based latent thermal energy storage demonstrating the feasibility of the developed solution for solar space heating applications.

Finally, this brief overview on the experimental setups and methodologies for the characterization of latent thermal energy storages for solar applications illustrated how this kind of systems should be designed to achieve reliable and accurate results. Differently from other applications, there is the need to use of weather station (temperature, wind, humidity) and pyranometer to fully characterize the weather conditions. The PV systems also need power measurement while the accuracy of the thermal performance evaluation depends upon the accuracy of the selected temperature sensors (considering the whole measuring chain; see Table 3.1) and flow rate instruments (in the case of solar thermal). Considering the common size of this kind of storages, the use of PT-100 sensors and standard size thermocouples can be suggested.

3.3.2 Building Applications: Active Systems

The use of latent thermal energy storages in building application can involve active or passive elements or both. The terms active refers to the fact that the PCM-based latent thermal energy storage is charged and discharged by a fluid (mainly air) flowing through the PCM, which comes or is coming, from or to, the ambient to be conditioned.

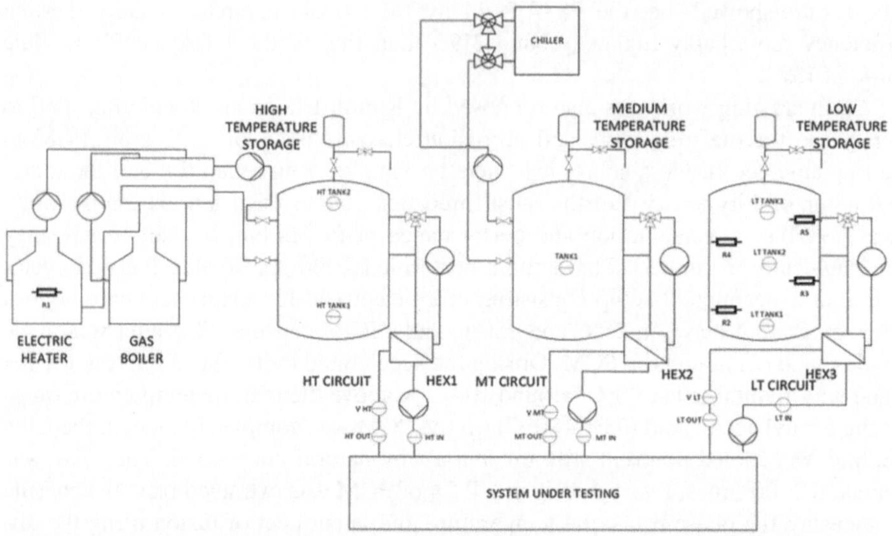

FIGURE 3.10 Test stand for hybrid heat pump and thermal storage test. (Adapted from Palomba et al. [60]).

Differently, the term passive refers to those applications in which the PCM is commonly embedded in the walls, floors, roofs windows, etc. and it acts as a thermal barrier to minimize the thermal losses or gains of the indoor environments.

Focusing on the active systems, this section only considers the studies in which the PCM-based solution is the main component or it is part of the whole air conditioning system. For example, Palomba et al. [60] experimentally analyzed the performance of a hybrid cascade heat pump and thermal-electric energy storage for residential building. Figure 3.10 shows the experimental test stand, which was used to test the system that can be considered a reference for this kind of experiments.

The test stand shown in Figure 3.10 consists of three independent sensible thermal storages set at three different thermal levels (i.e. high, medium and low temperature) to simulate: heat source (1.5 m^3 water tank), ambient heat sink (1.5 m^3 water tank), and heat source for evaporation (0.75 m^3 water/glycol mixture tank). The system is equipped with data acquisition and control; it also permits to simulate the PV generation by means of a DC power supply that can be remotely controlled while the power can be recorded. The temperatures are measured with thermocouples and magnetic flow meters measure the fluid volumetric flow rates. Figure 3.11 reports a schematic of the hybrid cascade heat pump with thermal/electric storage, including all the measuring devices, that was connected to the test stand of Figure 3.10.

The system performance was estimated by measuring the heat flow rates of each component, and, in particular, the following parameters were evaluated: the thermal heat flow rate to the heat source circuit of the sorption chiller; the heat rejection of the sorption chiller; the evaporation heat flow rate of the sorption chiller; the condensation heat flow rate of the vapor compression heat pump; the heat flow rate at the evaporator of the vapor compression heat pump; the charging and discharging

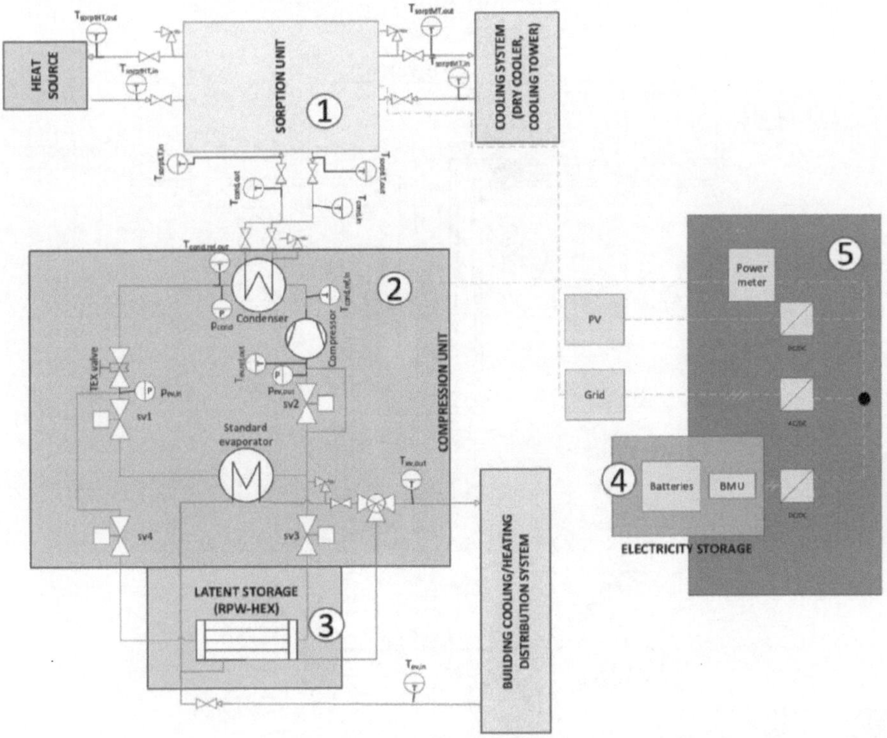

FIGURE 3.11 Hybrid cascade heat pump with thermal/electric storage. (Adapted from Palomba et al. [60]).

heat flow rates of the PCM storage; the electric energy input for the operation of the compressor of the heat pump, and the auxiliary power for the sorption module. This method allowed for an accurate experimental characterization of the performance of the individual components and of the entire system.

This example highlights that if complex systems need to be tested, the auxiliary test stands must fulfill the energy and power requirements in order to guarantee constant boundary conditions. This translates in large sensible thermal storages which ensure stability and instant power according to their size. In fact, an undersized auxiliary thermal storage can lead to unstable operating conditions of the simulated heat source; on the other hand, an oversized auxiliary thermal storage can reduce the test stand capabilities in simulating transient conditions (day and night) of the heat source.

Liu et al. [61] experimentally evaluated a vertical earth-to-air heat exchanger system which integrated an annular PCM to mitigate the thermal fluctuation of the ambient air temperature. This experimental campaign involved a large equipment as shown in Figure 3.12: as it can be seen, an annular PCM storage was wrapped around the pipe. The system pre-cool the outdoor air before it is introduced in the building. The PCM, air and soil temperatures were measured using PT100, the air flow rate was also measured. The collected data sets were used to validate a numerical model.

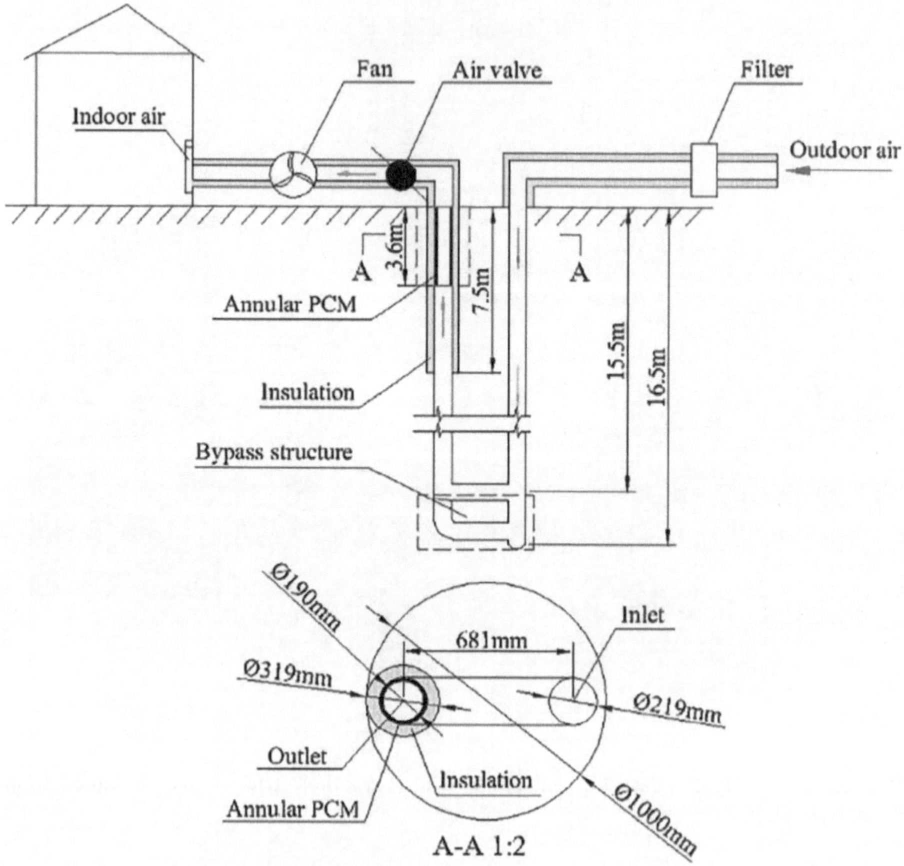

FIGURE 3.12 Schematic of the vertical earth-to-air heat exchanger application to air conditioning. (Adapted from Liu et al. [61]).

Encapsulated PCMs for building applications have also been studied by several authors. For example, Tyagi et al. [62] presented the thermal performance of PCM-based thermal energy storage consisting in 3750 spherical shape high density polyethylene balls filled with calcium chloride hexahydrate. The authors designed and built a prototype room and distributed several thermocouples to assess the performance of the system. Morovat et al. [63] proposed a study on the use of an active energy storage consisting of PCM encapsulated panels to be installed in the ceiling plenum of an office, a mechanical room or in other suitable locations.

The use of plate-encapsulated PCM in air heat exchangers for ventilation applications was also proposed by Kumirai et al. [64]. The authors compared two types of paraffins and a salt hydrate PCM that can be used in climatic conditions with summer ambient temperatures of 30–35°C. A horizontal air duct was designed to test the different plate-encapsulated PCM heat exchangers. The air enters in the wind tunnel from a temperature controlled test room in which both the dry bulb temperature and absolute air humidity are measured. The air passes through an electrical section

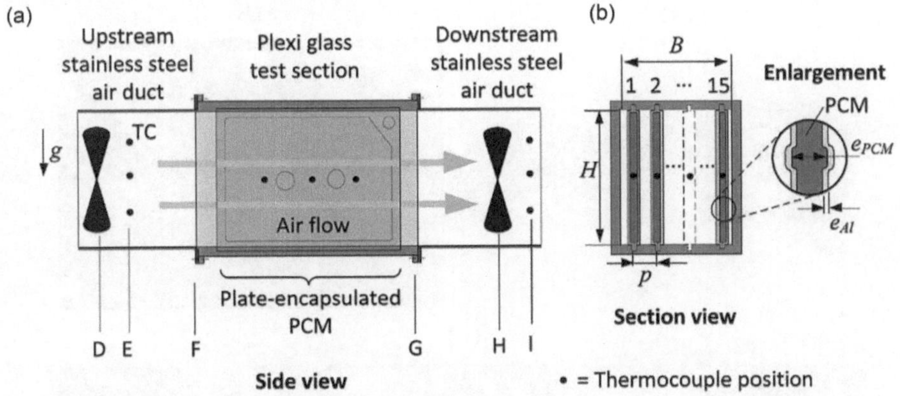

FIGURE 3.13 Plate-encapsulated PCM heat exchanger: side (a) and section (b) views. (Adapted from Kumirai et al. [64]).

equipped with Nickel Chromium heaters. The air temperature was measured before and after the heating section as well as the electrical power supplied; thus, from a thermal balance the air flow rate could have been estimated.

One of the main challenges when dealing with air flow driven by buoyancy forces is the temperature stratification, thus the authors implemented an interesting solution: a passive blade mixer. After the mixer five T-type thin thermocouples were used to measure the temperature at the inlet of the test section. The test section is shown in Figure 3.13.

The acrylic glass test section contained 15 vertically oriented commercial encapsulated PCM plates. The PCM temperatures were monitored by T-type thermocouples installed inside the capsules. At the outlet of the test section another passive blade mixer is used to eliminate any air stratification before the air temperature measurements, carried out using other five T-type thermocouples. The described experimental setup allowed for accurate measurements of the thermal performance of encapsulated PCM air heat exchangers.

Another interesting application of this kind of technology is the one proposed by Lee et al. [65] who implemented a PCM unit in a central air circulation system that also uses a roof ventilation layer. The PCM heat exchanger was used to store the cold energy of the cooled air by radiative cooling and the energy of the heated air by solar heat collections. The system was tested in an experimental house.

3.3.3 Building Applications: Passive Systems

PCMs can easily be embedded in the building envelope within walls [66–74], roofs [75–77], floors [78,79], ceilings [80], and window shutters [81–83]; they can also be mixed with concrete [78,84], plaster [85,86], and gypsum [68,87]. Figure 3.14 shows a few examples of the proposed solutions.

Among the cited works, there are interesting methodologies proposed to characterize the performance of the envisaged solutions. When dealing with large systems,

FIGURE 3.14 Examples of passive building solution. (a) Shape stabilized PCM mortar bricks (Adapted from Wang et al. [73]); (b) Smart gypsum composite boards with PCMs (Adapted from Khadiran et al. [87]); (c) Plug and play wall panels to test the performance of PCMs based walls (Adapted from Lee et al. [88]); (d) Concrete sandwich with PCM layer (Adapted from Marani and Madhkhan [84]); (e) Building roofs with PCMs (Adapted from Chung and Park [77]); (f) Window shutters with PCM (Adapted from Silva et al. [81]); (g) Concrete core slab with PCM (Adapted from Navarro et al. [78]).

as in the case of real scale building envelope components, there are a few challenges to be tackled and solved to achieve reliable results. The flexibility is one of the major challenges. Lee et al. [88] proposed a novel methodology which allows for faster installation and dismantling of the test walls. As shown in Figure 3.14c, this methodology is based on plug and play concept, in which the walls are fastened to the envelope of their institutional research building using clamps with rubber tips. The real scale lab was air conditioned by a central air conditioning system equipped with a 995 kW water chiller and sensible water thermal energy storage of around 3900 L. T-type thermocouples directly connected to the acquisition system were used and two heat flux meters with dimensions of 5.08 cm × 5.08 cm × 0.48 cm were installed on the interior surfaces of each panel. This methodology allows for a fair comparison between the current wall without PCMs and the solution implementing the latent thermal energy storage showing the great benefits achievable in terms of daily heat transfer reduction and peak heat flux reduction.

Another method widely used to evaluate the performance of walls, roofs, floors, etc. is represented by the direct comparison of different solutions (i.e. with and without PCM) using two independent test rooms/building. This technique was used by Singh and Bhat [68] who compared the performance of gypsum board with and without PCM using a two lab-scale test rooms.

A similar study was conducted by Lu et al. [79] to experimentally evaluate a PCM floor solution coupled with a solar water heating system. In this case, two full-scale identical buildings were set and used.

An interesting work was also carried out by Silva et al. [83], who designed and built a test cell (7.00 m × 2.35 m × 2.58 m, length × width × height) and the internal floor area of 5.17 m. The cell is subdivided in two sub-cells by an insulated internal wall; the aim of this work was to test the performance of an innovative window shutter developed by the authors. The cell was equipped with: two heat flux meters; ten PT100 probes; one anemometer; two hydrometers; one solar radiation probe; and one relative humidity and temperature sensor. The boundary of the test cell was insulated using sandwich panels that incorporate 4 cm to 8 cm of insulation material. However, the authors identified by means of an infrared camera (Testo®875i) several thermal bridges where potential heat losses occurred that could not be overlooked.

This study permits to highlight another possible challenge of this kind of experiments. In fact, another non-negligible issue when dealing with the application of PCM to walls is represented by the scale effect; in many cases, at lab scale, it is not possible to test real size walls and only small pieces of wall are built and tested. However, the parasitic side effects can affect the performance of the system and must be carefully considered. This issue might remain also in the case of similar small scale prototypes tested contemporary: parasitic heat flow rate and non-uniform temperature distribution can lead to inaccurate estimations of energy savings. In order to overcome this problem, dedicated test facilities need to be designed and used.

At lab scale, the reference standard for the determination of the thermal transmission properties of walls is the EN ISO 8990 [89]. The guarded hot box facility comprises a cold box, an inner box (metering box) surrounded by a guarded box, the thermal guard aim is to minimize the lateral heat flow in the specimen and the heat flow through the metering box walls, as specified by the cited standard. Bahrar et al.

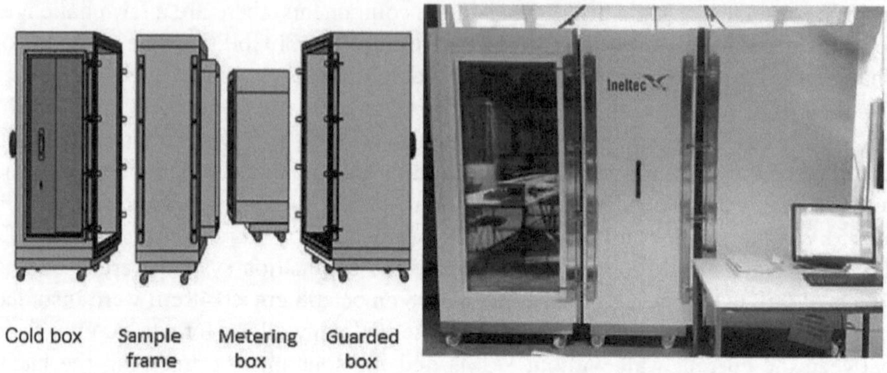

FIGURE 3.15 Guarded hot box test facility. (Adapted from Bahrar et al. [67]).

[67] used a guarded hot box test facility to characterize the effects of microencapsulated PCMs in textile reinforced concrete panel (Figure 3.15).

According to the EN ISO 8990 [89], the authors used a set of temperature sensors located in each box to measure the air temperature distribution; other surface temperature sensors (PT100) were installed on each side of the test panel and two heat flux sensors were used to evaluate the heat flow rate. Both boxes were equipped with fans to homogenize the air temperature and the air velocities were measured to estimate the convection heat transfer coefficient. The tests duration was eight days.

Zhang et al. [66] proposed an interesting method to experimentally evaluate structural insulated panels outfitted with PCMs. The authors designed a dynamic wall simulator which consists of a cubic box made up of six removable wall panels of dimensions of 1.219 m × 1.219 m. Six incandescent light bulbs with variable inputs were equidistantly placed at the center of the box. The heating system could be controlled to simulate various daylight hours with varying solar irradiance. Two fans were used to circulate the air. The internal volume of the box simulated the ambient while the space of the laboratory was supposed to be the inside space of the building. The temperature of the air and surfaces and the wall heat flux were measured by means of T-type thermocouples and heat flux meters, respectively. Jin et al. [72] adopted a similar dynamic wall simulator to study the performance of PCM-based thermal shields for building walls applications. Gounni et al. [90] designed a smaller box to experimentally analyze the optimal PCM layer locations in wood walls. The results demonstrated that the best location for the PCM layer is close to the heat source.

Finally, Fioretti et al. [74] developed two original experimental setups to test a novel wall for refrigerated containers using PCM. The authors designed an indoor experiment to develop a PCM-added prototype panel to be compared against the reference panel (without PCM) under defined environmental conditions. The tests were run by locating a refrigerated test box inside a climatic chamber where a solar simulator reproduced the solar irradiation flux. PT100 sensors and a heat flux meter were used to measure the internal and external panel surfaces temperatures and the heat flux through the panel, respectively. The air temperature was measured by T-type

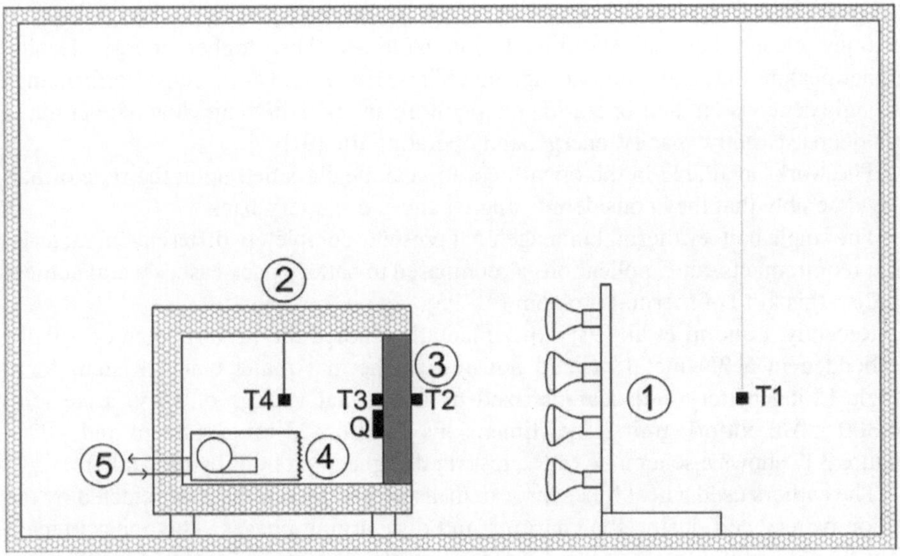

FIGURE 3.16 Indoor test facility. (Adapted from Fioretti et al. [74]).

thermocouples. A schematic of the indoor test facility is shown in Figure 3.16, which consisted of a solar simulator, a climatic test room with the refrigerated test box inside, a PCM-based panel, a cooling unit, and a thermostatic bath. The reported main instruments are air temperature inside the cold room, prototype panel external surface temperature, prototype panel internal surface temperature, refrigerated test box internal temperature, and heat flux.

The peculiarity of this work is that the authors also performed outdoor experiments using two identical cold rooms one of them fitted with a PCM external layer. Both rooms were instrumented to monitor the temperatures of the external/internal surfaces, the PCM surface layer and the internal cold room using PT100 resistance thermal detector. The heat flux of the south oriented wall of the cold rooms was measured by means of heat flux meters. A meteorological station was used to monitor the outdoor climatic conditions.

One concluding remark regards the fact that, in most of the current experiments relative to building applications, there is one final critical issue: the repeatability of the results. Most of the works demonstrates the repeatability by running multiple tests on the same PCM-based walls, roofs, gypsum boards but this kind of applications should include multiple tests on different samples to verify the repeatability also of the manufacturing process.

3.3.4 Thermal Management of Electrical Batteries

The thermal management of the electrical batteries is becoming more and more important because of the large use of the electrical storages based on Li-ion cells in many different either stationary or mobile technological fields: renewable energy systems, electrical vehicles, aeronautical and aerospace applications. As compared to other electrochemical systems, Li-ion batteries show higher energy density, higher peak power, and moderate ageing effects. However, Li-ion cells' performance strongly depends on their operating temperature and both high and low temperatures can decrease their capacity, energy, and operating life [91].

The works available in the open literature can be classified upon the type of battery assembly that they considered: single battery or battery pack.

The single battery thermal management presents completely different characteristics, requirements, and applications as compared to battery pack case; several authors studied this kind of thermal problem [91–95].

Recently, Landini et al. [91] experimentally studied the performance of a PCM embedded in a 3D metal printed housing as thermal management system for a single Li-ion battery cell characterized by a nominal voltage of 3.7 V, a capacity of 300 mAh, cut-off limit 2.5 V, dimensions 32 mm × 23 mm × 4.8 mm and 7.02 g. Figure 3.17 shows a schematic of the test rig designed and built by the authors.

The authors used a heat flux sensor to measure and record the heat rejected by the Li-ion battery cell during the charging and discharging phases. This measurement is of fundamental interest because it represents the instant value of the heat flux that passes through the aluminum housing and it is absorbed by the PCM. PT100 sensors were installed to measure the surface temperature of the Li-ion cell and the temperature of the PCM.

One of the main issues in this kind of measurement is represented by the presence of an avoidable thermal contact resistance between the Li-ion surface and the aluminum container. This implies that the surface temperature of the Li-ion surface changes as a function of this thermal contact resistance. Landini et al. [91] used a non-silicon based thermal paste to minimize the thermal contact resistance.

As a general rules, repeated tests with different amount of thermal paste or other thermal interface materials should be run to verify the reliability/repeatability of

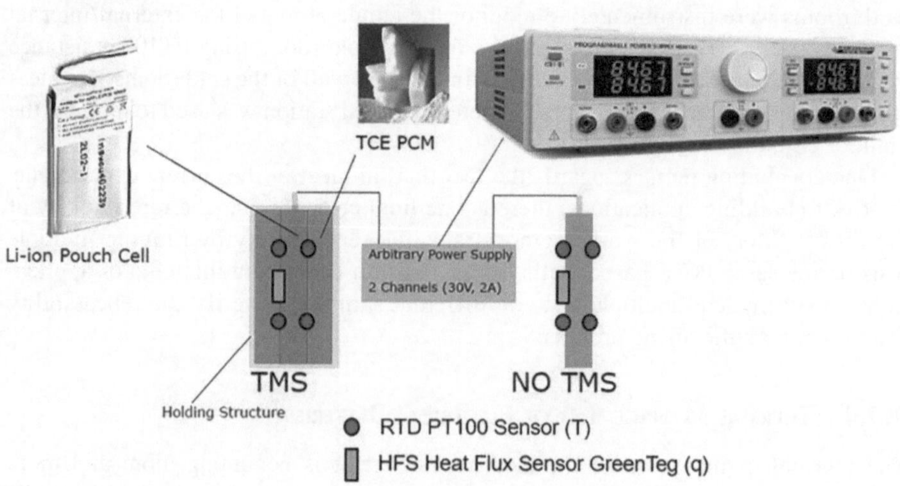

FIGURE 3.17 Schematic of the test facility. (Adapted from Landini et al. [91]).

the surface temperature measurement. This activity also allows for the identification of the best practice to minimize the thermal contact resistance in this kind of experiments.

Landini et al. [91] run a complete characterization of the Li-ion by studying a different discharge rate (DR) that is defined as the constant current which discharges the entire nominal capacity of the battery in one hour. For example, a Li-ion cell of capacity equal to 1Ah is discharged at a current of 1 A under DR = 1C. In general, it can be stated that DR = 1C refers to stationary electrical energy storage system discharge regime while higher DR such as 3C or 5C can be representative of an electrical vehicle discharge load [91].

Another possible issue when dealing with battery cell thermal characterization is the initial state of the battery; the authors usually perform the experiments using new battery cells but a calibration cycle should be conducted before the experiments to ensure that the battery cells will exhibit the real behavior. Celik et al. [93] conducted a calibration procedure using a battery autotest system. The steps of this calibration procedure are the followings: (1) complete charge with a constant current of 0.5C; (2) idle time of 1 hour; (3) discharge at 1C to the cut-off limit; (4) idle time of 3 hour; (5) fully charge at 0.5C; and (6) idle time 3 hour. After this calibration procedure, the cell was ready for testing under specified conditions.

Zhang et al. [92] performs experiments to validate a numerical model of a single battery cell with a capacity of 25 Ah. The authors located the electrical battery inside a climatic chamber, they used a battery simulation testing system to charge and discharge the battery; five T-type thermocouples were located on each side of the battery which is immersed in the PCM. Hence, in this case, there is not any thermal contact resistance; however, the authors did not measure the heat generated by the battery during the charging/discharging processes, which was estimated by means of the lumped parameter model proposed by Bernardi et al. [96].

A direct visualization of the phase change process was proposed by Sun et al. [94] who designed a thermal management system that comprises a cylindrical battery, metal fins, PCM, and a nylon transparent housing. During the tests, however, a customized electrical heater was used to simulate the battery, which supplied a constant heat flow rate. The authors performed 20 cycles to prove the repeatability of the tests and the stability of the PCM. Figure 3.18 shows the visualization of the melting process through the nylon housing. A similar experiment was also run by Yang et al. [95].

As a concluding remark, it should be noted that many works involve electrical heaters rather than real battery cells to simulate the charging/discharging processes; however, Landini et al. [91] demonstrated that the heat dissipated by the battery cells depends upon several parameters being a function of the time. This is also confirmed by Hémery et al. [97], who performed an experiment to verify how the electrical heaters could realistically simulate the heat power generation of the battery. The results showed that at low C-rate, the curves of heater wall temperature did not match those of the real battery, because the electrical heater could not simulate the temperature fluctuations induced by the heat of reaction. At higher C-rates, the electrical heater capabilities improved but, in general, the simulating temperature profiles are not fully consistent with the real battery profiles.

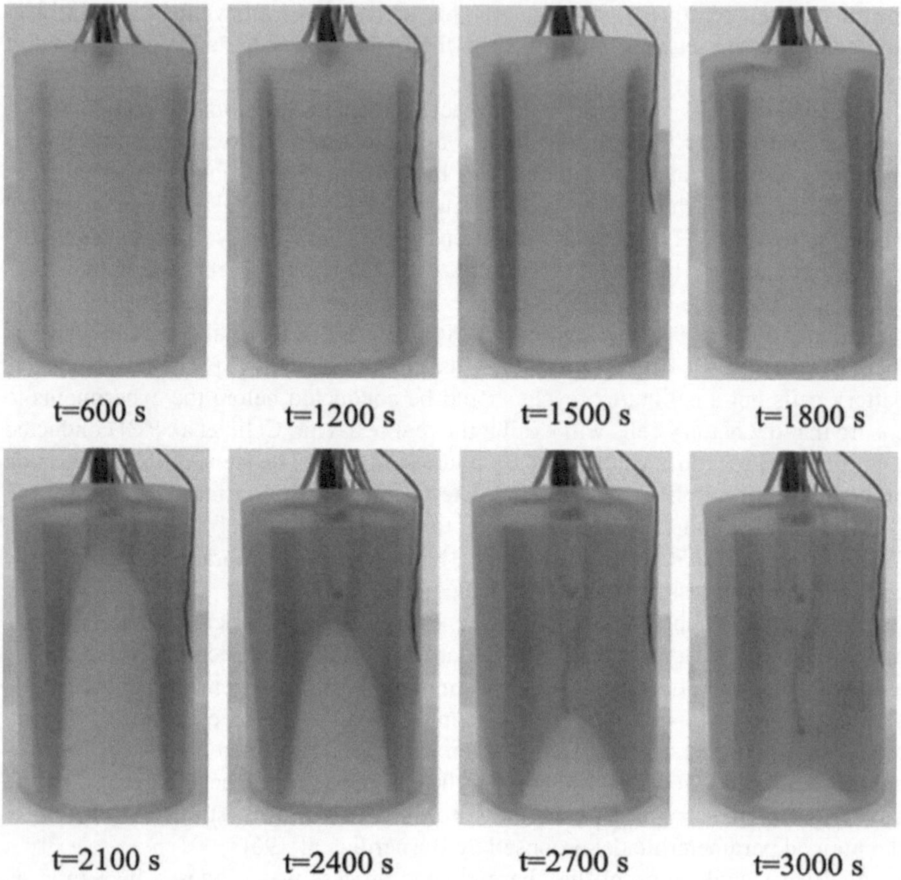

FIGURE 3.18 Visualization of the melting process. (Adapted from Sun et al. [94]).

Thus, it is strongly recommended to use real electrical batteries in the experiments as well as to insert heat flux sensor to monitor the actual heat flux dissipated.

Battery packs are of extreme interest because they are applied in several different stationary and mobile technologies. Several authors investigated the potential use of PCM to develop efficient thermal management system for battery pack assembly [98–106].

Hémery et al. [97] experimentally simulated a battery pack of 27 cells, which were simulated by as many electrical heaters for safety reasons. A Pt temperature sensor was installed on the wall of each casing to monitor their individual thermal response. The authors proposed two kind of tests: the first involved the active air cooling while the second investigated a semi-passive thermal management method using PCM.

As shown in Figure 3.19, in the semi-passive test, the electrical heaters are inserted inside a finned can containing PCM and sealed. The 27 cans are setup in staggered rows in the test section, without gap distance. Two cooling plates are located on the

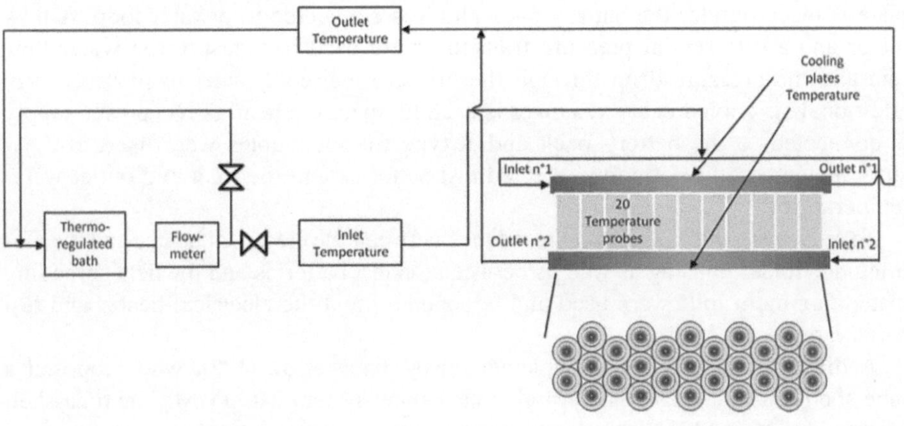

FIGURE 3.19 Semi-passive cooling system with PCM. (Adapted from Hémery et al. [97]).

top and bottom of the cans to circulate the water. The water temperatures and flow rate are measured by means of Pt probes and a volumetric flow meter.

Zhang et al. [107] conducted an experimental campaign on an original hybrid thermal management system using PCM and bottom liquid cooling techniques for a large-sized power battery module, including 106 test batteries in 18650 format.

Also in this case the test batteries used an electrical heater to simulate the heat generation rate at different C-rate.

Figure 3.20 shows a simplified schematic of the test apparatus used by Zhang et al. [107] to test the battery pack. As it can be seen the test batteries are located in vertical position, the liquid PCM is poured inside a Bakelite container. A cold

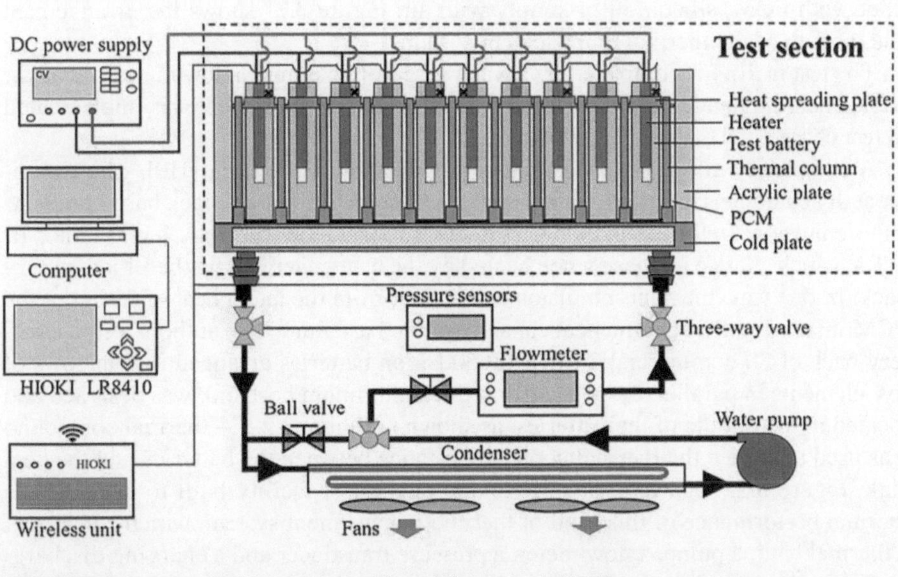

FIGURE 3.20 Test facility. (Adapted from Zhang et al. [107]).

plate is placed under the battery pack and it is connected to a water loop. A flow meter and a differential pressure transducer are used to measure the water flow rate and the pressure drop through the minichannel cold plate, respectively. An additional air cooled heat exchanger is used to reject the heat. A DC power supply is connected to the battery pack and K-type thermocouples were used to measure the temperature of a few selected test batteries and the inlet and outlet water temperatures.

To reduce the thermal contact resistance between the electrical heater and the aluminum cylinder housing as well as between the test batteries and the heat spreading plate, aluminum foils were used and wrapped around the electrical heater and test battery, respectively.

A different approach was implemented by Jiang et al. [108], who proposed a tube-shell heat exchanger as thermal management system with PCM. The tube-shell prototype contained five aluminum tubes on which baffles were mounted, arranged horizontally with gap distance. Each aluminum tube hosted five cells wrapped with expanded graphite/paraffin composite which were connected in series.

The authors monitored the temperature of the surface of 5 cells located in the center of each tube; another thermocouple monitored the temperature of the module. A DC electronic load was used to simulate the loads. Experiments either in natural or forced convection were run and compared against those obtained for a similar battery pack without PCM and baffles.

Hekmat and Molaeimanesh [109] studied a different kind of Li-ion batteries; five high-capacity prismatic cells with the sizes of 148 mm × 129 mm × 4 mm, a nominal voltage of 3.8 V and a nominal capacity of 5500 mAh. The cells were located vertically side by side in a glass box; on the surface of each cell a temperature sensor was placed. Once the cells were positioned inside the box, they were sealed and, thus, six separate zones were generated in which a cooling pipe was placed. Those zones were filled with PCM, silicon oil or simply with air. Figure 3.21 shows the arrangement and a photo of the thermal management system.

The test facility is completed by a water reservoir, a pump, a flow meter and an air cooled heat exchanger to reject the heat. Finally, a battery analyzer was implemented to test charge and discharge the batteries.

A completely different approach was proposed by Wang et al. [110], who experimentally characterize a thermal management system of a battery pack based on PCM nano-emulsion. In this case, the PCM is seeded inside the working fluid as nanoparticles, which change in phase when heated by the heat generated by the Li-ion battery pack. In this way, the nano-emulsion can benefit from the latent heat of fusion of the PCM increasing the specific heat capacity of the base fluid. The authors tested a battery pack of 20 commercial 2.6 Ah 18650 Li-ion batteries arranged in four rows of five elements in parallel. A special 6 mm thick aluminum heat sink was designed and located in the middle of the batteries, as shown in Figure 3.22. A thermal compound was used to reduce the thermal contact resistance between the batteries and the heat sink. Figure 3.22 also reports a schematic of the test facility built to evaluate the thermal performance of this kind of thermal management system, which consists of a thermal bath, a pump, a flow meter, a pressure transducer and a charging/discharging battery test module.

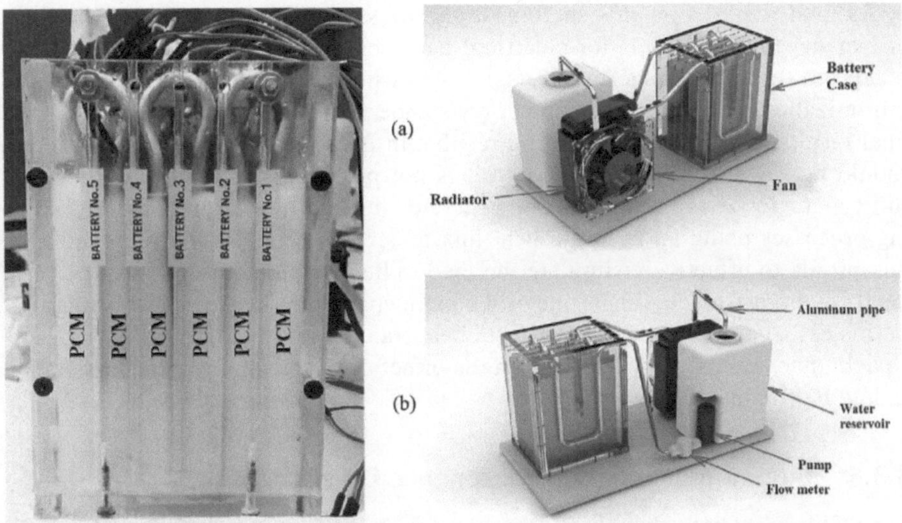

FIGURE 3.21 Module structure (left) and test facility (right). (Adapted from Hekmat and Molaeimanesh [109]).

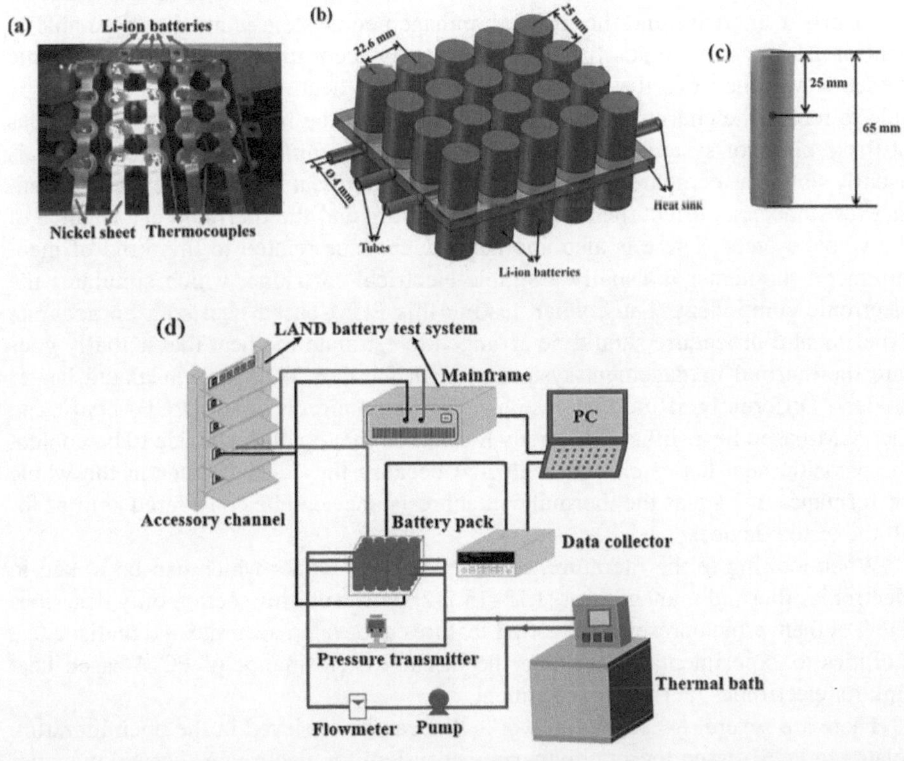

FIGURE 3.22 Experimental test section and test facility. (Adapted from Wang et al. [110]).

As final remarks on this section, it can be stated that the study of the thermal management systems for electrical batteries presents several issues mostly related to the difficulties (mainly safety) in using the real batteries and to the minimization of the thermal contact resistance. Both these issues can affect the final results and lead to misleading results; in general, the use of real batteries should be suggested, however, when it is not possible, one option might be to fully characterize the heat flow rate generated during the charging and discharging processes using an accurate heat flux meter and then use the recorded heat flux profile to effectively simulate the load to be dissipated. Regarding the thermal contact resistance, this is one of the main challenges of the heat transfer and, in this case, can lead to a non-uniform heat transfer on the different batteries of a pack; thus, a sensitivity analysis on the effect of the thermal contact resistance should be conducted.

3.3.5 Thermal Management of Electronic Devices

The PCMs seem to be thought to be coupled with an electronic device to be cooled; however, as demonstrated by Righetti et al. [111], the design of thermal management system can be tricky and lead to misleading results. Considering the operation of a PCM-based heat sink coupled with an electronic device, the heat generated is stored inside the PCM which melts, thus, when the solid is fully melted, the PCM temperature starts to increase and the thermal management system is not anymore able to control the junction temperature which increases accordingly. When the electronic device is switched off, the PCM should reject the heat solidifying in order to be able to repeat the charging cycle. This brief description highlights the main issues of these kinds of system: the solidification (discharging/heat rejection) process is usually slower as compared to the melting (charging/heat storing); the thermal contact resistance can affect the junction temperature and the thermal performance of the whole system. There is also another hidden issue related to this kind of measurement: the heater is usually a simple electrical cartridge, which simulates the electronic component. Thus, when testing this PCM-based heat sink an accurate experimental procedure should be arranged to estimate the heat that actually goes into the thermal management system even in the case of highly insulated heater holders. Differently, if the tests use a real printed circuit board (PCB) board and the PCM-based heat sinks are directly located on the top of the device to be cooled, the parasitic heat losses can be neglected because they do not enter in the whole performance as long as the thermal contact resistance can be considered similar for all the tested devices.

When looking to the literature, there are a lot of works which can be linked to electronics thermal management [1,12–15,112]; however, this section only describes a few of them which present interesting features to develop accurate and realistic test facilities to experimentally evaluate the thermal performance of PCM-based heat sink for electronics thermal management.

There are several experimental works that can be retrieved in the open literature related to PCM-based thermal management system for electronics thermal management [113–124].

Miers and Marconnet [116] proposed an experimental investigation of composite PCM heat sinks for passive thermal management. The authors designed a simple though accurate experimental setup where they tested different PCM-based heat sinks with realistic power profiles, controlled interfacial loading, and *in situ* temperature measurement. The performance was evaluated considering two main parameters: (1) the time it took the test heater chip to reach the cut-off temperature of 95°C, and (2) the period of a full melt-regeneration cycle. Figure 3.23 shows the thermal setup developed by Miers and Marconnet [116].

The thermal test setup consists of a chip with an integrated array of resistors and temperature sensors mounted on a PCB breakout board used to simulate a processor with a high degree of control over the heating patterns and provide *in situ* temperature measurements. A thin TIM (thermal interface material) pad is used between the PCM heat sink and the chip. A polyether ether ketone (PEEK) insulation block was used as interface with the vertical loading fixture to control the interfacial pressure (10 psi, around 0.69 bar). This simple setup allows for a direct and reliable comparison between different PCM heat sinks because, keeping constant the interfacial pressure with the same TIM, the thermal contact resistance remains the same and thus the temperature measured on the chip surface is always representative of the junction temperature.

Gharbi et al. [123] designed a similar test facility to compare different heat sink configurations (Figure 3.24); it consists of a PCM container, a heater, a power supply, a temperature data logger, a personal computer, and a digital camera. The PCM container has one wall in contact with a copper slab that it is used to spread the heat

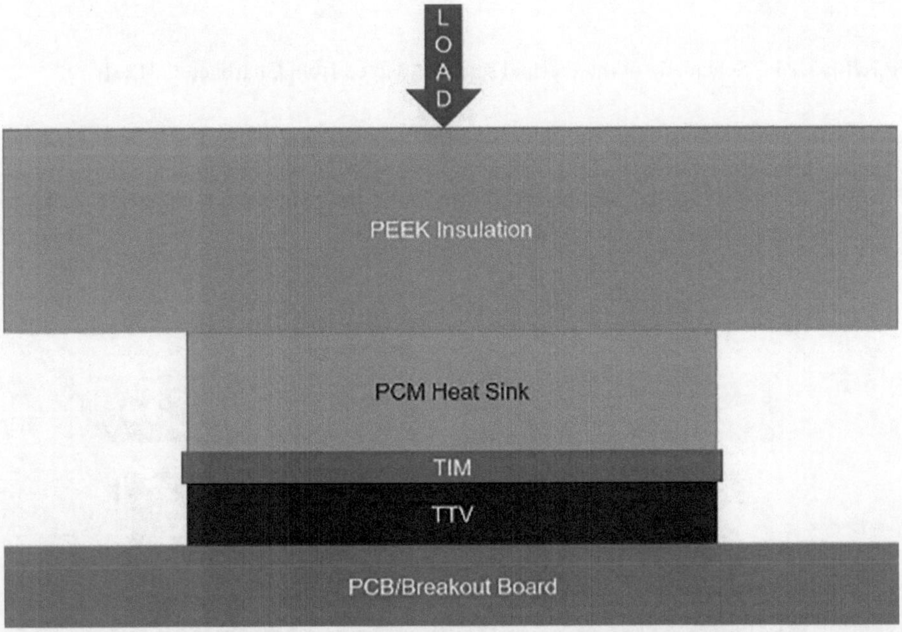

FIGURE 3.23 Schematic of the thermal setup. (Adapted from Miers and Marconnet [116])

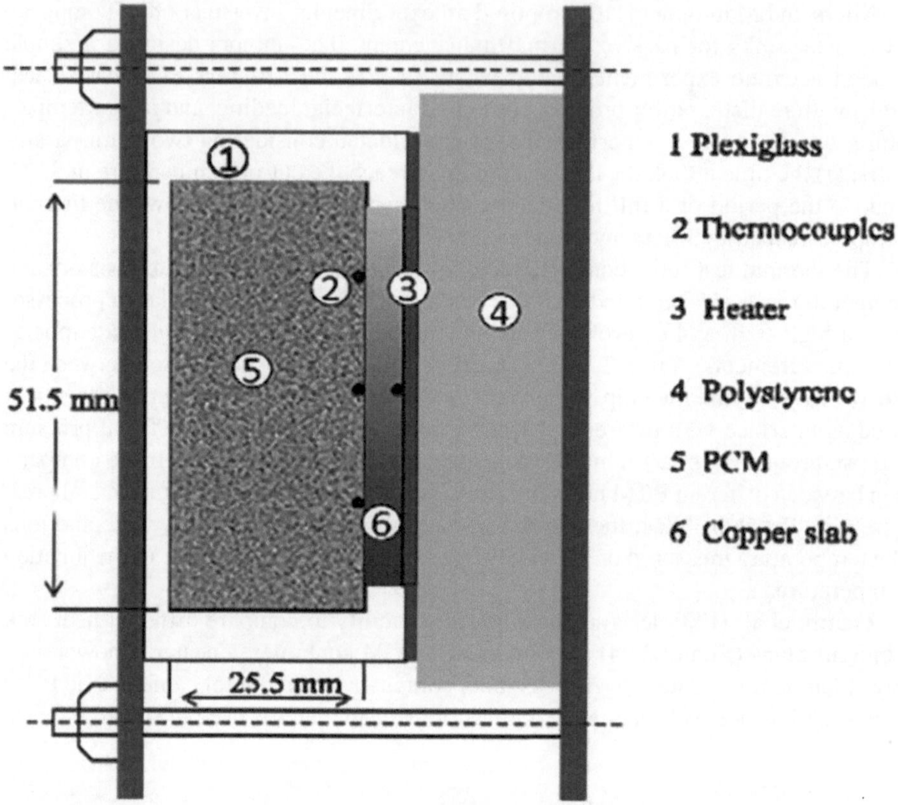

FIGURE 3.24 Schematic of the thermal setup. (Adapted from Gharbi et al. [123]).

generated by the heater with the same size that is insulated with a thick layer of polystyrene. The other walls of the PCM container were made of transparent Plexiglas sheets to allow for the photographic observation of the melting process. A few examples of this visualizations are reported in Figure 3.25.

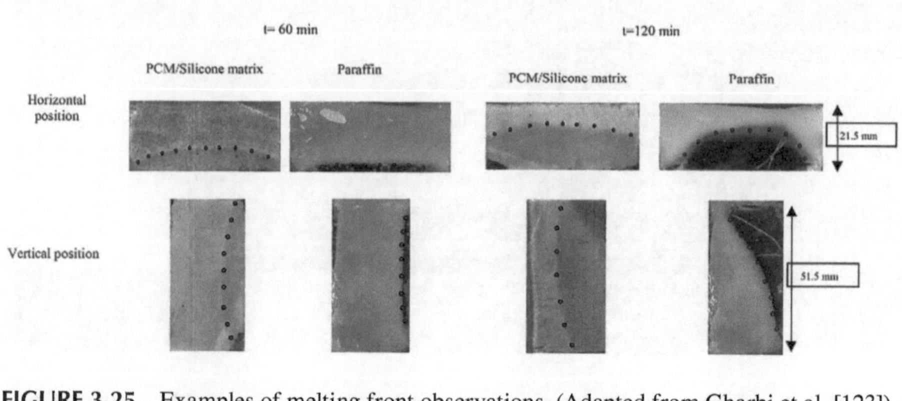

FIGURE 3.25 Examples of melting front observations. (Adapted from Gharbi et al. [123]).

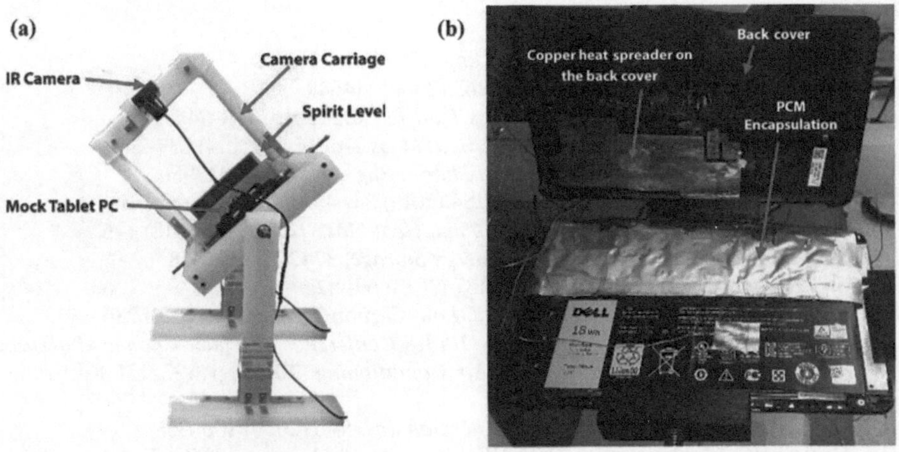

FIGURE 3.26 Experimental test facility. (Adapted from Ahmed et al. [117])

Ahmed et al. [117] investigated the performance of a tablet PC with a PCM thermal energy storage unit encapsulated in thin aluminized laminated film under continuous operation. Figure 3.26 shows the test stand designed and built to isolate the experimental tablet PC from potential external thermal interference by minimizing physical contacts.

The test stand held the experimental tablet PC while permitting to rotate it to any angle. A thermal camera is mounted on a guide to adjust the focus. The experimental tablet PC was used to assess the effects of integrating PCM-based units into a system comparable to a real-life tablet PC. It consisted of the original Windows tablet parts as a substrate (battery, screen, and casing). The most important part of the experimental tablet PC was the mock PCB. The heat dissipation from the tablet's electronics was applied using a 25.4 mm × 25.4 mm square Kapton heater with a rated power density of 10 W/in^2 at 28 VAC. Nine T-type-thermocouples were used to monitor the heat spreading on the front surface, back cover, and internal components. During each experiment, the PCM heat sink unit was placed on the PCB and the back cover was attached, then the experimental tablet PC was placed on the test stand using 3D printed clamps. At the beginning of each experiment, NUC (non-uniformity correction) operation of the IR camera was done to avoid non-uniformity in the IR image data. Cardboard shades were used all around the setup to avoid any unwanted light coming toward the tablet PC to avoid reflections from the back cover.

In this particular experiment, the thermal contact resistance could vary because of the flexibility of the PCM-based unit, which did not allow the control of the pressure exerted on the substrate. The contact was ensured by the 3D printed clamps that held the PCM-based unit.

As a final remarks, this brief overview permitted to show that, in the case of electronic thermal management, the experimental setup can be really simple but a few major points need to be considered to collect reliable results.

REFERENCES

1. V. Joshi, M.K. Rathod, *Appl. Therm. Eng.* 178 (2020) 115518.
2. A. Kumar, S.K. Saha, *Appl. Energy.* 263 (2020) 114649.
3. S. Wu, T.X. Li, et al., *Int. J. Heat Mass Transfer.* 102 (2016) 733–744.
4. B. Kamkari, H. Shokouhmand, *Int. J. Heat Mass Transf.* 78 (2014) 839–851.
5. M.Y. Yazici, M. Avci, O. Aydin, *Appl. Therm. Eng.*, 159 (2019) 113956.
6. C. Li, H. Yu, Y. Song, *Energy Build.* 184 (2019) 34–43.
7. W. Li, H. Wan, P. Zhang, et al., *Int. Comm. Heat Mass Transf.* 96 (2018) 1–6.
8. V. Safari, H. Abolghasemi, et al, *J. Energy Storage.* 37 (2021) 102458.
9. S. Seddegh, X. Wang, et al., *Energy.* 137 (2017) 69–82.
10. R.M. Lazzarin, S. Mancin, et al., *Int. J. Low-Carbon Technol.* 13 (2018) 286–291.
11. R. Lazzarin, S. Mancin, M. Noro, in: *11th IIR Conference on Phase Change Materials and Slurries for Refrigeration and Air Conditioning*, 2016: pp. 122–131 Karlsruhe, Germany, May 18–20, 2016.
12. G. Righetti, L. Doretti, et al., *Int. J. Thermofluids.* 5–6 (2020) 100035.
13. G. Righetti, G. Savio, et al, 25th IIR International Congress of Refrigeration, 2019, August 24–30, Montreal, CA.
14. G. Righetti, G. Savio, et al., *Int. J. Therm. Sci.* 153 (2020), 106376.
15. S. Mancin, A. Diani, et al., *Int. J. Therm. Sci.* 90 (2015) 79–89.
16. G. Righetti, C. Zilio, et al. *Appl. Therm. Eng.*, 196 (2021) 117276.
17. Y. Du, G. Xiao, et al., *Int. J. Thermophys.* 35 (2014) 1577–1589.
18. G. Chen, G. Sun, D. Jiang, Y. Su, *Int. J. Heat Mass Transf.* 152 (2020) 119480.
19. N.S. Dhaidan, A.F. Khalaf, *Int. Comm. Heat Mass Transf.* 111 (2020) 104476.
20. A.K. Hassan, J. Abdulateef, et al., *Case Stud. Therm. Eng.*, 21 (2020) 100675.
21. R. Rybár, M. Beer, M. Kaľavský, *J. Energy Storage.* 21 (2019) 72–77.
22. X. Yang, Z. Niu, J. Guo, et al., *Int. Comm. Heat Mass Transf.* 117 (2020) 104775.
23. X.Y. Li, D.Q. Qu, L. Yang, K. di Li et al., *Appl. Energy*, 189 (2017) 211–220.
24. R.M. Saeed, J.P. Schlegel, et al., *Energy Convers. Manag.* 181 (2019) 120–132.
25. A. Raul, M. Jain, S. Gaikwad, S.K. Saha, *Appl. Therm. Eng.* 143 (2018) 415–428.
26. Y. Wang, K. Yu, X. Ling, *Energy Build.* 210 (2020) 109744.
27. A.K. Raul, P. Bhavsar, S.K. Saha, *J. Energy Storage.* 20 (2018) 279–288.
28. M. Kabbara, D. Groulx, A. Joseph, *Int. J. Therm. Sci.* 130 (2018) 395–405.
29. S. Seddegh, S.S.M. Tehrani, et al., *Appl. Therm. Eng.* 130 (2018) 1349–1362.
30. P. Zhang, F. Ma, X. Xiao, *Appl. Energy.* 173 (2016) 255–271.
31. M. Fadl, D. Mahon, P.C. Eames, *Int. J. Heat Mass Transf.* 173 (2021) 121262.
32. M. Fadl, P.C. Eames, *Energy.* 188 (2019) 116083.
33. J. Gasia, A. de Gracia et al., *Appl. Energy.* 235 (2019) 1389–1399.
34. H. Niyas, C.R.C. Rao, P. Muthukumar, *Solar Energy.* 155 (2017) 971–984.
35. A. Siahpush, J. O'Brien, et al., *Heat Transf. Eng.* 40 (2019) 1600–1618.
36. K. D'Avignon, M. Kummert, *Appl. Therm. Eng.* 99 (2016) 880–891.
37. M. Sajawal, T.U. Rehman, et al., *Case Stud. Therm. Eng.*, 15 (2019) 100543.
38. R.A. Kumar, B.G. Babu, M. Mohanraj, *Int. J. Green Energy.* 14 (2017) 1238–1255.
39. A. Saxena, N. Agarwal, E. Cuce, *J. Energy Storage.* 30 (2020).
40. S. Bouadila, M. Fteïti, et al., *Energy Convers. Manag.* 78 (2014) 904–912.
41. A.E. Kabeel, M. Abdelgaied, *Solar Energy.* 144 (2017) 71–78.
42. M.S. Yousef, H. Hassan, *J. Clean. Prod.* 209 (2019) 1396–1410.
43. K. Schwarzer, M.E.V. da Silva, *Solar Energy.* 82 (2008) 157–163.
44. G. Palanikumar, S. Shanmugan, et al., *Renew. Energy.* 178 (2021) 260–282.
45. M.H. Abokersh, M. El-Morsi, et al., *Energy.* 139 (2017) 1111–1125.
46. V.V. Bhagwat, S. Roy, et al., *Sustain. Energy Technol. Assess.* 45 (2021) 101171.
47. K. Chopra, A.K. Pathak, et al., *Energy Conv. Manag.* 203 (2020) 112205.

48. K. Chopra, V. v. Tyagi, et al., *Energy Convers. Manag.* 198 (2019) 111896.
49. M. Bilardo, G. Fraisse, et al., *Solar Energy.* 183 (2019) 425–440.
50. Y.W. Beng, Thermal Management of Concentrator Photovoltaics, 2009, Doctoral Dissertation, University of Warwick.
51. S. Maiti, S. Banerjee, et al., *Solar Energy.* 85 (2011) 1805–1816.
52. S. Sharma, A. Tahir, et al., *Solar Energy Mater. Solar Cells.* 149 (2016) 29–39.
53. A. Hasan, J. Sarwar, H. Alnoman, S. Abdelbaqi, *Solar Energy.* 146 (2017) 417–429.
54. Z. Li, T. Ma, J. Zhao, A. Song, Y. Cheng, *Energy.* 178 (2019) 471–486.
55. P. Sudhakar, R. Santosh, et al., *Renew. Energy.* 172 (2021) 1433–1448.
56. R. Simón-Allué, I. Guedea, R. Villén, G. Brun, *Solar Energy.* 190 (2019) 1–9.
57. M. Carmona, A. Palacio Bastos, J.D. García, *Renew. Energy.* 172 (2021) 680–696.
58. International Organization for Standards, ISO 9806:2013 Solar Energy - Solar thermal collectors - Test methods, ISO/TC 180 Solar Energy, 2013.
59. G. Raam Dheep, A. Sreekumar, *J. Energy Storage.* 23 (2019) 98–105.
60. V. Palomba, A. Bonanno, et al., *Energies.* 14 (2021) 2580.
61. Z. Liu, Z. (Jerry) Yu et al. *Energy Convers. Manag.* 186 (2019) 433–449.
62. V. v. Tyagi, A.K. Pandey, et al., *Energy Build.* 117 (2016) 44–52.
63. N. Morovat, A.K. Athienitis, et al., *Energy Build.* 199 (2019) 47–61.
64. T. Kumirai, J. Dirker, J. Meyer, *J. Build. Eng.* 22 (2019) 75–89.
65. H. Lee, A. Ozaki, M. Lee, *Build. Environ.* 124 (2017) 104–117.
66. Y. Zhang, X. Sun, M.A. Medina, *Appl. Therm. Eng.* 178 (2020) 115454.
67. M. Bahrar, Z.I. Djamai, et al., *Sustain. Cities Society.* 41 (2018) 455–468.
68. S.P. Singh, V. Bhat, *Energy Build.* 159 (2018) 191–200.
69. X. Sun, J. Jovanovic, et al., *Energy.* 180 (2019) 858–872.
70. G. Zhou, M. Pang, *Energy.* 93 (2015) 758–769.
71. X. Shi, S.A. Memon, et al., *Energy Build.* 71 (2014) 80–87.
72. X. Jin, M.A. Medina, X. Zhang, *Energy.* 73 (2014) 780–786.
73. X. Wang, H. Yu, et al., *Energy Convers. Manag.* 120 (2016) 81–89.
74. R. Fioretti, P. Principi, B. Copertaro, *Energy Convers. Manag.* 122 (2016) 131–141.
75. Q. Al-Yasiri, M. Szabó, *Case Stud. Constr. Mater.* 14 (2021) e00522.
76. H.J. Akeiber, S.E. Hosseini, et al., *Energy Convers. Manag.* 150 (2017) 48–61.
77. M.H. Chung, J.C. Park, *Energies.* 10 (2017) 195.
78. L. Navarro, A. de Gracia, et al., *Energy Build.* 116 (2016) 411–419.
79. S. Lu, Y. Zhao, K. Fang, Y. Li, P. Sun, *Energy Build.* 140 (2017) 245–260.
80. M. Alizadeh, S.M. Sadrameli, *Energy Build.* 188–189 (2019) 297–313.
81. T. Silva, R. Vicente, C. Amaral, A. Figueiredo, *Appl. Energy.* 179 (2016) 64–84.
82. T. Silva, R. Vicente, *Appl. Therm. Eng.* 84 (2015) 246–256.
83. T. Silva, R. Vicente, et al., *Energy Build.* 88 (2015) 110–121.
84. A. Marani, M. Madhkhan, *J. Mater. Res. Technol.* 12 (2021) 760–775.
85. M. Kheradmand, M. Azenha, J.L.B. de Aguiar, *Energy.* 94 (2016) 250–261.
86. A. Sarı, V.V. Tyagi, *J. Mater. Civil Eng.* 28 (2016) 04015137.
87. T. Khadiran, M.Z. Hussein, et al., Journal of Materials in Civil Engineering. 28 (2016) 04015137.
88. K.O. Lee, M.A. Medina, X. Sun, *Energy Build.* 86 (2015) 86–92.
89. International Standards Organization, EN ISO 8990 Calibrated and Guarded Hot Box, 1996.
90. A. Gounni, M. el Alami, et al., *J. Solar Energy Eng.* 140 Issue 4 (2018).
91. S. Landini, R. Waser, et al., *Appl. Thermal Eng.* 173 (2020) 115238.
92. W. Zhang, Z. Liang, G. Ling, L. Huang, *J. Energy Storage.* 41 (2021) 102849.
93. A. Celik, H. Coban, et al., *Int. J. Energy Res.* 43 (2019) 3681–3691.
94. Z. Sun, R. Fan, et al., *Int. J. Heat Mass Transf.* 145 (2019) 118739.
95. H. Yang, H. Zhang, Y. Sui, C. Yang *Appl. Therm. Eng.* 128 (2018) 489–499.

96. D. Bernardi, E. Pawlikowski, J. Newman, *J. Electrochem. Soc.* 132 (1895) 5–12.
97. C.V. Hémery, F. Pra, J.F. Robin, P. Marty, *J. Power Sources.* 270 (2014) 349–358.
98. A. Hussain, I.H. Abidi, et al., *Int. J. Therm. Sci.* 124 (2018) 23–35.
99. W. Wang, X. Zhang, C. Xin, Z. Rao, *Appl. Therm. Eng.* 134 (2018) 163–170.
100. Y. Azizi, S.M. Sadrameli, *Energy Convers. Manag.* 128 (2016) 294–302.
101. Z. Ling, F. Wang, X. Fang, X. Gao, Z. Zhang, *Appl. Energy.* 148 (2015) 403–409.
102. J. Yan, K. Li, H. Chen, Q. Wang, J. Sun, *Energy Convers. Manag.* 128 (2016) 12–19.
103. A. Hussain, C.Y. Tso, C.Y.H. Chao, *Energy.* 115 (2016) 209–218.
104. Y. Lv, X. Yang, X. Li, et al., *Appl. Energy.* 178 (2016) 376–382.
105. Z. Ling, J. Cao, W. Zhang, et al., *Appl. Energy.* 228 (2018) 777–788.
106. M. Pan, Y. Zhong, *Int. J. Heat Mass Transf.* 126 (2018) 531–543.
107. H. Zhang, X. Wu, Q. Wu, S. Xu., *Appl. Therm. Eng.* 159 (2019) 113968.
108. G. Jiang, J. Huang, M. Liu, M. Cao, *Appl. Therm. Eng.* 120 (2017) 1–9.
109. S. Hekmat, G.R. Molaeimanesh, *Appl. Therm. Eng.* 166 (2020) 114759.
110. F. Wang, J. Cao, Z. Ling, Z. Zhang, X. Fang, *Energy.* 207 (2020) 118215.
111. G. Righetti, C. Zilio, et al., *Therm. Eng.* 196 (2021) 117276.
112. Y.Z. Ling, X.S. Zhang, et al., *Renew. Energy.* 154 (2020) 636–649.
113. R. Gulfam, W. Zhu, et al., *Energy Convers. Manag.* 156 (2018) 25–33.
114. D. Moore, A. Raghupathy, W. Maltz, in: *Annual IEEE Semiconductor Thermal Measurement and Management Symposium*, Institute of Electrical and Electronics Engineers Inc., 2016: pp. 213–217, Budapest, Hungary.
115. F. Kaplan, C. de Vivero, et al., in: *2014 32nd IEEE International Conference on Computer Design, ICCD 2014*, Institute of Electrical and Electronics Engineers Inc., 2014: pp. 256–263, 19–22 October 2014, Seoul, South Korea.
116. C.S. Miers, A. Marconnet, *J. Heat Transf.* 143 (2021) 013001.
117. T. Ahmed, M. Bhouri, et al., *Int. J. Therm. Sci.* 134 (2018) 101–115.
118. R. Kandasamy, X.Q. Wang, et al., *Appl. Therm. Eng.* 27 (2007) 2822–2832.
119. Y. Tomizawa, K. Sasaki, et al., *Appl. Therm. Eng.* 98 (2016) 320–329.
120. R. Kandasamy, X.Q. Wang, et al., *Appl. Therm. Eng.* 28 (2008) 1047–1057.
121. F.L. Tan, C.P. Tso, *Appl. Therm. Eng.* 24 (2004) 159–169.
122. F.L. Tan, W. Shen, et al., in: *Proceedings of the Electronic Packaging Technology Conference, EPTC*, 2009: pp. 640–645, 9–11 December 2009, Singapore.
123. S. Gharbi, S. Harmand, et al., *Appl. Therm. Eng.* 87 (2015) 454–462.
124. Z. Deng, C. Zhang, Q. Sun, L. Wu, F. Yao, D. Xu, *J. Therm. Anal. Calorimet.* 144 (2021) 869–882.

4 Design Criteria for Advanced Latent Heat Thermal Energy Storage Systems

Chunrong Zhao and Kamel Hooman
The University of Queensland

CONTENTS

4.1 INTRODUCTION

The objective of this chapter is to put forward design criteria of phase change material (PCM)-based advanced thermal energy storage tanks. The problem at hand is a nonlinear, moving-boundary heat transfer problem, which is firstly studied by Stefan [1] toward the melting of polar ice cap. Only a handful of analytical solutions are presented for this complex problem. Existing approximate solutions are obtained by relying on simplifications (one-dimensional, semi-infinite domain, etc.) mainly ignoring convection effects. Nonetheless, natural convection during melting and solidification can be very important. Here, we review and list some of the approximate solutions for melting and solidification in storage tanks filled with pure PCM. Then we move on to consider the effect of thermal conductivity enhancers which need to be optimized to benefit from both conduction and convection.

DOI: 10.1201/9781003213260-4

4.2 GEOMETRIC IMPACT ON MELTING

Without loss of generality, we focus on natural convection during melting and offer results which can be extended to solidification as well. Here we consider storage tanks of different shapes.

4.2.1 RECTANGULAR ENCLOSURES

4.2.1.1 Top Heating

If a rectangular enclosure of H (height)$\times L$ (length) is heated from the top wall (while gravity is pulling down) then the melting process can be convection-free due to thermal stratification (see Figure 4.1). The dimensionless form of thermal energy equation is given by [2]

$$\frac{\partial^2 \theta}{\partial Y^2} = \text{Ste}\frac{\partial \theta}{\partial \tau} \tag{4.1}$$

with the interface condition as

$$\left.\frac{\partial \theta}{\partial Y}\right|_{Y=\gamma} + \frac{d\gamma}{d\tau} = 0 \tag{4.2}$$

where the dimensionless numbers are defined as: temperature $\theta = (T - T_m)/(T_w - T_m)$; melt front $Y = y/H$; melt fraction $\gamma = s/H$; time $\tau = \text{Ste}\,\text{Fo}$. Note that "$s$" denotes the melt front, Stefan number is $\text{Ste} = c_p(T_w - T_m)/h_{sf}$ and Fourier number $\text{Fo} = \alpha t/H^2$.

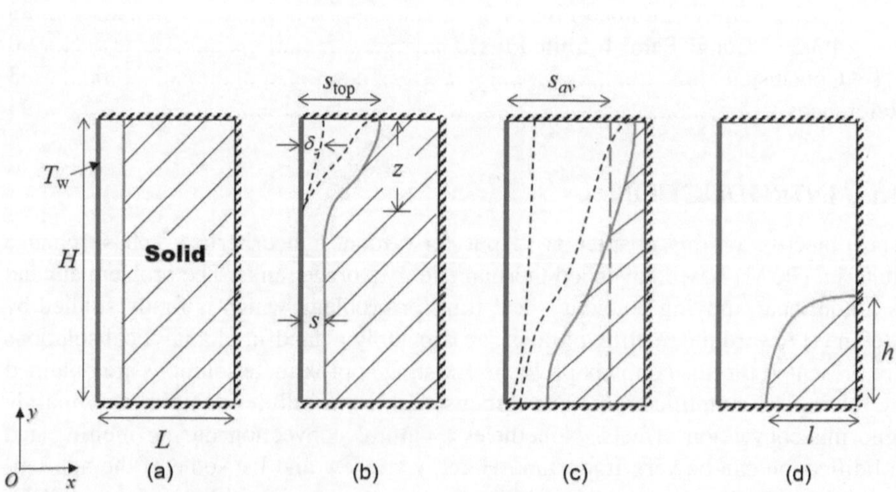

FIGURE 4.1 Four-regime model for isothermal lateral melting in a rectangular enclosure: (a) conduction stage; (b) transition stage; (c) convection stage and (d) "shrinking-solid" stage.

The boundary/initial conditions are

Initial condition	$\theta(Y,0)=0$
Boundary conditions	$\theta(0,\tau)=1, \theta(\gamma,\tau)=0$

With the effects of subcooling ignored from the initial condition, an assumption we will revisit in this chapter, using quasi-steady approximation (infinitesimal Stefan number), Eqs. (4.1) and (4.2) can be solved leading to

Melt front	$s(t)=\sqrt{2\text{Ste}\cdot\alpha t}$
Temperature field	$\theta(Y,\tau)=Y/2\text{Ste}\cdot\alpha t$
Heat flux	$q''=k(T_w-T_m)/\sqrt{2\text{Ste}\cdot\alpha t}$
Melting time	$t_{\text{melt}}=H^2/(2\alpha\text{Ste})$

4.2.1.2 Lateral Heating

Now consider the same enclosure this time heated from one side with all other walls perfectly insulated. Here, natural convection accelerates melting. The four-regime model, developed by Jany and Bejan [3], is introduced here as illustrated in Figure 4.1.

1. Conduction Stage

   In this stage, natural convection is too weak to contribute to melting. Similar to melting with top heating, the dimensionless solutions of melt layer thickness, liquid fraction and Nusselt number are

Melt front	$s(t)\sim H\tau_H^{1/2}$
Liquid fraction	$\gamma=s/L\sim\text{AR}\times\tau_H^{1/2}$
Nusselt number	$\text{Nu}=Q/\left[k(T_w-T_m)\right]\sim\tau_H^{-1/2}$

   where the aspect ratio of the rectangular enclosure is defined as $\text{AR}=H/L$.

2. Transition Stage

   As time goes on, the liquid layer at the top of the reservoir moves and transfers heat by natural convection within the region denoted by s_{top} and z in Figure 4.1. The scales for s_{top} and z for the top reservoir, the Nusselt number are

Top reservoir length	$s_{\text{top}}\sim H\left(\tau_H+\text{Ra}_H\tau_H^{5/2}\right)^{1/2}$
Top reservoir height	$z\sim H\times\text{Ra}_H\tau_H^2$
Nusselt number	$\text{Nu}\sim\tau_H^{-1/2}+\text{Ra}_H\tau_H^{3/2}$

where the height-based Rayleigh number is written as $\mathrm{Ra}_H = \left[g\beta(T_w - T_m)H^3\right]/(\alpha\nu)$. The transition stage lasts until $\tau_1 \sim \mathrm{Ra}_H^{-1/2}$ when $z \sim H$, and Nusselt number reaches a minimum value scaled with $\mathrm{Nu}_{\min} \sim \mathrm{Ra}_H^{1/4}$.

3. Convection Stage

As expected during this stage convective flow is strong to a point that heat is mainly transferred by convection. The curvature of solid–liquid interface is not parallel due to buoyancy-induced flow upward, therefore, a height-averaged melt front is proposed, which defined as $s_{\mathrm{av}}(t) = \dfrac{1}{H}\displaystyle\int_0^H s(y,t)dt$. The scales of s_{av} and Nusselt number are

Height-averaged melt length	$S_{\mathrm{av}} \sim H \times \mathrm{Ra}_H^{1/4}\tau_H$
Nusselt number	$\mathrm{Nu} \sim \mathrm{Ra}_H^{1/4}$

At the end of the third stage, which lasts $\tau_2 \sim \mathrm{Ra}_H^{-1/4} L/H$, one expects $s_{\mathrm{av}} \sim L$. The existence of this stage should satisfy with $\mathrm{AR} < \mathrm{Ra}_H^{1/4}$.

4. Shrinking Solid" Stage

During the final stage, the convective heat transfer is still dominant, while its strength weakens as time progresses. The shape of the left solid PCM is simplified as a right-angled triangle, where the length (l) is assumed to be fixed and the height (h) reduces as time progresses. The scales of h and Nusselt number are

Height of solid PCM	$H\left[1-(H/l)\,\mathrm{Ra}_H^{1/4}(\tau-\tau_2)^4\right]$
Nusselt number	$\mathrm{Nu} \sim \mathrm{Ra}_h^{1/4}$

where the time scale of complete melting is $\tau_3 \sim \tau_2 + (l/H)\mathrm{Ra}_H^{-1/4}$. Therefore, the time scale of this regime accounts for $1/(1+l/L)$ of the time for complete melting, which equals to at least 50% approximately when $l/L = 1$. This can be used as a criterion to optimize melting. Note that this fourth stage is the longest among all and a proper design would be to shorten this stage.

4.2.1.3 Basal Heating

For melting with basal heating, Benard–Rayleigh convection could be expected. Madruga and Curbelo [4] reported four distinct regimes, i.e. the conductive regime, the linear regime, the coarsening regime and turbulent regime for the range of melt layer height-based Rayleigh number up to 10^9. Favier et al. [5] conducted theoretical analysis of solid–liquid interface advances during the basal melting, where the thermal energy balance was raised as

$$\mathrm{Nu} \times \frac{k\left(T_w - T_m\right)}{s} \sim \rho h_{sf}\frac{ds}{dt} \tag{4.3}$$

where "s" denotes the melt layer height, and the Nusselt number is given by

$$\mathrm{Nu} \sim \begin{cases} 1, & \mathrm{Ra}_s \le 1708 \\ \mathrm{Ra}_s^m, & \mathrm{Ra}_s > 1708 \end{cases} \tag{4.4}$$

Thus, the solutions of Eq. (4.3) are

$$s \sim \begin{cases} [\alpha \mathrm{Ste} \cdot t]^{1/2}, & \mathrm{Ra}_s \le 1708 \\ \left[(2-3m)\mathrm{Ste} \cdot \mathrm{Ra}_H^m \dfrac{\alpha t}{H^{3m}},\right]^{1/(2-3m)} & \mathrm{Ra}_s > 1708 \end{cases} \tag{4.5}$$

where the exponent value is typically at the range of $1/4 \le m \le 1/3$ [6].

4.2.1.4 Inclined Enclosure

Comparing the cases above, the complete melting time for the same mass of PCM in a tank is approximated as

Top heating	$\tau_3 \sim (L/H)^2$
Lateral heating	$\tau_3 \sim \left[(L+l)/H\right]\mathrm{Ra}_H^{-1/4}$
Basal heating	$\tau_3 \sim \left[1/(2-3m)\right](L/H)^{2-3m}\mathrm{Ra}_H^{-m}$

For a more comprehensive analysis we use the mean convective enhancement factor ($\bar{\omega}$), proposed by [7], as the ratio of the time-averaged actual heat flux by natural convection to a hypothetical merely conductive heat flux. Assuming $\mathrm{Ste} \to 0$, to ensure all systems start from the same stored thermal energy, one has

$$\bar{\omega} = \frac{\overline{q''_{\mathrm{conv}}}}{q''_{\mathrm{cond}}} \sim \frac{Q/\tau_{3,\mathrm{conv}}}{Q/\tau_{3,\mathrm{cond}}} = \frac{\tau_{3,\mathrm{cond}}}{\tau_{3,\mathrm{conv}}} \tag{4.6}$$

Hence, for lateral and basal heating scenarios, one has

Lateral heating	$\bar{\omega} \sim (1+l/L)^{-1}\ \mathrm{AR}^{-1}\mathrm{Ra}_H^{1/4}$
Basal heating	$\bar{\omega} \sim (2-3m)\mathrm{AR}^{-3m}\mathrm{Ra}_H^m$

Moreover, the time-consuming "shrinking solid" regime of lateral heating is absent in the basal heating, which means the melting time of lateral heating could be around two times longer than that of basal heating. Kamkari et al. [8] experimentally investigated the effect of inclination angle (90°, 45° and 0°) on melting time, the melt fraction evolution and melting time versus inclination angle are displayed in Figure 4.2.

One notes that, with 45° as the inclination angle, the liquid fraction is linearly proportional to the melting time ($\gamma \sim 0.15\tau_H \mathrm{Ra}_H^{1/3}$), but the linear trend breaks down

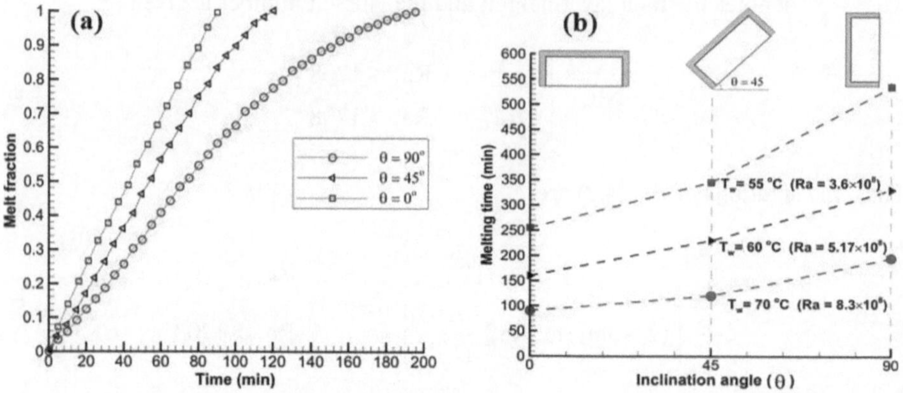

FIGURE 4.2 Melt fraction evolution under constant wall temperature at 70°C (a) and melting times versus inclination angles (b). (Source: Kamkari et al. [8].)

when liquid fraction exceeds 0.85. Moreover, the "shrinking solid" regime is significantly shortened with the inclination angle. This is in line with Yang et al.'s [9] observations.

4.2.1.5 Start Up

In both experimental and numerical analysis of PCM tanks, the phase change temperature has to be different from that of the heated or cooled wall during charge or discharge, respectively. What is missing in almost all of the theoretical works on PCMs is the initial response of the PCM before melting starts. Here, a simple model is presented to account for transient conduction in the PCM tank shown in Figure 4.1. The model is generic enough to be extended to basal and/or inclined heating as there is no flow (solid state). Consider the PCM near the heated wall. We start with a case when the wall is heated and maintained at a uniform and constant heat flux. This is a boundary condition which can be tested in laboratory using an electric heater for instance [10]. We rely on two extremes to obtain a temperature profile using scale analysis. The early stage of heat transfer is considered as the extreme where the total heat input is stored. Hence, the heat balance is given by

$$q'' \sim \rho c_p H \frac{\Delta T}{t} \tag{4.7}$$

Hence, the early response of the PCM is a linear function of time with isoflux heating, that is,

$$\Delta T \sim \frac{q'' H}{k} \frac{\alpha t}{H^2} \tag{4.8}$$

The late regime is assumed as the extreme when the total heat input is conducted out of the control volume between the heated wall and the surface of the control volume.

The heat leaving the control volume penetrates to a depth δ which in turn moves with time. That is,

$$q'' \sim k\frac{\Delta T}{\delta} \tag{4.9}$$

where δ signifies the temperature penetration in the solid PCM. Using scale analysis for the thermal energy equation leads to $\delta \sim \sqrt{\alpha t}$. Making use of this, we have

$$\Delta T \sim \frac{q''H}{k}\sqrt{\frac{\alpha t}{H^2}} \tag{4.10}$$

Hence, in the late regime, temperature response is proportional to square root of time. Using the dimensionless temperature profiles, one has the two, early and late regime, responses as linear function of Fo and square root of Fo, that is

$$\theta_H = \frac{\Delta T}{\left(q''H/k\right)} \sim \begin{cases} \text{Fo}, & \text{early regime} \\ \text{Fo}^{1/2}, & \text{late regime} \end{cases} \tag{4.11}$$

Now, considering the case of isothermal wall heating, as considered in the current paper, the wall heat flux is no longer constant, but the wall temperature is prescribed. For early stage, however, heat only penetrates to a depth of $\sqrt{\alpha t}$ and the wall heat flux is

$$q'' \sim k\frac{T_w - T_i}{\sqrt{\alpha t}} \tag{4.12}$$

Combining Eq. (4.7) with Eq. (4.12), and assuming again that the total heat transferred to the PCM is stored in the control volume during the early stage, one has

$$\frac{\Delta T}{T_w - T_i} \sim \frac{\sqrt{\alpha t}}{H} \tag{4.13}$$

Interestingly, the transient temperature profile during the early stage for an isothermally heated wall increases with time following a square root form.

Considering the late regime, the heat balance for the isoflux case, Eq. (4.10), is recovered noting that heat flux is no longer constant, but it follows Eq. (4.12) above. Combining Eqs. (4.10) and (4.12), and after some algebraic manipulations, one has

$$\Delta T \sim T_w - T_i \tag{4.14}$$

That is, the final response of the PCM is to reach a constant temperature in the limit after a long enough period. In all the above equations, the temperature difference scales with $\Delta T \sim T - T_i$, where the initial temperature is known and is different from that of melting. The dimensionless early and late stages are given, respectively, as functions of Fo, by

$$\theta_T = \frac{\Delta T}{T_w - T_i} \sim \begin{cases} \mathrm{Fo}^{1/2}, & \text{early regime} \\ 1, & \text{late regime} \end{cases} \tag{4.15}$$

With no melting, the final temperature would merge to the wall temperature but in presence of melting, the extreme for temperature would be that of phase change.

In the above formulation, for either isothermal or isoflux heating, an approximate solution can be developed. The temperature distribution would follow the early or late regime in the limits when time is close to zero or long enough (infinity), respectively. These two solutions would be patched away from those limits and as noted by Bejan [11], this approximate solution would come with a maximum error when either of the two extreme solutions breaks down at a certain time, here denoted as critical time. The critical time can be obtained by equating the temperature solution from early and late regime. Interestingly, for both cases of isoflux or isothermal wall heating, the critical time scale is given by

$$\mathrm{Fo} = 1 \tag{4.16}$$

That is, the two extremes merge when time reaches a critical scale

$$t_c \sim \frac{H^2}{\alpha} \tag{4.17}$$

Figure 4.3 schematically reveals the evolution of dimensionless temperature for both scenarios (where subscripts H and T denote isoflux and isothermal wall heating cases, respectively).

The above equations should be used to approximate the temperature response for a given point (here a plane parallel to the heating plate) before any near-wall melting is observed.

One notes a concave transient temperature when plotted versus time (or Fo if a non-dimensional plot is generated). Interestingly, however, it can be shown that once a thin near-wall region starts to undergo phase change, a convex temperature plot is to be expected instead. The reason can be explained using a heat balance equation for

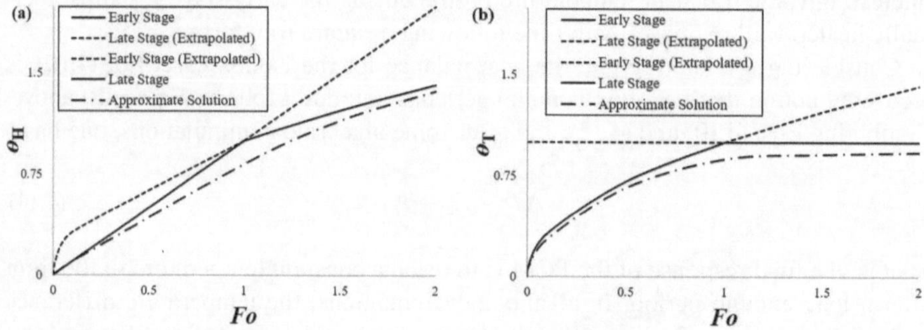

FIGURE 4.3 Schematics of temporal temperature variations for isoflux (a) or isothermal (b) scenarios.

a control mass of width H thicker than the melt layer ($H > s$). The heat transferred to tank is conducted through the liquid layer (of thermal conductivity k_l) and then transferred to a thin layer of thickness s over which the PCM undergoes phase change. The difference between the conducted and latent heat is then conducted across our arbitrarily selected control volume of width H (of thermal conductivity k_s). Using an order of magnitude analysis, this can be mathematically formulated as

$$k_l \frac{T_w - T_m}{s} - \frac{\rho h_{sf}}{t} s \sim k_s \frac{T_m - T}{H} \tag{4.18}$$

The above equation can be rearranged to read

$$T \sim T_m + (T_w - T_m)\left(\frac{k_l}{k_s}\frac{H}{s}\right)\left(\frac{\rho h_{sf} H^2}{k_l}\frac{s^2}{T_w - T_m}\frac{1}{t} - 1\right) \tag{4.19}$$

That is the temperature difference can be expressed as inverse-linearly proportional to time, which will be convex when plotted versus time. This trend has been observed in the data pertinent to experimental investigations for instance those published in (see Figure 3 of Mancin et al. [10]). While it is interesting to be able to explain the change in the temperature profile of a given plane above the heated wall (temporal variation), one can use this as a precursor for melting. That is, the temperature of a given plane will increase with time following a concave profile if that plane is close enough to the heated wall. Once that plane undergoes phase change, the temperature of other planes, further away from the heated wall, would follow a convex trend instead. This is mainly because of the change in the heating mechanism (directly from the heat source or through a melt front). By plotting the temperature of given planes in the tank, one would initially observe a concave temporal response. With time, a turning point would appear when the temperature profile tends to show a convex profile versus time. This would be a harbinger denoting phase change at the heated wall (within the accuracy limits imposed by temperature response delay).

4.2.2 Tubular, Cylindrical and Spherical Enclosures

In the absence of free convection, Peremans [12] modeled the PCM cuboids as different rectangles, as shown in Figure 4.4:

- 1D Transport: The cuboid is heated from one side, so that the heat can be diffused perpendicularly to the heated wall in one direction. The tubular PCM can also fit into this category when the ratio of the outer radius of annulus to the thickness of PCM layer tends to infinity.
- 2D Transport: Cylindrical PCM is characterized as a radial heat transport problem. Observing the system in the Cartesian systems, this translates to heat transport in two dimensions, like a cuboid PCM heated in two normal directions.

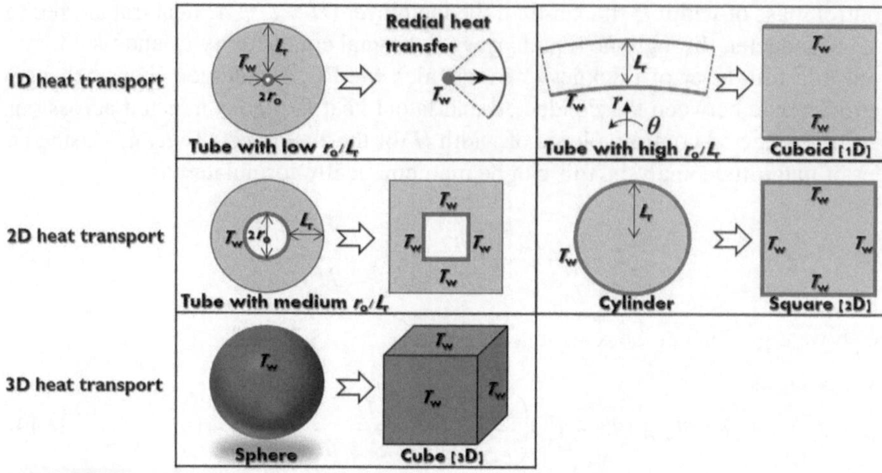

FIGURE 4.4 Classification of different container configurations according to heat transport dimensions.

- 3D Transport: A spherical PCM tank is characterized as a radial 3D heat transport problem in spherical coordinates. Similarly, a cuboid PCM is heated from all three directions and falls in this classification.

Using Buckingham π theorem, dimensional analysis was done for cuboid, tubular, cylindrical and spherical PCMs. However, explicit correlations were not presented while it was indicated that the Stefan number, is the only key parameter to affect the charging time for each geometry.

The dimensionless charge time, in the form of Fourier number, can be written as

$$\overline{t_{\text{ch}}} = \overline{\text{Fo}_{L_r}} = \frac{\alpha t_{\text{melt}}}{L_r^2} \tag{4.20}$$

where L_r is the characteristic length of heat transfer within PCM. Figure 4.5 exhibits the comparison of different configurations of PCM, consisting of 1D cuboid, 2D square, 3D cube, cylinder, sphere and three different tubes with various ratios of the outer radius of annulus to the thickness of PCM layer.

One notices that the shortest charge time of PCM is in a spherical enclosure, conversely, the tubular PCM with low $\overline{r_o} = r_o/L_r$, where r_o denotes the internal radius of the annulus and L_r is the difference between external and internal radii, offers the longest time for complete melting. Since the heat transfer distance is equivalent, the ratio of heating area to the PCM volume, $A_{\text{heating}}/V_{\text{PCM}}$, should be taken into consideration. For the spherical and cubic PCMs, the ratio is $3/L_r$, and the value becomes $2/L_r$ for cylindrical and squared PCMs. The ratio reduces to $1/L_r$ for 1D cuboidal and high $\overline{r_o}$ tubular PCMs. For tubular PCMs, as the $\overline{r_o}$ decreasing, the ratio keeps diminishing until reaching zero.

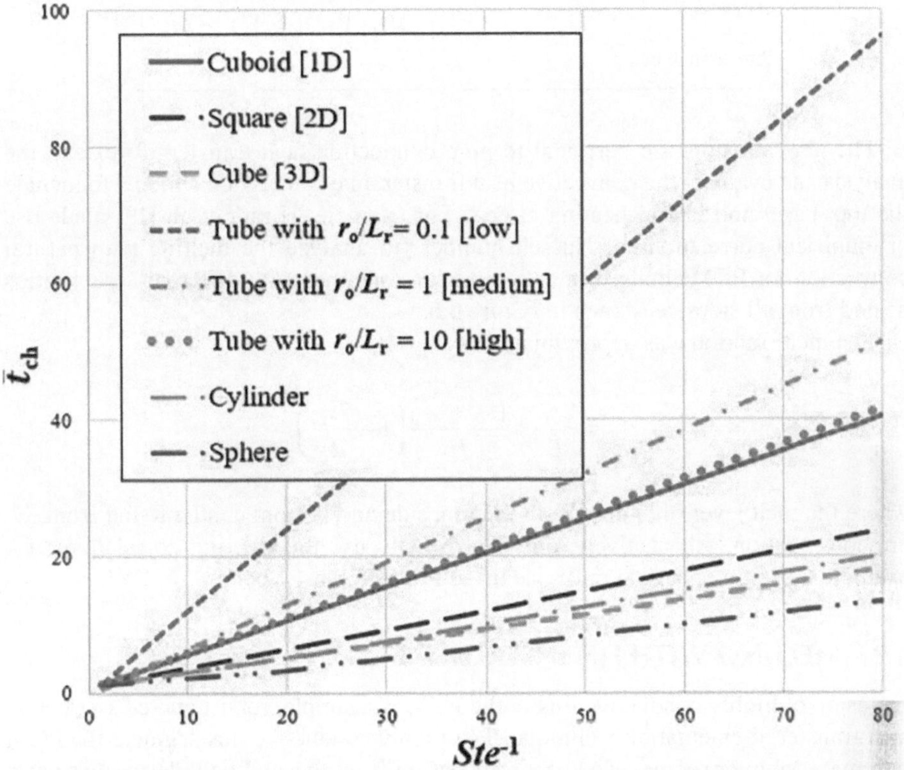

FIGURE 4.5 Dimensionless charge times of different cuboidal, tubular, cylindrical and spherical PCMs with respect to Stefan number.

There are limited analytical reports under convection-free assumption regarding tubular, cylindrical and spherical configurations due to the mathematical complexity. One example is raised for the outward melting inside a horizontally oriented tube [2], where the melt front, temperature field and melting time are

Melt front	$2s^2 \ln(s/r_i) = s^2 - r_i^2 + 4\alpha \text{Ste} \int_0^t \theta \, dt$
Temperature field	$\theta = \ln(r/s)/\ln(r_i/s), \; r_i \le r \le s$
Melting time	$t_{\text{melt}} = \left[r_i^2 - r_o^2 + 2r_o^2 \ln(r_o/r_i)\right]/(4\alpha \text{Ste})$

Another case is for inward melting within a spherical enclosure with the radius of $\bar{r}_o$ [2], the instantaneous melt front location and temperature distribution are

Melt front	$2(s/r_o)^3 - 3(s/r_o)^2 + 1 = 6\tau_{r_o}$
Temperature field	$\theta = (1 - s/r)/(1 - s/r_o), s \le r_o$

The above results are pertinent to pure conduction heat transfer. To extend the analysis and evaluate the convective heat transfer, one can rely on a model to include the top, basal and lateral heating effects. For example, Hirata et al. [13] made use of empirical correlations of Nusselt numbers to analyze the melting with natural convection for PCMs housed in a rectangular container with different aspect ratios heated from all sides, as shown in Figure 4.6.

The melt fraction was approximated as

$$\gamma = \frac{2s_L}{L} + \frac{(s_B + s_T)}{H}\left(1 - \frac{2s_L}{L}\right) \tag{4.21}$$

where the melt layer thickness from the top, side and bottom can be found from the previous section. Alternatively, one can directly use the existing correlations, for example those for spherical tanks, as listed in the Table 4.1 below.

4.3 MELTING WITH FIN INSERTS

Inclusion of highly conductive fins in the PCM is a simple, robust and cost-effective heat transfer augmentation technique. The highly conductive fins segment the PCM into many smaller sectors. The heat transported from thermal boundaries dissipates

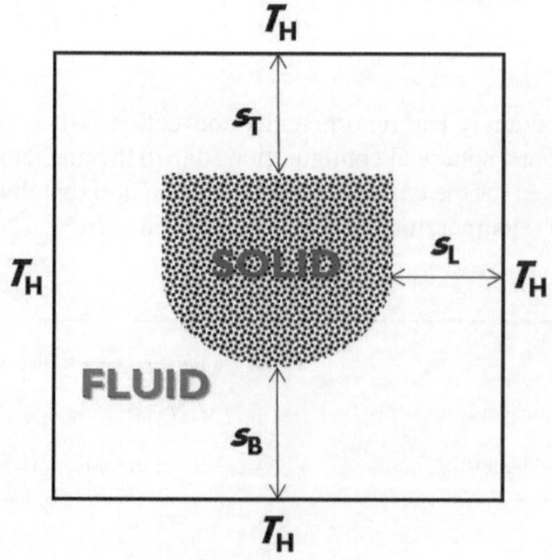

FIGURE 4.6 Analytical model of melting in a rectangular box isothermally heated from all sides.

TABLE 4.1
Correlations of Liquid Fraction for Melting in Spheres

References	Correlations of Liquid Fraction	Ranges of Application
Ho and Viskanta [14]	$\gamma = 0.782\left(\frac{\tau_R \mathrm{Ra}_R^{1/5}}{1+\mathrm{Ste}_i}\right)^{0.685}$	$4\times10^5 < \mathrm{Ra}_R < 1.4\times10^6$; $0.05 < \mathrm{Ste}/(1+\mathrm{Ste}_i) < 0.16$
Assis et al. [15]	$\gamma = 1-\left(1-\frac{\mathrm{Fo\,Ste}^{1/3}\mathrm{Gr}_R^{0.25}}{1.9}\right)^{3/2}$	$0.02 \le \mathrm{Ste} \le 0.1$; $\mathrm{Pr} \ge 1$
Archibold et al. [16]	$\gamma = 1-\left(1-\frac{\mathrm{Fo\,Ste}^{1/3}\mathrm{Gr}_R^{0.25}}{2.9}\right)^{2}$	$0.047 < \mathrm{Ste} < 0.1$; $7.2 < \mathrm{Pr} < 9.1$ $1.2\times10^4 < \mathrm{Gr}_R < 1\times10^5$;
Archibold et al. [17]	$\gamma = 1-\left(1-\frac{\mathrm{Fo\,Ste}^{0.37}\mathrm{Gr}_R^{0.25}}{2.8}\right)^{2.35}$	$0.048 < \mathrm{Ste} < 0.194$; $8.73 < \mathrm{Pr} < 9.05$ $1.32\times10^4 < \mathrm{Gr}_R < 2.06\times10^5$;
Archibold et al. [18]	$\gamma = 1-\left(1-\frac{\mathrm{Fo\,Ste}^{0.33}\mathrm{Gr}_R^{0.27}\,\mathrm{Pr}^{0.37}\,\chi^{0.72}\zeta^{-0.02}}{9.5}\right)^{1.8}$	$0.048 < \mathrm{Ste} < 0.145$; $8.9 < \mathrm{Pr} < 35$; $1.32\times10^4 < \mathrm{Gr}_R < 4.21\times10^5$; $0.6709 \le \chi = 1 - k_{\mathrm{PCM}}/k_{\mathrm{shell}} \le 0.9945$; $0.00259 \le \zeta = T_{\mathrm{in}}/T_m \le 0.0259$
Fan et al. [19]	$\gamma = 1.181 - 1.188\exp\left(-\frac{\mathrm{Fo\,Ste}^{0.33}\mathrm{Gr}_R^{0.41}}{2.022}\right)$	N/A

quickly through the fins and then to the PCM. Breaking down the PCM to smaller volumes will enhance the heat transfer while the fins hinder the convective flow structures which could have significantly improved the heat transfer [20]. Hence, a conduction-convection balance to maximize the heat transfer must be sought.

According to the literature [21,22] the fin number density, fin thickness and fin size will affect the performance of the system. Among these, the fin thickness seems to deliver only marginal effect on the performance enhancement (which is to be expected for thin highly conductive fins tested in different labs). The fins can be considered as either "short" lumped fins or "long" semi-infinite fins.

4.3.1 "Short" Lumped Fins

These are fins with uniform and invariant temperature (close to T_w); i.e. no conduction resistance to heat transfer through the fins with immediate temperature response. Short fins, attached to a heated wall, are tested in a rectangular PCM-filled tank with different orientation (see Figure 4.7).

- Basal Heating

  In this case, boundary layers will develop along the fins leaving a stratified region between the adjacent vertical boundary layers. To minimize the

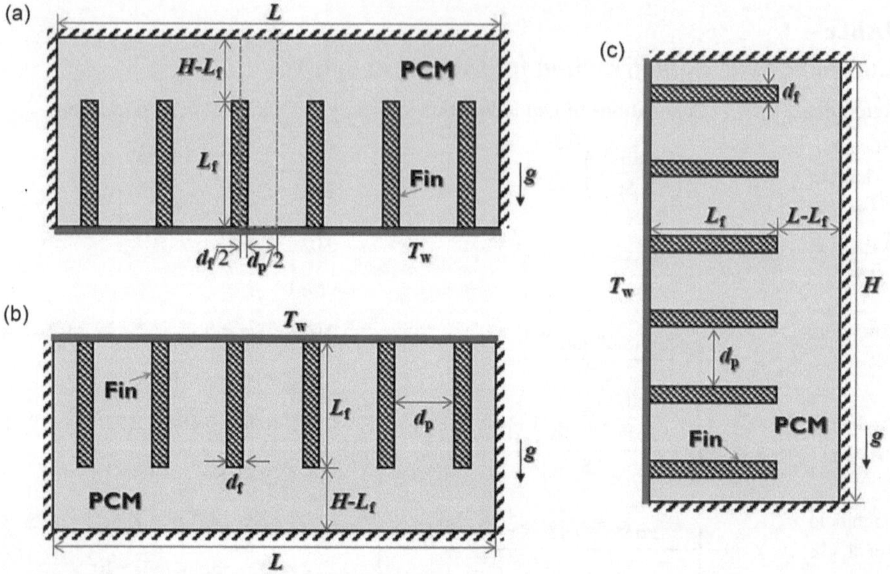

FIGURE 4.7 Schematic of finned rectangular PCM tank heated from the bottom (a), top (b) and side (c) walls.

melting time, an optimal fin spacing can be achieved by setting the two adjacent fins close enough to allow for the two hydrodynamic boundary layers to merge and form a rising plume. That is,

$$d_{p,\text{opt}} \sim 2L_f \text{Ra}_{L_f}^{-0.25}\,\text{Pr}^{0.5} \tag{4.22}$$

If the fin length reaches the top wall of the enclosure, namely $L_f \sim H$, one can expect the optimal fin spacing gets much smaller due to the sluggish convection. With fixed d_f/d_p (0.3), Shatikian et al. [23] reported that reducing the fin spacing always led to a faster melting rate. However, there can be a limit, which will be demonstrated in the following top heating part.

Those authors assess fin performance using the fin efficiency. The melt fraction was then expressed as an exponential function of the fin aspect ratio. Note that the numerical value of $\exp(-C \times \text{AR}_{\text{fin}})$ decreases with the fin aspect ratio and it tends to unity when the fin length approaches zero.

Assuming that the fin temperature is constant, equals to T_w, Lacroix and Benmadda [24] obtained the following optimal fin spacing:

$$d_{p,\text{opt}} = -4.173 \times 10^{-8}\,\text{Ra}_H + 1.4376 \tag{4.23}$$

where the $L_f/H = 0.75$ and the correlation was valid for $2.10 \times 10^6 \le \text{Ra}_H \le 8.57 \times 10^6$.

Here, considering constant heat flux as the thermal boundary, under $L_f = H$, Shatikian et al. [25] correlated melt fraction with a parameter combination, $\tau_{d_p/2}\mathrm{Ra}_H^{-1/4}\mathrm{Ra}_{d_p/2}^{1/6}$, for various fin length and thickness under the same $\mathrm{AR}_{\mathrm{fin}}$ constraint. Note that the temperature difference in those dimensionless parameters was replaced by $q_w''d_p/(2k_{\mathrm{PCM}})$. The Nusselt number correlation was summarized as

$$\mathrm{Nu} = \frac{q_w''d_p}{2k_{\mathrm{PCM}}\Delta T} = 5 + \frac{4}{\tau_{d_p/2}\mathrm{Ra}_H^{-1/4}\mathrm{Ra}_{d_p/2}^{1/6} + 0.015} \tag{4.24}$$

It indicates that increasing fin length and fin spacing can enhance the heat transfer rate.

However, according to Saha and Dutta [26], Eq. (4.24) does not monotonically hold as, in the limits, with lower and higher aspect ratios, the error is considerable. Therefore, three distinct correlations, based on different length scales, were proposed for rectangular enclosures with different aspect ratios ($\mathrm{AR} = H/d_p$), namely, shallow ($\mathrm{AR} \ll 1$), rectangular ($\mathrm{AR} \sim 1$) and tall ($\mathrm{AR} \gg 1$) enclosures, respectively (see Table 4.2).

We have conducted computational fluid dynamics simulation for the same problem to note that the melt front away from the fin tip in y direction (namely s_y) is linearly proportional to that from the side of the fin s_x (in *x*-direction) as shown in Figure 4.8. This is in line with He et al.'s [27] observation for a transient plume rising above a heated plate.

- Top Heating

  For the configuration shown in Figure 4.7b with $L_f = H$, Akhilesh et al. [28] define the PCM volume fraction

$$\varepsilon = \frac{1}{1 + d_f/d_p} \tag{4.25}$$

and the Number of fin-PCM units in the enclosure

$$N = \frac{L}{d_p + d_f} \tag{4.26}$$

TABLE 4.2
Correlations of Different Enclosures with Their Applicability Ranges

Enclosure Type	Correlations	Applicable Range
Shallow	$\mathrm{Nu} = 1.2\times10^{-3}\dfrac{H}{d_p}\dfrac{\mathrm{Ra}_H^{1/4}}{\tau_{d_p/2}}$	$3.6\times10^{-2} \le \dfrac{H}{d_p} \le 0.395$
Rectangular (medium AR)	$\mathrm{Nu} = 7.0\times10^{-2}\left(\dfrac{\mathrm{Ra}_H^{1/4}}{\tau_{d_p/2}\mathrm{Ra}_{d_p/2}^{1/6}}\right)^{3/4}$	$0.395 \le \dfrac{H}{d_p} \le 1.939$
Tall	$\mathrm{Nu} = 4.0\times10^{-5}\dfrac{H}{d_p}\dfrac{\mathrm{Ra}_{d_p/2}^{1/2}}{\tau_{d_p/2}}$	$1.939 \le \dfrac{H}{d_p} \le 5$

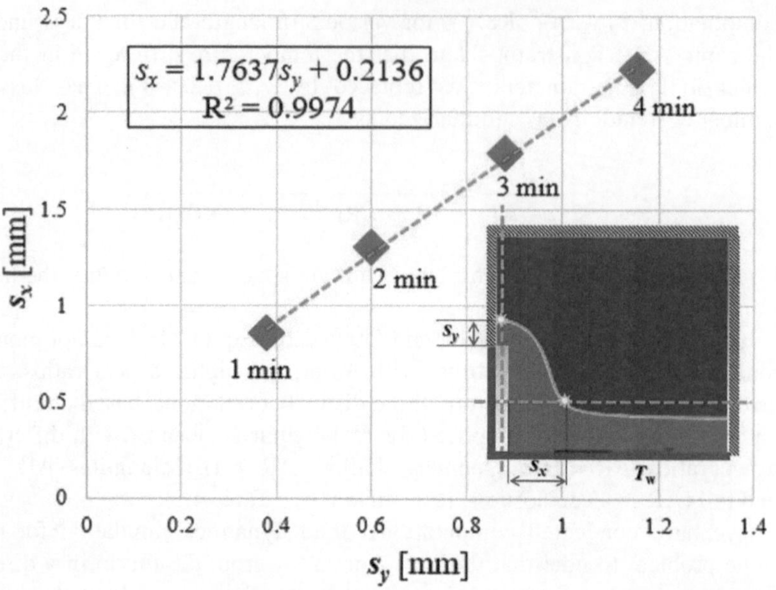

FIGURE 4.8 Relationship between vertical and horizontal melt fronts for isothermal basal melting with vertical fins.

where L denotes the length of the storage enclosure. Accordingly, one can get the number of fins per unit system length (fin number density)

$$n_f = \frac{N}{L} = \frac{1-\varepsilon}{d_f} \tag{4.27}$$

It is tempting to assume that the fastest melting rate coincides with infinite fin number density ($N \to \infty$). However, the simulation results indicate out that there is a critical value for n_f that prolongs the time to reach a set point temperature (SPT) beyond which the SPT time is more or less insensitive to n_f as the PCM has already been fully melted (see Figure 4.9).

The critical fin thickness scales with

$$d_{f,\text{crit}}^2 \sim \frac{k_{\text{PCM}} H\left(T_{\text{SPT}} - T_m\right)}{q''}\left[\frac{(1-\varepsilon)^2}{\varepsilon}\left(1+\left(\frac{1}{2}+\frac{(1-\varepsilon)\left(\rho c_p\right)_{\text{fin}}}{\varepsilon\left(\rho c_p\right)_{\text{PCM}}}\right)\text{Ste}\right)\right] \tag{4.28}$$

Similarly, Levin et al. [29] numerically investigated the effects fin volume fraction and fin number density on the melting time, as presented in Figure 4.10.

- Lateral Heating

  For the isothermal lateral heating case, Sharifi et al. [30] reported that increasing fin length and reducing fin spacing enhances the melting rate (see Figure 4.7c). Moreover, two melting behaviors, rapid and slow melting, were characterized. Rapid melting assumes $T_{\text{fin}} = T_w$, with liquid fraction

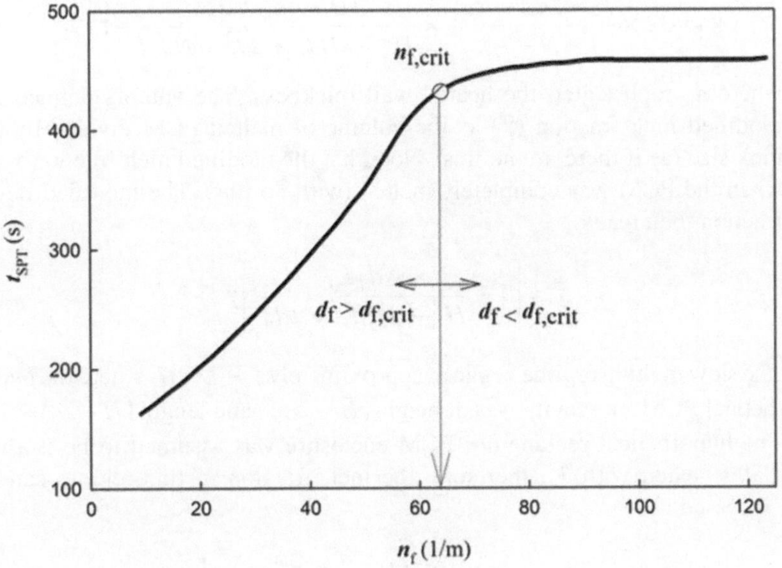

FIGURE 4.9 Time taken for the PCM system temperature to reach the set point temperature versus number of fins per system length.

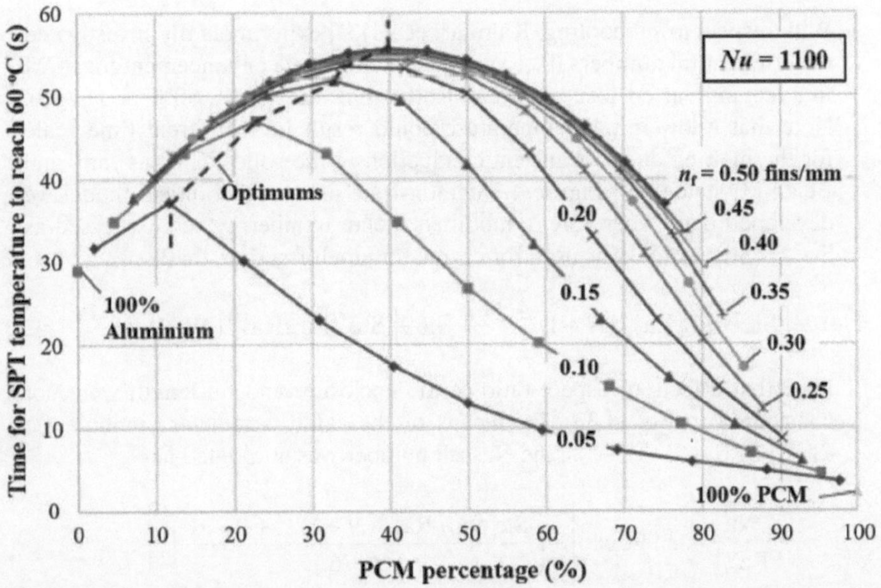

FIGURE 4.10 SPT time as a function of PCM percentage (ε) and number of fins per system length (n_f). (Source: Levin et al. [29].)

$$\gamma = \sqrt{2} \times \frac{H - 2d_w - Nd_f}{(N+1)} \times \frac{3H - 4d_w + 2Nd_f}{H^2 - 3Hd_w + 2d_w^2 - Nd_f L_f} \tau_{d_p/2}^{1/2} \tag{4.29}$$

where d_w represented the heated wall thickness. The authors defined the modified melt fraction ($\tilde{\gamma}$) as the volume of melted PCM divided by the tank size (as if there are no fins). Note that the modified melt fraction $\tilde{\gamma} = 1$ when the PCM was completely melted (with no fins). The modified liquid fraction then reads

$$\tilde{\gamma} = \left[1 - \frac{Nd_f L_f}{H^2 - 3Hd_w + 2d_w^2}\right]\gamma \tag{4.30}$$

The slow melting regime began at approximately $\gamma \sim L_f / H$, where the hypothetical PCM sub-cavity was of height $(H - 2d_w)$ and length $(H - L_f - d_w)$. The hypothetical rectangular PCM enclosure was assumed to be isothermally heated with T_w, therefore, the melt fraction in this regime can be expressed as

$$\gamma = \left[\frac{1.98}{2 + \text{Ste}} \times \frac{H^{5/4}(H - 2d_w)^{3/4}}{H^2 - 3Hd_w + 2d_w^2 - Nd_f L_f}\right]\tau_H \text{Ra}_H^{1/4} \tag{4.31}$$

The modified melt fraction was expressed as

$$\tilde{\gamma} = \left[1 - \frac{Nd_f L_f}{H^2 - 3Hd_w + 2d_w^2}\right]\gamma \tag{4.32}$$

With respect to subcooling, Kamkari et al. [31] experimentally investigated the effect of fin numbers (0, 1 and 3) on melting rate enhancement for PCM in a rectangular enclosure with an isothermal wall (55°C, 60°C and 70°C). Note that a low initial temperature could result in a different time scale for the melting due to transient conduction in the solid PCM, as Jany and Bejan [3] stated and Figure 4.3 demonstrate using a theoretical model we developed here. Therefore, a modified Stefan number, which expressed as $\text{Ste}^* = (\text{Ste}_i + \text{Ste})$, was used for correlating melt fraction, that is,

$$\gamma = -0.21\left[\text{Ste}^{*2}\text{Fo}_H \text{Ra}_H^{1/4}(N+1)^{1/2}\right]^{1.9} + 0.69\left[\text{Ste}^{*2}\text{Fo}_H \text{Ra}_H^{1/4}(N+1)^{1/2}\right]^{1.2} \tag{4.33}$$

Note that effects of aspect ratio of the enclosure and fin length were not considered in Eq. (4.33). By means of the same parameter combination ($\text{Ste}^{*2}\text{Fo}_H \text{Ra}_H^{1/4}(N+1)^{1/2}$), the Nusselt number was correlated as

$$\frac{\overline{\text{Nu}}}{\text{Ra}_H^{1/4}} = 2.996\exp\left[-\left(\frac{\text{Ste}^{*2}\text{Fo}_H \text{Ra}_H^{1/4}(N+1)^{1/2} + 0.731}{0.476}\right)^2\right]$$
$$+ 0.284\exp\left[-\left(\frac{\text{Ste}^{*2}\text{Fo}_H \text{Ra}_H^{1/4}(N+1)^{1/2} - 0.418}{1.795}\right)^2\right] \tag{4.34}$$

The transient surface-averaged Nusselt number was based on the height of the enclosure and the heat transfer coefficient was calculated as $\bar{h}(t) = Q_{\text{total}}(t)/\left(A_w\left(T_w - T_m\right)\Delta t\right)$, where the total heat transfer area was $\left(H + 2N \times L_f\right)W$. Moreover, the melt fronts of different fin numbers and wall temperatures were exhibited in Figure 4.11.

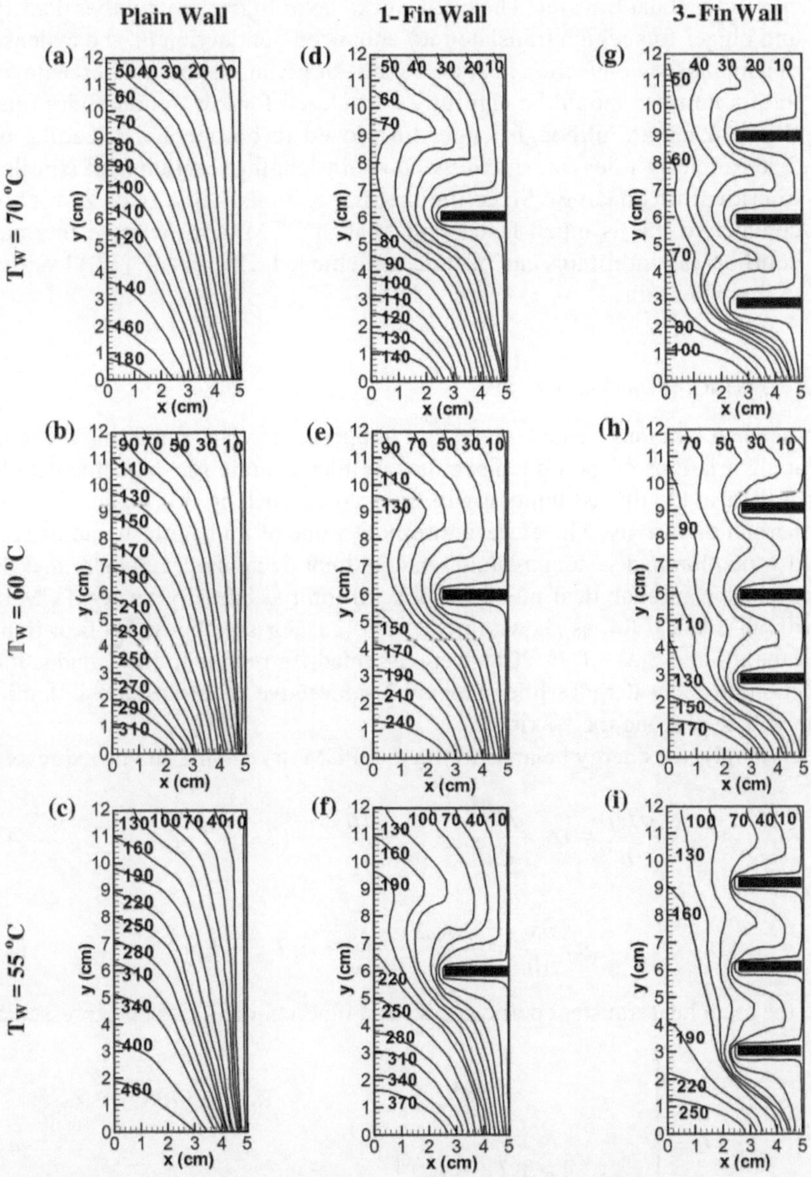

FIGURE 4.11 (a–i) Advances of melt front in the rectangular enclosure with different fin numbers and wall temperatures. (Source: Kamkari and Shokouhmand [31].)

To summarize, one could roughly ascertain the fin size and fin number density based on the requirement of storage capacity, cost and charging/discharging rates for convection-free cases (top heating for instance). For such cases, reducing fin spacing could benefit the melting rate enhancement up until the fin spacing reaches a critical value beyond which the improvement is marginal. For lateral and basal heating, the convection strength can be considerable and inclusion of fins highly deteriorates the convective heat transfer. The cited works favor higher fin number density and bigger fins which translate into enhanced conduction (at the expense of suppressed convection heat transfer). For given fin volume percentage, the fin number should be carefully considered for low conductivity fins. With fin length, although longer fin proved to be beneficial leading to faster melting rates, a critical value of fin length is anticipated (similar to critical fin spacing). Since the thermal response time scale of highly conductive fins is much faster than that of PCM, approximate thermal equilibrium within fins can be quickly achieved (Zhao et al. [32]) leading to inefficient fins.

4.3.2 "Long" Semi-Infinite Fins

The preceding citations treated the fin as a lumped capacitance ignoring the temporal/spatial variations of the fin temperature. While accurate for short fins with high thermal diffusivity, the assumption will breakdown for long fins and/or those with poor thermal diffusivity. Therefore, a threshold value of fin length should be sought beyond which lumped system assumption could not be applied. Lamberg and Siren [33] developed an analytical model for a semi-infinite laterally heated PCM storage with an internal fin, as shown in Figure 4.12a. For simplicity, 1D heat transfer within the fin and Ste $\to 0$ in PCM were assumed. In region 1, only conduction in x-direction was considered; while in region 2, convective heat transfer was dominant transferring heat along the y-axis.

Accordingly, the energy balance for fin and PCM in region 2 can be expressed as

$$\frac{\partial T_{\text{fin}}}{\partial t} = \alpha_{\text{fin}} \frac{\partial^2 T_{\text{fin}}}{\partial x^2} - \frac{2h}{(\rho c_p)_{\text{PCM}} d_f} (T_{\text{fin}} - T_m) \tag{4.35}$$

$$\rho_{\text{PCM}} h_{sf} \frac{\partial s_y}{\partial t} = \frac{k_{\text{PCM}} \Delta T}{x} \frac{\partial s_y}{\partial x} + h(T_{\text{fin}} - T_m) \tag{4.36}$$

where the local heat transfer coefficient, h, as a function of s_y, can be expressed as

$$h = \begin{cases} \dfrac{k_{\text{PCM}}}{s_y}, & \text{Ra}_{s_y} \le 1708 \\ 0.072 \left[\dfrac{g\beta(T_{\text{fin}} - T_m)}{\alpha \nu} \right]^{1/3} k_{\text{PCM}}, & \text{Ra}_{s_y} \le 1708 \end{cases} \tag{4.37}$$

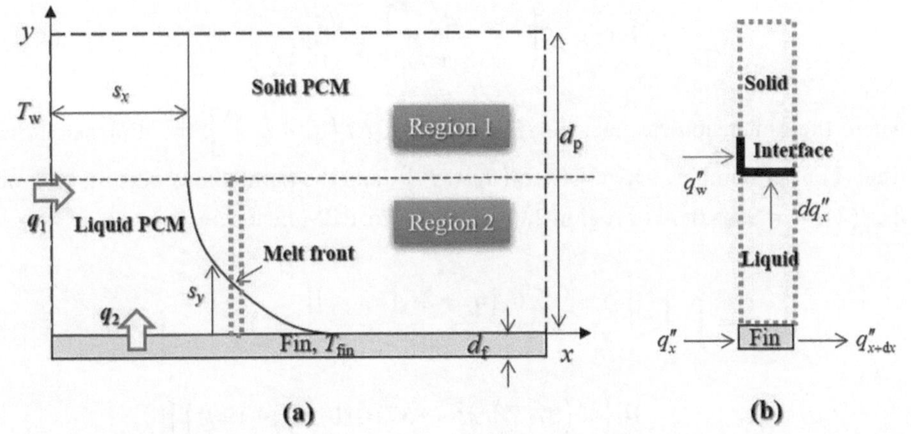

FIGURE 4.12 PCM storage with a semi-infinite fin (a) and energy flows in the arbitrary differential element (b).

Properties were evaluated at $(T_w + T_m)/2$ in Eq. (4.37). The fin temperature reads

$$T_{\text{fin}} = T_m + \Delta T \times e^{-mx} \times f(x,t) \tag{4.38}$$

where $m = \sqrt{2h/k_{fin}d_f}$ and the function $f(x,t)$ was

$$f(x,t) = \frac{1}{2}\left\{\left[1 - \text{erf}\left(\frac{x}{2\sqrt{\alpha_{\text{fin}}t}} - m\sqrt{\alpha_{\text{fin}}t}\right)\right] + e^{2mx}\left[1 - \text{erf}\left(\frac{x}{2\sqrt{\alpha_{\text{fin}}t}} + m\sqrt{\alpha_{\text{fin}}t}\right)\right]\right\} \tag{4.39}$$

Meanwhile, the melt front location, s_y, in the y direction was

$$s_y = h(T_{\text{fin}} - T_m)x\frac{-bx + \sqrt{(bx)^2 + 2abt}}{ab} \tag{4.40}$$

with $a = k_{\text{PCM}}\Delta T$ and $b = \rho_{\text{PCM}}h_{sf}$. The solution is invalid for $x \to 0$ therefore Neumann's exact solution is to be applied instead when x approaches zero.

Equations (4.38)–(4.40) were cross-verified numerically, where the analytical predicted melt front s_y was close to numerical results for $x > 0.025$ m. In addition, the fin temperatures remained unchanged after 280 s while the values were close to the initial temperature for $x > 0.2$ m. These phenomena indicate the existence of a critical time instant and fin length, which are required to be obtained for fin-PCM system design.

Mostafavi et al. [34] further conducted optimizations of fin volume fraction, $1 - \varepsilon = d_f/(d_p + d_f)$, in terms of maximizing total absorbed heat $(\bar{q}_1(\tau) + \bar{q}_2(\tau)$ for a given finite time during melting (see Figure 4.12). In region 1 (wall-to-PCM heat transfer),

$$\bar{q}_1(\tau) = 2\left(1 - \frac{d_f}{d_p + d_f}\right)\frac{\sqrt{\tau}}{\sqrt{\pi}\,\mathrm{erf}(\lambda)} \tag{4.41}$$

where the dimensionless heat flux $\bar{q} = q\Big/\left[c_{p,\mathrm{PCM}}\Delta T\left(d_p + d_f\right)^2\right]$, the dimensionless time (Fourier number) was $\tau = \alpha_{\mathrm{fin}} t\big/\left(d_p + d_f\right)^2$, and the parameter λ was the root of $\lambda\,\mathrm{erf}(\lambda)e^{\lambda^2} = \mathrm{Ste}/\sqrt{\pi}$. For region 2, (wall-to-fin-to-PCM heat transfer),

$$\bar{q}_2(\tau) = \int_0^{\tau}\int_0^{\bar{L}_f}\left[\frac{\theta_f}{\bar{s}_y} + \mathrm{Ste}\frac{\theta_f\left(\theta_f + 2\bar{s}_y\left(\theta_f'/\bar{s}_y'\right)\right)}{6\bar{s}_y} - \mathrm{Ste}^2\frac{\theta_f\left(40\left(\theta_f'/\bar{s}_y'\right)^2\bar{s}_y^2 + 85\bar{s}_y\theta_f\left(\theta_f'/\bar{s}_y'\right) + 19\theta_f^2\right)}{360\bar{s}_y}\right]d\bar{x}\,d\tau \tag{4.42}$$

where the dimensionless parameters were $\bar{L}_f = L_f\big/\left(d_p + d_f\right)$, $\bar{x} = x\big/\left(d_p + d_f\right)$, $\bar{s}_y = s_y\big/\left(d_p + d_f\right)$, $\theta_f = (T_f - T_m)/(T_w - T_m)$. The dimensionless melt front location was given as

$$\bar{s}_y = \sqrt{2\mathrm{Ste}\int_0^{\tau}\theta_f\left(1 - \frac{\mathrm{Ste}}{3}\theta_f + \frac{7\mathrm{Ste}^2}{45}\theta_f^2\right)d\tau} \tag{4.43}$$

Accordingly, effects of time available for heat transfer ($\bar{t}$) and thermal properties ($k_{\mathrm{fin}}/k_{\mathrm{PCM}}$) on the optimal fin volume fraction, $\left[d_f/\left(d_p + d_f\right)\right]_{\mathrm{opt}}$ (that maximizes the total heat absorption), were investigated with a sample of results presented in Figure 4.13.

One realizes that higher thermal conductivity ratio (fin to PCM) and longer time available for heat transfer would lead to a smaller optimal fin volume fraction. Keep

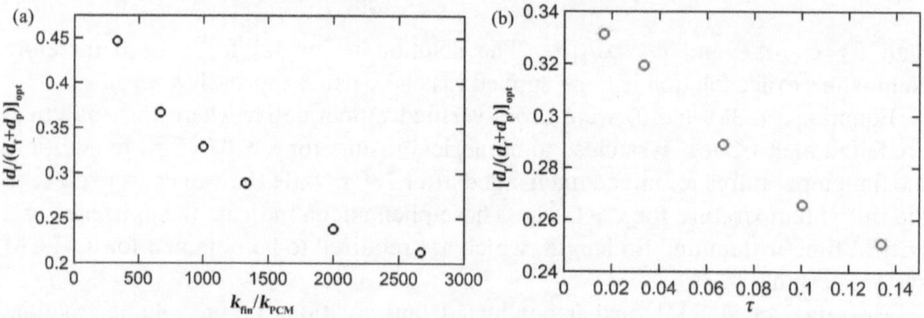

FIGURE 4.13 Effects of thermal properties (a) and time available for heat transfer (b) on the optimal fin volumetric percentage $\left[d_f/\left(d_p + d_f\right)\right]_{\mathrm{opt}}$. (Source: Mostafavi et al. [34].)

in mind that these results are developed by neglecting natural convection effects therefore their applicability is limited.

In the interest of brevity, we would not present results pertinent to non-straight fins, but the reader may consult Zhao et al. [35] for a recent survey of the literature on the topic. A particular feature to those fins is their potential to induce localized melting. That is, islands of melted PCM can be observed downstream the melt front putting the melt front between the heated wall and those locally melted patchy islands. This can be attributed to high conductivity of the fins transferring the heat faster from the heated wall to the bulk of PCM compared with the slower conduction-dominated melt front movement. Zhao et al. [36] provides a criterion to predict and evaluate the phenomenon.

Adjacent fins can be modeled as a porous medium; see Bejan [37] or Hooman and Gurgenci [38]. While porosity can be directly calculated using geometrical details, to determine permeability, form drag coefficient and effective thermal conductivity one must rely on experimental data while theoretical models can also be developed similar to Zhao et al. [39], Hooman and Dukhan [40] and Xiao et al. [41,42]. Volume-averaged models to simulate fin-PCM compounds are computationally less expensive than continuum approaches. Moreover, such models lend `themselves better to theoretical analysis. For instance, basal heating of a finned tank can be modeled as Rayleigh–Darcy–Benard convection for which the onset of convection coincides with a minimum Rayleigh–Darcy number value of about 40. This leads to a maximum (approximate) porosity value of $\varepsilon = 1 \Big/ \left\{1 + \left[\left(k_f/k_s\right)\left(L_f/d_p\right)\left(\mathrm{Ra}_{d_p}/4000\right)\right]^{0.33}\right\}$. With this, fin-to-fin distance can be obtained and a full design can follow using appropriate constrains. Analytical studies focusing on melting in porous media are reviewed by Zhao et al. [35], and this chapter will not cover the same ground. Insertion of a porous matrix in a PCM tank can remarkably accelerate the charging/discharging rates, meanwhile, it will inevitably increase the tank size, add to the cost, suppress the natural convection of liquid PCM and brings in material compatibility issues. An alternative to expensive foams has been the use of additively manufactured periodic structures; see Righetti et al. [43], Zhao et al. [44] and Opolot et al. [45] where the latter quantitatively evaluated the effects of thermal contact resistance for such systems.

4.4 CONCLUSION

Melting characteristics of PCM in different configurations (planar, tubular, cylindrical and spherical) were reviewed and analyzed. In particular, top, lateral, basal and inclined heating of a PCM in the rectangular enclosure was clarified mathematically in detail. Two important points are listed here:

- For melting in the same rectangular configuration, the strength of natural convection is negligibly small for top heating and strongest for basal heating. With lateral heating, the “shrinking solid” stage is the most time-consuming, almost half of the total melting time.
- Criteria for assuming 1D or 2D heat transfer are presented. In the absence of such simplifications, one must rely on 3D models.

Introduction of highly conductive fins can augment the heat transfer, but the fins must be designed carefully. Fin shape, length, thickness, material, as well as fin number density can significantly influence the performance of the storage system.

REFERENCES

1. J. Stefan, Über die Theorie der Eisbildung, insbesondere über die Eisbildung im Polarmeere. *Annalen der Physik und Chemie*. 42 (1891) 269–286.
2. S. Kakac, Y. Yener, C.P. Naveira-Cotta. *Heat conduction*. CRC Press, Boca Raton, FL, 2018.
3. P. Jany, A. Bejan. Scaling theory of melting with natural convection in an enclosure. *International Journal of Heat and Mass Transfer*. 31(6) (1988) 1221–1235.
4. S. Madruga, J. Curbelo. Dynamic of plumes and scaling during the melting of a Phase Change Material heated from below. *International Journal of Heat and Mass Transfer*. 126 (2018) 206–220.
5. B. Favier, J. Purseed, L. Duchemin. Rayleigh-Bénard convection with a melting boundary. *Journal of Fluid Mechanics*. 858 (2019) 437–473.
6. S. Grossmann, D. Lohse. Scaling in thermal convection: a unifying theory. *Journal of Fluid Mechanics*. 407 (2000) 27–56.
7. J. Vogel, J. Felbinger, M. Johnson. Natural convection in high temperature flat plate latent heat thermal energy storage systems. *Applied Energy*. 184 (2016) 184–196.
8. B. Kamkari, H. Shokouhmand, F. Bruno. Experimental investigation of the effect of inclination angle on convection-driven melting of phase change material in a rectangular enclosure. *International Journal of Heat and Mass Transfer*. 72 (2014) 186–200.
9. X.H. Yang, X.Y. Wang, Z. Liu, Z.X. Guo, K. Hooman. Thermal performance assessment of a thermal energy storage tank: effect of aspect ratio and tilted angle. *International Journal of Energy Research*. 45(7) (2021) 11157–11178.
10. S. Mancin, A. Diani, L. Doretti, K. Hooman, L. Rossetto. Experimental analysis of phase change phenomenon of paraffin waxes embedded in copper foams. *International Journal of Thermal Sciences*. 90 (2015) 79–89.
11. A. Bejan. Simple methods for convection in porous media: scale analysis and the intersection of asymptotes. *International Journal of Energy Research*. 27(10) (2003) 859–874.
12. B. Peremans. Optimal Parameter and Fin Design for PCM-based Thermal Energy Storage. Ph.D. thesis, 2020.
13. T. Hirata, Y. Makino, Y. Kaneko. Analysis of natural convection melting inside isothermally heated horizontal rectangular capsule. *Wärme-und Stoffübertragung*. 28(1) (1993) 1–9.
14. C.J. Ho, R. Viskanta. Heat transfer during inward melting in a horizontal tube. *International Journal of Heat and Mass Transfer*. 27(5) (1984) 705–716.
15. E. Assis, L. Katsman, G. Ziskind, R. Letan. Numerical and experimental study of melting in a spherical shell. *International Journal of Heat and Mass Transfer*. 50 (9–10) (2007) 1790–1804.
16. A.R. Archibold, M.M. Rahman, D.Y. Goswami, E.L. Stefanakos. Parametric investigation of the melting and solidification process in an encapsulated spherical container. *Energy Sustainability*. 44816 (2012) 573–584.
17. A.R. Archibold, J. Gonzalez-Aguilar, M.M. Rahman, D.Y. Goswami, M. Romero, E.K. Stefanakos. The melting process of storage materials with relatively high phase change temperatures in partially filled spherical shells. *Applied Energy*. 116 (2014) 243–252.
18. A.R. Archibold, M.M. Rahman, D.Y. Goswami, E.K. Stefanakos. Analysis of heat transfer and fluid flow during melting inside a spherical container for thermal energy storage. *Applied Thermal Engineering*. 64(1–2) (2014) 396–407.

19. L.W. Fan, Z.Q. Zhu, Y. Zeng, Q. Ding, M.J. Liu. Unconstrained melting heat transfer in a spherical container revisited in the presence of nano-enhanced phase change materials (NePCM). *International Journal of Heat and Mass Transfer.* 95 (2016) 1057–1069.
20. J. Vogel, M. Johnson. Natural convection during melting in vertical finned tube latent thermal energy storage systems. *Applied Energy.* 246 (2019) 38-52.
21. B. Kamkari, H. Shokouhmand. Experimental investigation of phase change material melting in rectangular enclosures with horizontal partial fins. *International Journal of Heat and Mass Transfer.* 78 (2014) 839–851.
22. C.R. Zhao, M. Opolot, M. Liu, F. Bruno, S. Mancin, K. Hooman. Numerical study of melting performance enhancement for PCM in an annular enclosure with internal-external fins and metal foams. *International Journal of Heat and Mass Transfer.* 150 (2020) 119348.
23. V. Shatikian, G. Ziskind, R. Letan. Numerical investigation of a PCM-based heat sink with internal fins. *International Journal of Heat and Mass Transfer.* 48(17) (2005) 3689–3706.
24. M. Lacroix, M. Benmadda. Analysis of natural convection melting from a heated wall with vertically oriented fins. *International Journal of Numerical Methods for Heat & Fluid Flow.* 8(4) (1998) 465–478.
25. V. Shatikian, G. Ziskind, R. Letan. Numerical investigation of a PCM-based heat sink with internal fins: constant heat flux. *International Journal of Heat and Mass Transfer.* 51(5–6) (2008) 1488–1493.
26. S.K. Saha, P. Dutta. Heat transfer correlations for PCM-based heat sinks with plate fins. *Applied Thermal Engineering.* 30(16) (2010) 2485–2491.
27. S.Y. He, F. Sabri, K. Hooman. Transient natural convection: scale analysis of dry cooling towers. *Journal of Thermal Analysis and Calorimetry.* 139(4) (2020) 2891–2897.
28. R. Akhilesh, A. Narasimhan, C. Balaji. Method to improve geometry for heat transfer enhancement in PCM composite heat sinks. *International Journal of Heat and Mass Transfer.* 48(13) (2005) 2759–2770.
29. P.P. Levin, A. Shitzer, G. Hetsroni. Numerical optimization of a PCM-based heat sink with internal fins. *International Journal of Heat and Mass Transfer.* 61 (2013) 638–45.
30. N. Sharifi, T.L. Bergman, A. Faghri. Enhancement of PCM melting in enclosures with horizontally-finned internal surfaces. *International Journal of Heat and Mass Transfer.* 54(19–20) (2011) 4182–4192.
31. B. Kamkari, H. Shokouhmand. Experimental investigation of phase change material melting in rectangular enclosures with horizontal partial fins. *International Journal of Heat and Mass Transfer.* 78 (2014) 839–851.
32. C.R. Zhao, Y.B. Sun, J.Y. Wang, S.Y. He, K. Hooman. Fin Design Optimization to Enhance PCM Melting Rate inside a Rectangular Enclosure. *Applied Energy* (Under review).
33. P. Lamberg, K. Siren. Analytical model for melting in a semi-infinite PCM storage with an internal fin. *Heat and Mass Transfer.* 39(2) (2003) 167–176.
34. A. Mostafavi, M. Parhizi, A. Jain. Theoretical modeling and optimization of fin-based enhancement of heat transfer into a phase change material. *International Journal of Heat and Mass Transfer.* 145 (2019) 118698.
35. C.R. Zhao, M. Opolot, M. Liu, J. Wang, F. Bruno, S. Mancin, K. Hooman. Review of analytical studies of melting/solidification enhancement with fin and/or foam. *Applied Thermal Engineering.* 207 (2022) 118154.
36. C.R. Zhao, M. Opolot, M. Liu, J. Wang, F. Bruno, S. Mancin, K. Hooman. Periodic Structures for Melting Enhancement: observation of critical cell size and localized melting. *International Journal of Heat and Mass Transfer.* (under review).
37. A. Bejan. *Convection Heat Transfer.* John Wiley & Sons, Hoboken, NJ, 2013.

38. K. Hooman, H. Gurgenci. Porous medium modeling of air-cooled condensers. *Transport in Porous Media.* 84(2) (2010) 257–273.
39. C.R. Zhao, M. Opolot, M. Liu, F. Bruno, S. Mancin, K. Hooman. Phase change behaviour study of PCM tanks partially filled with graphite foam. *Applied Thermal Engineering.* (2021) 117313.
40. K. Hooman, N. Dukhan. A theoretical model with experimental verification to predict hydrodynamics of foams. *Transport in Porous Media.* 100(3) (2013) 393–406.
41. T. Xiao, X.H. Yang, K. Hooman, T.J. Lu. Analytical fractal models for permeability and conductivity of open-cell metallic foams. *International Journal of Heat and Mass Transfer.* 177 (2021) 121509.
42. T. Xiao, J.F. Guo, X.H. Yang, K. Hooman, T.J. Lu. On the modelling of heat and fluid transport in fibrous porous media: analytical fractal models for permeability and thermal conductivity. *International Journal of Thermal Sciences.* 172 (2022) 107270.
43. G. Righetti, G. Savio, R. Meneghello, L. Doretti, S. Mancin, Experimental study of phase change material (PCM) embedded in 3D periodic structures realized via additive manufacturing. *International Journal of Thermal Sciences.* 153 (2020) 106376.
44. C.R. Zhao, M. Opolot, M. Liu, F. Bruno, S. Mancin, R. Flewell-Smith, K. Hooman. Simulations of melting performance enhancement for a PCM embedded in metal periodic structures. *International Journal of Heat and Mass Transfer.* 168 (2021) 120853.
45. M. Opolot, C.R. Zhao, M. Liu, S. Mancin, F. Bruno, K. Hooman. Investigation of the effect of thermal resistance on the performance of phase change materials. *International Journal of Thermal Sciences.* 164 (2021) 106852.

5 Multi-Scale Modeling in Solid–Liquid Phase Change Conjugate Heat Transfer for Thermal Energy Storage Applications

Qinlong Ren, Li Chen, Yu-Bing Tao, and Wen-Quan Tao
Xi'an Jiaotong University

CONTENTS

DOI: 10.1201/9781003213260-5

5.1 INTRODUCTION

Solar energy becomes a promising alternative for overcoming the shortage of fossil energy and world environmental pollution during the past decades. However, the intermittent nature of solar energy limits its flexible large-scale applications. Thermal energy storage (TES) offers an indispensable bridge between the solar energy conversion and utilization at different peak periods [1]. Generally, the TES technologies are classified into sensible heat storage systems, latent heat storage (LHS) systems, and chemical heat storage systems. Based on the advantages of high energy storage density and nearly constant operating temperature, LHS system is attractive for storing solar thermal energy through solid–liquid phase change [2,3]. Under this situation, the heat transfer capability of LHS system plays an essential role in its energy storage efficiency. Nevertheless, the thermal conductivity of phase change materials (PCMs) is usually low, impeding the corresponding heat absorption and release rates in the LHS system. Facing this scientific challenge, effectively enhancing the solid–liquid phase change conjugate heat transfer is an inevitable task for the practical applications of LHS units in the renewable solar thermal energy system.

To ameliorate the thermal behavior of LHS system, several heat transfer enhancement technologies are developed in recent years including adding nanoparticles, using extended fins, and inserting porous media [4,5]. When nanoparticles such as carbon nanotubes, graphite, graphene, metal, and metal oxide are dispersed into the PCMs, the thermophysical properties of nano-enhanced PCMs including thermal conductivity, latent heat capacity, melting temperature, density, and viscosity are obviously altered [6]. Specifically, due to the nanoparticle clustering phenomenon, the nano-enhanced PCMs usually exhibit higher thermal conductivity than that predicted by the effective medium theories [7], making them convincing for improving thermal performance of LHS system. However, the existence of nanoparticles also increases the viscosity of PCMs [8], significantly weakening the convective heat transfer. Besides, the thermal stability of nano-enhanced PCMs is also challenging because the dispersed nanoparticles confront aggregation and sedimentation during charging and discharging cycles of LHS system, limiting the lifetime and robustness of nano-enhanced PCMs [9]. Alternatively, due to the simple fabrication and low construction cost, high thermal conductive extended fins with numerous geometries like longitudinal, circular/annular, plate, pin, and biomimetic tree shape are widely inserted into the LHS system to consolidate its heat transfer capability [10]. Even though the extended fins enhance the heat transfer area with improved energy storage rate, it also attenuates the energy storage capacity of LHS system to some extent. Furthermore, despite that the prolonged fin length in LHS system strengthens the heat transfer depth, the fin efficiency is commonly decreased due to its reduced inside temperature [11]. Under this circumstance, the extended fin configurations need to be optimized in consideration of the conjugate heat transfer occurring on the interface between fins and PCMs to maximize the thermal response of LHS system. On the other hand, diversified porous media with high thermal conductivity such as metal foams, metal fiber felts, expanded graphite, and graphite foam are integrated into the LHS system to accelerate its thermal response behavior [12]. Due to the interconnected high-speed conductive heat transfer channels inside porous media,

the effective thermal conductivity of porous medium-PCM composites is obviously improved [13]. Nevertheless, when the porous media exist in the LHS system, the convective heat transfer of PCMs is depressed due to the porous medium skeleton induced limited space for fluid flow development. Similar to the use of extended fins, the energy storage capacity of porous medium-PCM composites is reduced with their decreased porosity [14]. Besides, the gas void cavities are generated by the notable volume variation of PCMs during melting and solidification, affecting the conjugate heat transfer process inside the LHS system [15]. Furthermore, the contact thermal resistance between porous media and PCMs also influences their conjugate heat transfer rate. Hence, the morphology and thermophysical property of porous media require careful design when they are applied to enhance the thermal response ability of LHS system. The aforementioned discussions show that the heat transfer enhancement techniques for LHS system always exhibit the outstanding advantages simultaneously accompanied by several drawbacks. Therefore, heat transfer optimization becomes necessary for the LHS system to balance its energy storage efficiency and energy storage capacity in order to modulate various trade-off effects.

Due to the advantages of high flexibility and low cost compared with experimental measurements, multi-scale modeling is a powerful tool to investigate diverse heat and mass transfer processes in engineering fields [16–18]. In multi-scale modeling, the physical nature at different scales is fully considered and utilized such as molecular dynamics (MD) simulation at atomic scale, lattice Boltzmann method (LBM) at mesoscopic scale, and finite volume method (FVM) at macroscopic scale to solve the various governing equations in different regions. Then, the complicated transport phenomena are coupled through particular mathematical schemes at the interfacial boundaries between MD-LBM [19], MD-FVM [20], and LBM-FVM [21,22]. Due to its atomic characteristic, MD simulation is convincing for exploring the microscopic phenomena and thermophysical properties of PCMs including melting/freezing point, thermal conductivity, latent heat, and viscosity [23]. The melting/freezing point of PCMs is essentially computed through interface method [24] and void method [25] because of its significant influence on solid–liquid phase change during TES. Moreover, the thermal conductivity of PCMs could be calculated by equilibrium molecular dynamics (EMD) simulation [26,27] and non-equilibrium molecular dynamics (NEMD) simulation [28,29]. Therefore, the thermophysical properties of PCMs achieved by distinct MD simulations offer important foundations for investigating the latent heat TES process with different heat transfer enhancement technologies at mesoscopic and macroscopic scales.

Based on its mesoscopic kinetic nature and parallel characteristic, LBM is becoming a powerful numerical method to simulate complicated heat transfer and fluid flow phenomena [30,31]. The existing lattice Boltzmann models for solid–liquid phase change heat transfer could be generally classified into three categories: (1) the phase-field method [32,33], (2) the enthalpy-based method [34], and (3) the immersed boundary method [35]. For the phase-field-based LBM, the solid–liquid interface is implicitly tracked by an auxiliary parameter which smoothly varies across the diffusive phase interface. Consequently, extremely fine grids are required in the interfacial region. In addition, when the immersed boundary-based LBM is applied to simulate the solid–liquid phase change heat transfer, the solid–liquid interface is explicitly tracked by the Lagrangian nodes, while the temperature and fluid flow fields are solved on the Eulerian nodes. However, the interpolation process of physical variables between Lagrangian nodes and Eulerian nodes dramatically burdens the computational load. Alternatively, the enthalpy-based LBM becomes the most robust and efficient approach for investigating solid–liquid phase change conjugate heat transfer. Originally, the enthalpy-based LBM is developed by Jiaung et al. for heat conduction with solid–liquid phase change [36]. Even though their simulation of solid–liquid phase change heat conduction at a constant melting temperature with mushy zone is demonstrated to be consistent with analytical or previous numerical results, iteration steps for the non-linear latent heat source are necessary. To avoid the iteration of latent heat source term, Eshraghi and Felicelli derived an implicit enthalpy-based LBM scheme for heat conduction with solid–liquid phase change through solving a linear system of equations [37]. However, numerically solving a linear system of equations at each time step also decreases the computational efficiency. Through modifying the equilibrium distribution function for temperature, Huang et al. successfully developed a two-dimensional (2D) single-relaxation-time (SRT) enthalpy-based LBM for solid–liquid phase change heat transfer without iteration of latent heat source or solving a linear system of equations [38]. Under this situation, the computational efficiency of LBM for simulating solid–liquid phase change is obviously improved. However, the SRT enthalpy-based LBM suffers from the drawback of numerical instability, and the specific heat and thermal conductivity could not be individually decoupled from the relaxation time, limiting its application for handling coupled Dirichlet–Neumann conditions in the conjugate heat transfer. To overcome these shortcomings, Huang et al. further developed a 2D multiple-relaxation time (MRT) enthalpy-based LBM to reduce the numerical diffusion during solid–liquid phase change, and the specific heat and thermal conductivity are also decoupled with each other to guarantee the continuity of heat transfer at the interface between different materials [39]. Afterwards, several different MRT enthalpy-based LBM schemes are developed for solid–liquid phase change heat transfer including axisymmetric cylindrical model [40], axisymmetric spherical model [41], and three-dimensional (3D) Cartesian model [42] as well as its representative elementary volume (REV) scale model in porous media [43]. Particularly, the enthalpy-based LBM considering contact thermal resistance is also developed making it more appropriate to investigate the solid–liquid phase change conjugate heat transfer in the practical LHS system [44]. Based on the tremendous efforts, the enthalpy-based LBM is widely used to optimize the thermal performance of LHS system especially for pore-scale clarifying the conjugate heat transfer inside complicated porous medium-PCM composites through reconstructing their diverse morphologies by X-ray computed tomography [45] or numerical methods [46–49]. Furthermore, the graphic processor units (GPUs) computing becomes attractive in recent years, and the enthalpy-based LBM is also successfully fitted onto GPUs to achieve parallel simulation with high computational acceleration ratio [50–51], making LBM more powerful for designing the thermal performance of LHS system.

Besides, with the continuous development during the past half century, FVM is robust for computational fluid dynamics (CFD) and heat transfer problems [52]. In particular, FVM is also applied to model solid–liquid phase change heat transfer in solar TES systems [53]. Based on the IDEAL algorithm [54], Li et al. derived an FVM with consideration of non-Darcy effect, local natural convection, and thermal non-equilibrium for solid–liquid phase change conjugate heat transfer in porous medium-PCM composite [55]. In addition, Ling et al. developed a sharp-interface-based FVM with SIMPLER algorithm to simulate solid–liquid phase change heat transfer [56]. The solid–liquid interface is captured by combing volume-of-fluid (VOF) and level-set (LS) methods (VOSET), and the interaction between solid and liquid phases is handled by an immersed boundary method (IBM). Furthermore, as a commonly used FVM-based CFD commercial software, ANSYS Fluent is also widely used for designing the thermal behavior of LHS system with different heat transfer enhancement technologies for practical engineering applications [57–59]. From the aforementioned discussions, multi-scale modeling is an effective technique for optimizing the thermal response of LHS system. In this chapter, we highlight the multi-scale numerical methods for solid–liquid phase change conjugate heat transfer.

5.2 MULTI-SCALE NUMERICAL METHODS AND COUPLING SCHEMES

Multi-scale modeling takes the advantages of various numerical methods at different scales to investigate the complicated physical phenomena with high efficiency. In this section, the MD simulation for computing thermophysical properties of PCMs is firstly discussed. Then, the enthalpy-based MRT LBM in 2D and 3D Cartesian coordinates is elaborated, respectively. Besides, the reconstructing methodologies for porous media and the principle of GPUs computing are further presented. Then, the FVM is briefly introduced, and the coupling scheme between LBM and FVM is finally analyzed.

5.2.1 Molecular Dynamics Simulation

MD simulation is a powerful tool for the prediction of thermophysical properties of PCMs and the solid–liquid interface interactions during melting and solidification processes [60–62]. The microphysical model of PCMs is often constructed by common modeling tools, such as Materials Studio [63] or visual molecular dynamics (VMD) software [64]. Then, the MD simulations can be performed with software including Material studio or large-scale atomic/molecular massively parallel simulator (LAMMPS) package.

5.2.1.1 Common Used Force Field

The molecular force field is a functional model describing the potential energy interaction between the atoms in PCM, which is also known as potential energy function. In the classical MD, the potential energy function plays a vital role because it is closely related to the micro-motion of atoms, which directly affects the accuracy of simulation results. For the MD simulation of PCMs, it is important to select the suitable force field. In this section, the commonly used force fields for PCM are introduced.

a. CHARMM Potential

CHARMM is used to describe the interaction of organic systems, such as paraffin-based PCM. The pair styles compute Lennard Jones (LJ) and Coulomb interactions with additional switching or shifting functions that ramp the energy and/or force smoothly to zero between an inner and outer cutoff, which can be written as follows [61,62]:

$$E_{vdw} = \begin{cases} \mathrm{LJ}(r) & r < r_{\mathrm{in}} \\ S(r) \times \mathrm{LJ}(r) & r_{\mathrm{in}} < r < r_{\mathrm{out}} \\ 0 & r > r_{\mathrm{out}} \end{cases} \tag{5.1}$$

$$E_{\mathrm{coul}} = \begin{cases} C(r) & r < r_{\mathrm{in}} \\ S(r) \times C(r) & r_{\mathrm{in}} < r < r_{\mathrm{out}} \\ 0 & r > r_{\mathrm{out}} \end{cases} \tag{5.2}$$

$$\mathrm{LJ}(r) = 4\varepsilon\left[\left(\frac{\sigma}{r}\right)^{12} - \left(\frac{\sigma}{r}\right)^{6}\right] \tag{5.3}$$

$$C(r) = \frac{Cq_i q_j}{\varepsilon r} \tag{5.4}$$

$$S(r) = \frac{\left[r_{\mathrm{out}}^2 - r^2\right]^2 \left[r_{\mathrm{out}}^2 + 2r^2 - 3r_{\mathrm{in}}^2\right]}{\left[r_{\mathrm{out}}^2 - r_{\mathrm{in}}^2\right]^3} \tag{5.5}$$

where E_{vdw} is the van der Waals energy and E_{coul} is the electrostatic energy. $S(r)$ is the energy switching function. In addition, r_{in} and r_{out} represent the start distance and the end distance of the switching function, respectively.

$$E_{\mathrm{bond}} = K_b(b - b_0)^2 \tag{5.6}$$

$$E_{\mathrm{angle}} = K_\theta(\theta - \theta_0)^2 + K_{\mathrm{ub}}(r - r_{\mathrm{ub}})^2 \tag{5.7}$$

The angle style uses an additional Urey-Bradley angle term based on the distance between the first and third atoms in the angle.

$$E_{\mathrm{dihedral}} = K_\chi(1 + \cos(n\chi - \delta)) \tag{5.8}$$

b. LJ Potential

For CPCM containing pure PCM and nanoparticles, it is difficult to describe the complicated interactions with a specific potential function. A more general LJ potential energy is usually used to describe the interactions with relative less accuracy, which can be written as follows:

$$U_{\text{non-bond,LJ}}(r_{ij}) = \begin{cases} 4\varepsilon_{ij}\left(\dfrac{\sigma_{ij}^{12}}{r_{ij}^{12}} - \dfrac{\sigma_{ij}^{6}}{r_{ij}^{6}}\right) & r_{ij} < r_{\text{cut}} \\ 0 & r_{ij} \geq r_{\text{cut}} \end{cases} \tag{5.9}$$

where ε, r, and σ represent the well depth (kcal/mol), the distance between atoms (Å) and the characteristic length (Å) for atom i and j, respectively; ε_{ij} and σ_{ij} are all parameters related to atomic types and follow Lorentz–Berthelot mixing rules [29], which are, respectively, determined by:

$$\varepsilon_{ij} = \sqrt{\varepsilon_i \times \varepsilon_j} \tag{5.10}$$

$$\sigma_{ij} = \frac{\sigma_i + \sigma_i}{2} \tag{5.11}$$

5.2.2 Lattice Boltzmann Method

Based on its kinetic characteristics, LBM has been developed as an effective approach to simulate the solid–liquid phase change heat transfer with convenient interface tracking capability. The LBM-based simulation of LHS system could be classified into pore-scale method and REV-scale method [34]. In this section, the pore-scale lattice Boltzmann modeling of solid–liquid phase change conjugate heat transfer is highlighted.

5.2.2.1 Governing Equations for Solid–Liquid Phase Change Conjugate Heat Transfer

The enthalpy-based MRT LBM has already been developed as a major tool for investigating the solid–liquid phase change conjugate heat transfer in the LHS system. The macroscopic governing equation for the solid–liquid phase change heat transfer is given as [65]:

$$\frac{\partial(\rho_f h)}{\partial t} + \nabla\cdot(\rho_f c_{pf} \boldsymbol{u} T_f) = \nabla\cdot(\lambda_f \nabla T_f) \tag{5.12}$$

where ρ_f, c_{pf}, and λ_f are density, specific heat, and thermal conductivity of PCMs, respectively; T_f and $\boldsymbol{u}$ are temperature and fluid flow velocity vector of PCMs, respectively; Besides, t is the time, and the enthalpy h is defined as:

$$h = c_{pf}(T_f - T_m) + f_l h_{sl} \tag{5.13}$$

where T_m is the melting temperature of PCMs, which is also used as the reference temperature to define the enthalpy value; h_{sl} is the latent heat of PCMs, and f_l is its liquid fraction during melting and solidification. The fluid flow of PCMs at liquid state is governed by the continuity equation and Navier–Stokes equations as:

$$\nabla \cdot \boldsymbol{u} = 0 \tag{5.14}$$

$$\rho_f \left(\frac{\partial \boldsymbol{u}}{\partial t} + \boldsymbol{u} \cdot \nabla \boldsymbol{u} \right) = -\nabla p + \mu \nabla^2 \boldsymbol{u} + \rho_f \boldsymbol{g} \beta \left(T_f - T_m \right) \tag{5.15}$$

where p is the pressure, μ is the dynamic viscosity, g is the gravitational acceleration, and β is the thermal expansion coefficient. Besides, when the extended fins or porous media are inserted into the PCMs to accelerate the heat transfer of LHS system, the corresponding heat conduction occurring inside the solid materials is governed as [66]:

$$\frac{\partial \left(\rho_s c_{ps} T_s \right)}{\partial t} = \nabla \cdot \left(\lambda_s \nabla T_s \right) \tag{5.16}$$

where T_s is the temperature of extended fins or porous media; ρ_s, c_{ps}, and λ_s are density, specific heat, and thermal conductivity of extended fins or porous media, respectively. As aforementioned, conjugate heat transfer occurs at the interface between PCMs and extended fins or porous media, and the Dirichlet–Neumann conditions need to be satisfied in order to guarantee the continuity of heat transfer when the contact thermal resistance is neglected:

$$T_f = T_s \tag{5.17}$$

$$-\lambda_f \nabla T_f = -\lambda_s \nabla T_s \tag{5.18}$$

The governing Eqs. (5.12)–(5.16) in the 2D and 3D Cartesian coordinates could be derived from the Boltzmann equation through Chapman–Enskog expansions. The procedure of solving partial differential equations (PDEs) using LBM includes collision and streaming steps of distribution function, boundary condition tackling, and computation of macroscopic variables. The MRT enthalpy-based LBM for solid–liquid phase change conjugate heat transfer with fluid flow in the 2D and 3D Cartesian coordinates is discussed in this section. Furthermore, the details about MRT enthalpy-based LBM for axisymmetric cylindrical or spherical coordinates could be found in Refs. [40,41].

5.2.2.2 MRT Enthalpy-Based LBM in 2D Cartesian Coordinate

The D2Q9 MRT enthalpy-based LBM is commonly used for simulating solid–liquid phase change conjugate heat transfer in 2D Cartesian coordinate [39]. The evolution equation for the distribution function of total enthalpy h is given by [66]:

$$g_i \left(\boldsymbol{x} + \boldsymbol{e}_i \Delta t, t + \Delta t \right) = g_i \left(\boldsymbol{x}, t \right) - \boldsymbol{M}^{-1} \boldsymbol{S} \left[\boldsymbol{m}(\boldsymbol{x}, t) - \boldsymbol{m}^{\text{eq}} (\boldsymbol{x}, t) \right] \tag{5.19}$$

where g_i is the distribution function of enthalpy in velocity space; $\boldsymbol{x}$ is the vector of location on the Cartesian grids; Δt is the time step; $\boldsymbol{e}_i$ is discrete velocity defined by:

$$\boldsymbol{e}_i=\begin{cases}\boldsymbol{e}_0=(0,0)\\ \boldsymbol{e}_i=c\left(\cos\left[(i-1)\pi/2\right],\sin\left[(i-1)\pi/2\right]\right), & i=1,2,3,4\\ \boldsymbol{e}_i=\sqrt{2}c\left(\cos\left[(2i-9)\pi/4\right],\sin\left[(2i-9)\pi/4\right]\right), & i=5,6,7,8\end{cases} \tag{5.20}$$

where c is the lattice speed. $\boldsymbol{m}(\boldsymbol{x},t)$ is the distribution function in momentum space given by:

$$\boldsymbol{m}(\boldsymbol{x},t)=\boldsymbol{M}g_i(\boldsymbol{x},t)=(m_0,m_1,m_2,m_3,m_4,m_5,m_6,m_7,m_8)^T \tag{5.21}$$

The transformation matrix $\boldsymbol{M}$ is defined as:

$$\boldsymbol{M}=\begin{bmatrix}1&1&1&1&1&1&1&1&1\\-4&-1&-1&-1&-1&2&2&2&2\\4&-2&-2&-2&-2&1&1&1&1\\0&1&0&-1&0&1&-1&-1&1\\0&-2&0&2&0&1&-1&-1&1\\0&0&1&0&-1&1&1&-1&-1\\0&0&-2&0&2&1&1&-1&-1\\0&1&-1&1&-1&0&0&0&0\\0&0&0&0&0&1&-1&1&-1\end{bmatrix} \tag{5.22}$$

Similarly, $\boldsymbol{m}^{\mathrm{eq}}(\boldsymbol{x},t)$ is the equilibrium distribution function in momentum space defined as:

$$\boldsymbol{m}^{\mathrm{eq}}(\boldsymbol{x},t)=\boldsymbol{M}g_i^{\mathrm{eq}}(\boldsymbol{x},t)=\left(m_0^{\mathrm{eq}},m_1^{\mathrm{eq}},m_2^{\mathrm{eq}},m_3^{\mathrm{eq}},m_4^{\mathrm{eq}},m_5^{\mathrm{eq}},m_6^{\mathrm{eq}},m_7^{\mathrm{eq}},m_8^{\mathrm{eq}}\right)^T \tag{5.23}$$

In MRT enthalpy-based LBM for solid–liquid phase change heat transfer, the equilibrium distribution function $\boldsymbol{m}^{\mathrm{eq}}(\boldsymbol{x},t)$ is given by:

$$\boldsymbol{m}^{\mathrm{eq}}(\boldsymbol{x},t)=\begin{pmatrix}h,-4h+2c_{pf,\mathrm{ref}}T_f+3c_{pf}T_f\dfrac{\boldsymbol{u}^2}{c^2},4h-3c_{pf,\mathrm{ref}}T_f-3c_{pf}T_f\dfrac{\boldsymbol{u}^2}{c^2},\\ c_{pf}T_f\dfrac{u}{c},-c_{pf}T_f\dfrac{u}{c},c_{pf}T_f\dfrac{v}{c},-c_{pf}T_f\dfrac{v}{c},c_{pf}T_f\dfrac{u^2-v^2}{c^2},c_{pf}T_f\dfrac{uv}{c^2}\end{pmatrix}^T \tag{5.24}$$

where $\boldsymbol{u}$ is the vector of fluid flow velocity, and u and v are its distributed velocities in the horizontal x coordinate and vertical y coordinate, respectively. $c_{pf,\mathrm{ref}}$ is the reference specific heat, and it is defined by the harmonic mean value of PCM specific heat c_{pf} and solid material specific heat c_{ps} to satisfy the Dirichlet–Neumann conditions of Eqs. (5.17) and (5.18) for conjugate heat transfer at PCM-solid material interface:

$$c_{pf,\mathrm{ref}} = \frac{2c_{pf}c_{ps}}{c_{pf} + c_{ps}} \tag{5.25}$$

Specifically, the relaxation matrix $\boldsymbol{S}$ in momentum space is given as:

$$\boldsymbol{S} = \mathrm{diag}\left(s_0, s_e, s_\varepsilon, s_j, s_q, s_j, s_q, s_e, s_e\right) \tag{5.26}$$

The relaxation parameters in the diagonal matrix $\boldsymbol{S}$ should satisfy the following relation:

$$s_0=1,\ s_j = 1/\tau_T,\ \text{and}\ 0 < s_{e,\varepsilon,q} < 2 \tag{5.27}$$

where τ_T is the relaxation time of enthalpy distribution function:

$$\tau_T = \begin{cases} 3\dfrac{\lambda_f}{\rho_{\mathrm{ref}}c_{p,\mathrm{ref}}} + 0.5, & \text{PCMs} \\[2ex] 3\dfrac{\lambda_s}{\rho_{\mathrm{ref}}c_{p,\mathrm{ref}}} + 0.5, & \text{Solid materials} \end{cases} \tag{5.28}$$

where ρ_{ref} is a harmonic mean value of PCM density ρ_f and density of solid materials ρ_s. Besides, to reduce the numerical diffusion during solid–liquid phase change heat transfer, the relaxation parameters should also be kept as [39]:

$$\left(\frac{1}{s_e} - \frac{1}{2}\right)\left(\frac{1}{s_j} - \frac{1}{2}\right) = \frac{1}{4} \tag{5.29}$$

The collision step for enthalpy distribution function is completed in momentum space as:

$$\boldsymbol{m}(\boldsymbol{x}, t + \Delta t) = \boldsymbol{m}(\boldsymbol{x}, t) - \boldsymbol{S}\left[\boldsymbol{m}(\boldsymbol{x}, t) - \boldsymbol{m}^{\mathrm{eq}}(\boldsymbol{x}, t)\right] \tag{5.30}$$

The post-collision distribution function in velocity space is inversely transformed as:

$$g_i(\boldsymbol{x}, t + \Delta t) = \boldsymbol{M}^{-1}\boldsymbol{m}(\boldsymbol{x}, t + \Delta t) \tag{5.31}$$

Then, the streaming process is handled as:

$$g_i\left(\boldsymbol{x} + \boldsymbol{e}_i\Delta t, t + \Delta t\right) = g_i\left(\boldsymbol{x}, t + \Delta t\right) \tag{5.32}$$

After handling the thermal boundary conditions with different numerical schemes [67–69], the macroscopic enthalpy h could be computed as:

$$h = \sum_i g_i \tag{5.33}$$

The transient liquid fraction f_l of PCMs could be calculated as:

$$f_l = \begin{cases} 0, \ h \le h_s \\ \dfrac{h - h_s}{h_l - h_s}, \ h_s < h < h_l \\ 1, \ h \ge h_l \end{cases} \tag{5.34}$$

The corresponding temperature T_f of PCMs could be further given as:

$$T_f = \begin{cases} T_m - \dfrac{h_s - h}{c_{pf}}, \ h \le h_s \\ T_m, \ h_s < h < h_l \\ T_m - \dfrac{h - h_l}{c_{pf}}, \ h \ge h_l \end{cases} \tag{5.35}$$

Besides, the temperature T_s of solid materials is computed by:

$$T_s = \frac{h}{c_{ps}} + T_m \tag{5.36}$$

To simulate the convective heat transfer caused by fluid flow, the details about 2D MRT LBM for incompressible flow could be found in Refs. [70,71].

5.2.2.3 MRT Enthalpy-Based LBM in 3D Cartesian Coordinate

The D3Q7 MRT enthalpy-based LBM is derived for simulating solid–liquid phase change conjugate heat transfer in the 3D Cartesian coordinate [42]. Similarly, the collision step of enthalpy distribution function is also carried out in momentum space as presented in Eq. (5.30). For D3Q7 MRT enthalpy-based LBM, the enthalpy distribution function $\boldsymbol{m}$ in momentum space and its corresponding transformation matrix $\boldsymbol{M}$ are defined as:

$$\boldsymbol{m}(\boldsymbol{x},t)=\boldsymbol{M}\left(g_0,g_1,g_2,g_3,g_4,g_5,g_6\right)^T \tag{5.37}$$

$$\boldsymbol{M}=\begin{bmatrix} 1 & 1 & 1 & 1 & 1 & 1 & 1 \\ 0 & 1 & -1 & 0 & 0 & 0 & 0 \\ 0 & 0 & 0 & 1 & -1 & 0 & 0 \\ 0 & 0 & 0 & 0 & 0 & 1 & -1 \\ 6 & -1 & -1 & -1 & -1 & -1 & -1 \\ 0 & 2 & 2 & -1 & -1 & -1 & -1 \\ 0 & 0 & 0 & 1 & 1 & -1 & -1 \end{bmatrix} \tag{5.38}$$

The equilibrium distribution function $\boldsymbol{m}^{\text{eq}}(\boldsymbol{x},t)$ is given as:

$$\boldsymbol{m}^{\text{eq}}(\boldsymbol{x},t)=\left(h,\frac{c_{pf}T_f u_x}{c},\frac{c_{pf}T_f u_y}{c},\frac{c_{pf}T_f u_z}{c},6h-21\omega_T c_{pf,\text{ref}}T_f,0,0\right)^T \tag{5.39}$$

where u_x, u_y, and u_z are respectively the velocities in three orthogonal directions of Cartesian coordinate; ω_T is a constant set to be 0.25 in this model, and $c_{pf,\text{ref}}$ is the reference specific heat defined by the harmonic mean value of PCM specific heat c_{pf} and solid material specific heat c_{ps} as presented in Eq. (5.25). In addition, the relaxation matrix $\boldsymbol{S}$ in momentum space for D3Q7 MRT enthalpy-based LBM is:

$$\boldsymbol{S}=\text{diag}\left(\sigma_0,\sigma_1,\sigma_2,\sigma_3,\sigma_4,\sigma_5,\sigma_6\right) \tag{5.40}$$

The relaxation parameters in the diagonal matrix $\boldsymbol{S}$ should satisfy the following relation to reduce the numerical diffusion during solid–liquid phase change:

$$\sigma_0=1,\ \sigma_1=\sigma_2=\sigma_3=\frac{1}{\tau_T},\ \sigma_4=\sigma_5=\sigma_6=2-\frac{1}{\tau_T} \tag{5.41}$$

The relaxation time τ_T of enthalpy distribution function in D3Q7 model is given as:

$$\tau_T=\begin{cases} 4\dfrac{\lambda_f}{\rho_{\text{ref}}c_{p,\text{ref}}}+0.5, & \text{PCMs} \\ 4\dfrac{\lambda_s}{\rho_{\text{ref}}c_{p,\text{ref}}}+0.5, & \text{Solid materials} \end{cases} \tag{5.42}$$

Similarly as the 2D MRT enthalpy-based LBM for solid–liquid phase change conjugate heat transfer, the macroscopic enthalpy h could be achieved by Eq. (5.33) after the streaming of post-collision distribution functions in velocity space and tackling the boundary conditions using appropriate numerical schemes. Then, the liquid fraction of PCM f_l, PCM temperature T_f, and solid material temperature T_s is further calculated according to Eqs. (5.34)–(5.36), respectively.

5.2.2.4 Numerical Reconstruction of Porous Media

High thermal conductive porous media are widely inserted into the LHS system to ameliorate its thermal response during charging and discharging processes because of their interconnected heat transfer channels. A significant advantage of MRT enthalpy-based LBM is its powerful capability in pore-scale modeling of the solid–liquid phase change conjugate heat transfer in porous medium enhanced LHS system. Under this circumstance, numerical methodologies for reconstructing porous medium morphology become essential for accurately simulating the melting and solidification behaviors. The quartet structure generation set (QSGS) is a famous approach to numerically reconstruct the porous foam geometries [72]. The metal foam-PCM composite is effectively reconstructed using QSGS as displayed in Figure 5.1 with respect to different porosity and pore size, and its reconstruction procedure is as follows [14]: (1) Randomly locate the center of a pore in the domain and expand it into the neighboring cells until reaching to the pore diameter $\left(d = d_p \pm 0.2d_p\right)$, where d is the reconstructed pore diameter and d_p is the average pore diameter. (2) Repeat the growing step 1 for other pores and the region of different pores could be overlapped. (3) Stop the growing process if the total area fraction of the pores reaches to $(1 - \varepsilon_{\text{ave}}) \pm 0.001$, where ε_{ave} is the average porosity of metal foams. (4) To reconstruct the metal foam with non-uniform porosity, steps 1–3 are used to generate five different layers with porosity $\left(\varepsilon = \varepsilon_{\text{ave}} \pm 0.02, 0.04\right)$, respectively.

Besides the porous foams, porous fibers are also promising for enhancing the heat transfer performance in the LHS system due to their flexibility in modulating the consolidated heat transfer directions. The random generation-growth method is generally used to reconstruct the morphology of metal fibers [73,74]. Specifically, the growth direction possibility of woven metal fibers follows the Gaussian distribution as [75]:

$$F_g(\theta) = \frac{1}{\sqrt{2\pi}} \exp\left[-\frac{(\theta - \theta_0)^2}{2\sigma_d^2}\right] \tag{5.43}$$

where θ_0 is the major growth direction of woven fibers, and σ_d is the distribution standard deviation of fiber growth direction which indicates the anisotropic degree of woven fibers. As shown in Figure 5.2, the woven metal fibers with different diameter,

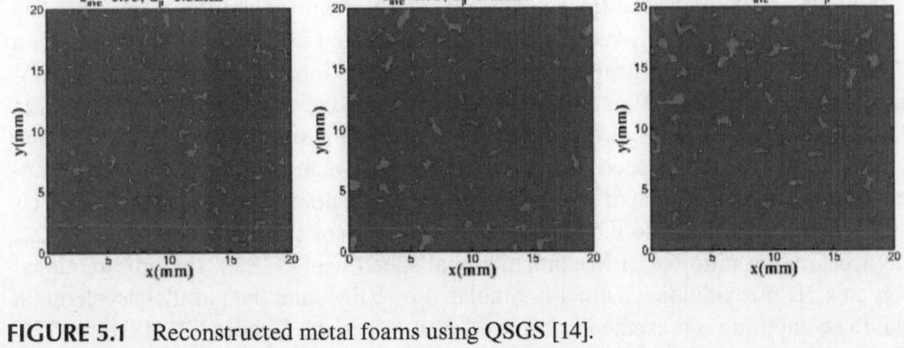

FIGURE 5.1 Reconstructed metal foams using QSGS [14].

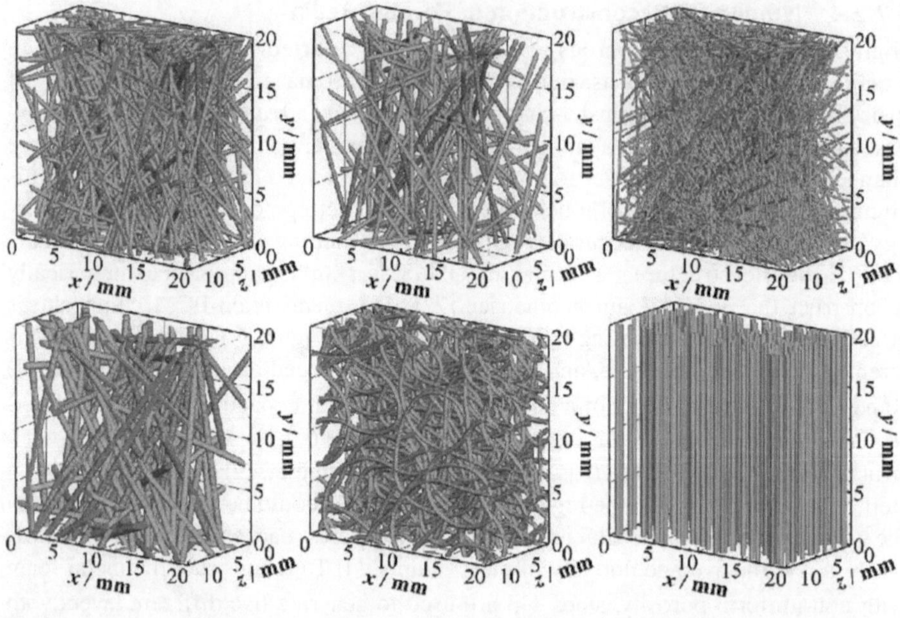

FIGURE 5.2 Reconstructed woven fibers with different diameter, porosity, curvature, and anisotropic degree [75].

porosity, curvature, and anisotropic degree are successfully reconstructed using the random generation-growth method [75].

5.2.2.5 Graphic Processor Units (GPUs) Computing

Based on the rapid hardware development of GPUs, GPU computing becomes attractive during the past decade due to its fantastic capability for tackling parallel modeling task. Due to its highly parallel kinetic nature, LBM has been successfully fitted onto GPUs to efficiently simulate several heat transfer and fluid flow problems on a CUDA programming platform [76–78]. During GPU computing, the CPU and GPUs work as host and devices, respectively. The sequential time steps and initial conditions of LBM are executed on CPU, while GPUs are responsible for all the parallel tasks [79]. Firstly, the initial conditions are set up in host memory and then move the data to the memory of GPUs. A kernel is a function executed in the concurrent threads on GPUs. When the kernels are started for LBM simulation, the collision and streaming steps of distribution functions, tackling of boundary conditions, and computing the macroscopic variables at different lattice grids are simultaneously carried out. Significantly, the kernels need to be synchronized between CPU and GPUs to guarantee the computational logic of LBM. Finally, the data achieved on the threads of GPUs needs to be moved back to the host memory on CPU for further printing out. To test the acceleration ratio on LBM computational speed using GPUs, the unidirectional flow in a 3D microfluidic channel is simulated by LBM, and the parallel acceleration of GPU computing is presented in Table 5.1 [80]. When the Quadro GV 100 type GPU

TABLE 5.1
Computational Acceleration of GPU-Based LBM

Lattice Grid Numbers	100 * 100 * 100	150 * 150 * 150	200 * 200 * 200
CPU intel core i7 8770 s/LBM time step	10.742	34.525	83.174
GPU Quadro GV 100 s/LBM time step	0.025	0.0664	0.1493
Acceleration ratio	429.68	519.95	557.09

is used for parallel computing, the LBM computation could be accelerated by 557.09 times at a grid number of 200 × 200 × 200 in comparison to serial CPU LBM code.

5.2.3 Finite Volume Method

When FVM is used to simulate the solid–liquid phase change heat transfer inside the LHS system, the enthalpy-based energy equation as shown in Eq. (5.12) is transformed into a temperature-based energy equation with a latent heat source term as:

$$\frac{\partial(\rho_f T_f)}{\partial t} + \nabla \cdot (\rho_f \boldsymbol{u} T_f) = \nabla \cdot \left(\frac{\lambda_f}{c_{pf}} \nabla T_f \right) - \frac{\rho_f h_{sl}}{c_{pf}} \frac{\partial f_l}{\partial t} \tag{5.44}$$

Therefore, the continuity equation of Eq. (5.14), momentum equation of Eq. (5.15), and energy equation of Eq. (5.44) are simultaneously solved, and they could be expressed by the following general governing equations:

$$\frac{\partial(\rho_f \phi)}{\partial t} + \frac{\partial(\rho_f u_x \phi)}{\partial x} + \frac{\partial(\rho_f u_y \phi)}{\partial y} + \frac{\partial(\rho_f u_z \phi)}{\partial y}$$
$$= \frac{\partial}{\partial x}\left(\Gamma_\phi \frac{\partial \phi}{\partial x}\right) + \frac{\partial}{\partial y}\left(\Gamma_\phi \frac{\partial \phi}{\partial y}\right) + \frac{\partial}{\partial z}\left(\Gamma_\phi \frac{\partial \phi}{\partial z}\right) + S_\phi \tag{5.45}$$

where ϕ is a general variable representing PCM temperature T_f and fluid flow velocities u_x, u_y, and u_z. The general diffusion coefficient Γ_ϕ and source term S_ϕ for continuity equation, momentum equation, and energy equation are summarized in Table 5.2.

Based on the FVM with integration over each control volumes, the general governing equation for solid–liquid phase change heat transfer in 2D Cartesian coordinate could be discretized into the following linear system of equations using IDEAL algorithm [54]:

$$a_P \phi_P = a_E \phi_E + a_W \phi_W + a_N \phi_N + a_S \phi_S + b \tag{5.46}$$

TABLE 5.2
Expression of General Diffusion Coefficient and Source Term in Various Governing Equations

Governing equations	ϕ	Γ_ϕ	S_ϕ
Continuity equation	1	0	0
Momentum equation	u	μ	$-\nabla p+\rho_f g\beta\left(T_f-T_m\right)$
Energy equation	T_f	$\frac{\lambda_f}{c_{pf}}$	$-\frac{\rho_f h_{sl}}{c_{pf}}\frac{\partial f_l}{\partial t}$

where coefficients a_E, a_W, a_N, a_S, a_P, and source term b are given as:

$$a_E = D_e A(P_{\Delta e}) = D_e A\left(|P_{\Delta e}|\right) + \left[\left|-F_e,\ 0\right|\right] \tag{5.47}$$

$$a_W = D_w A(P_{\Delta w}) = D_w A\left(|P_{\Delta w}|\right) + \left[\left|-F_w,\ 0\right|\right] \tag{5.48}$$

$$a_N = D_n A(P_{\Delta n}) = D_n A\left(|P_{\Delta n}|\right) + \left[\left|-F_n,\ 0\right|\right] \tag{5.49}$$

$$a_S = D_s A(P_{\Delta s}) = D_s A\left(|P_{\Delta s}|\right) + \left[\left|-F_s,\ 0\right|\right] \tag{5.50}$$

$$b = S_c \Delta V + a_p{}^0 \phi_p{}^0 \tag{5.51}$$

$$a_P = a_E + a_W + a_N + a_S - S_p \Delta V \tag{5.52}$$

$$a_P{}^0 = \frac{\rho_P \Delta V}{\Delta t} \tag{5.53}$$

The detail information of parametric coefficients in Eqs. (5.46)–(5.53) could be found in the Ref. [52]. Besides, several other FVM schemes using combined VOF and level-set approach (VOSET) or phase-field method are also developed for solid–liquid phase change heat transfer in the LHS system as presented in Refs. [56,81].

5.2.4 LBM-FVM Coupling Schemes

The coupling schemes of LBM-FVM are essential for carrying out multi-scale numerical modeling of heat transfer and fluid flow with economic computational load. In the multi-scale simulation method called "solving problems regionally and coupling at the interfaces", the computational domain is divided into several sub-regions and different numerical methods (or models) are used to simultaneously simulate transport processes in different sub-regions, while the information in different regions is exchanged at their interfaces as shown in Figure 5.3. Based on this, Chen et al. derived the expressions of LBM density distribution functions up to second order from the macroscopic velocity and density using Champan–Enskog expansion [21], and they apply the coupling simulation strategy using FVM and LBM to more complicated fluid flow. Furthermore, they also derived the expressions of LBM temperature (concentration) distribution functions up to second order from the macroscopic temperature (concentration) using Champan–Enskog expansion, and they applied the coupling simulation strategy to convective heat (species) transfer processes [82].

Besides, a general expression of the density distribution functions up to second order for any scalar that obeys the convective-diffusion (CD) equation is also proposed. For fluid flow usually encountered, the Chapman–Enskog multi-scale expansion technique is used to derive the distribution function up to $f(2)$, which is expressed as follows together with $f(0)$ and $f(1)$:

$$f_i^{(0)} = f_i^{(\mathrm{eq})} \tag{5.54}$$

$$f_i^{(1)} = -\tau_f \Delta t (c_{i\alpha} - u_\alpha)(c_{i\beta} - u_\beta) f_i^{(0)} c_s^{-2} \partial_{x\alpha}^{(1)} u_\beta \tag{5.55}$$

$$f_i^{(2)} = -\Delta t \tau_f \upsilon (c_{i\beta} - u_\beta) f_i^{(0)} c_s^{-2} \left[\frac{1}{\rho} (\partial_{x\beta}^{(1)} u_\alpha + \partial_{x\alpha}^{(1)} u_\beta) \partial_{x\alpha}^{(1)} \rho + \left(\partial_{x\alpha}^{(1)}\right)^2 u_\beta \right] \tag{5.56}$$

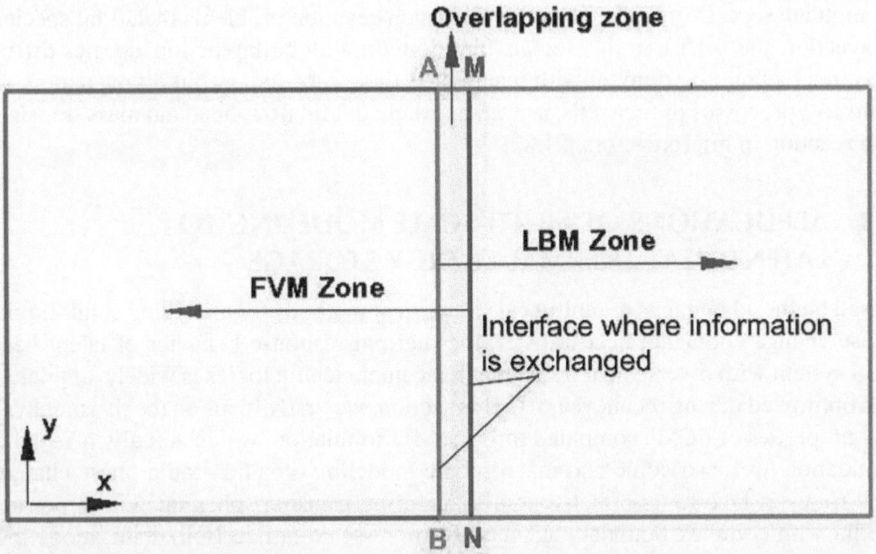

FIGURE 5.3 FVM–LBM coupled computational domain: One region is simulated using FVM and the other is simulated using LBM. Information is exchanged at their interfaces.

where $f_i^{(1)}$ and $f_i^{(2)}$ refer to the first-order and second-order distribution functions. Δt, τ_f, c, c_s are the time step, relaxation time, lattice speed, and sound speed in the LBM, respectively. The detail of the deriving process could be found in Ref. [83,84]. Several typical fluid flow problems are simulated including backward-facing step flow, flow around a circular cylinder, and lid-driven cavity flow [83,84] using the MRCA-DMDV, namely multi (M) regions (R) coupling (C) approach (A)–different (D) numerical methods (M) with different (D) basic variables (V), and good agreements are obtained between the coupling simulation results and numerical results in literature. For heat transfer or mass transport, the following distribution functions based on the Chapman–Enskog expansion $g(0)$, $g(1)$, and $g(2)$ are given:

$$g_i^{(0)} = g_i^{\mathrm{eq}} \tag{5.57}$$

$$g_i^{(1)} = -\tau_g \Delta t\left[(c_{i\alpha} - u_\alpha)C^{-1} g_i^{(eq)} \partial_{x\alpha}^{(1)} C + 0.5(c_{i\alpha} - u_\alpha)Cc_{i\beta}\,\partial_{x\alpha}^{(1)} u_\beta - 0.5\rho^{-1} Cc_{i\beta}\,\partial_{x\beta}^{(1)}\rho\right] \tag{5.58}$$

$$g_i^{(2)} = -\tau_g \Delta t\left[DC^{-1} g_i^{(eq)} \partial_{x\alpha}^{(1)} \partial_{x\alpha}^{(1)} C + 0.5Cc_{i\beta}\rho^{-1} v\left[\rho \partial_{x\alpha}^{(1)} \partial_{x\alpha}^{(1)} u_\beta + (\partial_{x\beta}^{(1)} u_\alpha + \partial_{x\alpha}^{(1)} u_\beta)\partial_{x\alpha}^{(1)}\rho\right]\right] \tag{5.59}$$

where τ_g is the relaxation time for temperature (concentration) distribution function. The detailed derivation can be found in Ref. [21]. The MRCA-DMDV is applied to simulate several typical convective-diffusion-reaction problems including species convection and diffusion in a rectangular domain with bulk reaction, species diffusion in a rectangular domain with reaction on the top boundary, fluid flow and mass transport processes in fuel cells, as well as coupled fluid flow, heat and mass transfer, and reaction in microreactors [21,82].

5.3 APPLICATIONS OF MULTI-SCALE MODELING TO LATENT HEAT THERMAL ENERGY STORAGE

Based on the advantages of multi-scale numerical methods for analyzing solid–liquid phase change conjugate heat transfer, the thermal response behavior of latent heat TES system with diverse heat transfer enhancement technologies is widely simulated and optimized during recent years. In this section, we firstly focus on the thermophysical properties of PCMs computed through MD simulation, which actually offers the foundation for mesoscopic and macroscopic modeling of solid–liquid phase change heat transfer. Due to its effectiveness in handling transport phenomenon in porous media with complex boundary geometries, pore-scale lattice Boltzmann modeling for solid–liquid phase change conjugate heat transfer in the LHS system is further reviewed. Finally, the macroscopic FVM for investigating the thermal behavior of LHS system is also discussed to promote practical engineering applications.

5.3.1 Thermophysical Properties of PCMs

MD simulations are successfully used to calculate several thermophysical properties of PCMs including thermal conductivity, specific heat, melting enthalpy, melting point, density, and thermal expansion coefficients, which serve as significant supplements for experimental measurements. In this section, we mainly focus on the MD simulations for specific heat and melting enthalpy of PCMs because these parameters significantly influence the charging and discharging processes of LHS system.

5.3.1.1 Specific Heat Capacity and Melting Enthalpy

The specific heat capacity determines the sensible heat storage density and melting enthalpy determines the latent heat storage density of PCMs, which are significant for thermal performance of LHS system. To calculate the specific heat capacity and melting enthalpy using MD simulation, PCM is heated from solid state to liquid state in NPT ensemble, and the corresponding energy, volume, and temperature are recorded. Then, the specific heat capacity and melting enthalpy can be obtained [85]:

$$c_p = \left(\frac{\partial H}{\partial T}\right)_P = \frac{1}{k_B T^2}\left\langle \delta\left(E_K + U + pV\right)^2 \right\rangle \approx \frac{\Delta H}{\Delta T \cdot \rho \cdot V} \tag{5.60}$$

where H is the total enthalpy of the system (kJ/mol); k_B is the Boltzmann constant, and E_K and U are the kinetic energy (kJ/mol) and the potential energy (kJ/mol) of the system, respectively. p, V, and T are the pressure (Pa), volume (m^3), and temperature (K) of the system, respectively. The melting enthalpy ΔH is defined as the difference of the total enthalpy of PCMs during the phase transition (J/g). As displayed in Figure 5.4, the specific heat c_p and melting enthalpy ΔH for NaCl-single walled carbon nanotube (SWCNT) composite are achieved by MD simulation with respect to different mass fraction of SWCNT [29]. The results indicate that the specific heat capacity c_p is enhanced with an increased mass fraction of SWCNT, while the corresponding melting enthalpy ΔH is attenuated.

5.3.2 Pore-Scale Modeling of LHS System

Based on the MRT enthalpy-based LBM for solid–liquid phase change conjugate heat transfer, pore-scale investigation of thermal behavior in porous medium-PCM composite is carried out. Under this circumstance, the morphology of porous media could be further optimized to improve the latent heat TES performance. Ren et al. presented a pore-scale investigation of melting process in a latent heat TES unit filled with metal foams by pore-scale lattice Boltzmann modeling under GPU acceleration [14]. The QSGS method is used to reconstruct the diversified morphology of metal foams. As shown in Figure 5.5, at a specific porosity of metal foams, the melting rate of PCMs is accelerated with a decreased pore size due to the enhanced specific surface area of metal foams, which also makes the temperature distribution inside LHS unit more uniform. In addition, due to the trade-off effect between increasing the heat transfer rate by using metal foam with lower porosity and the reduced amount of

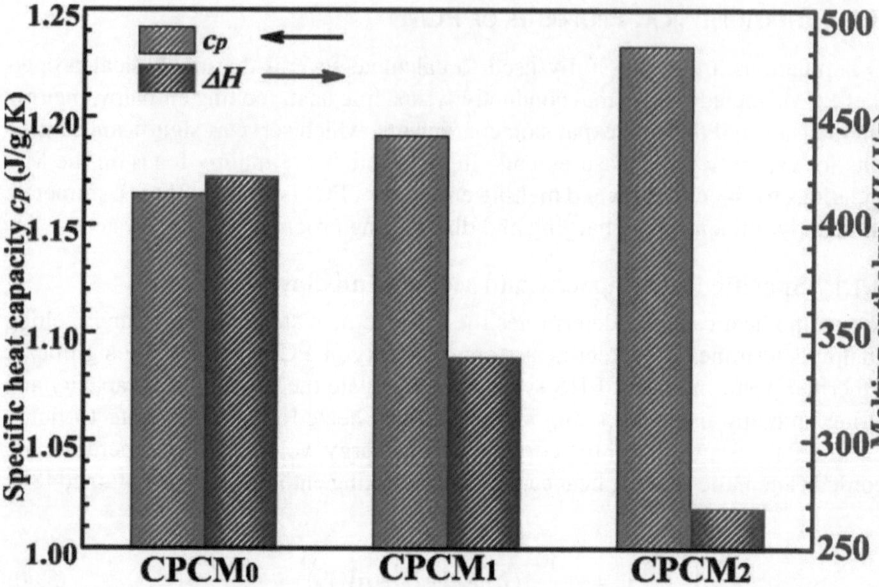

FIGURE 5.4 The melting enthalpy ΔH and specific heat capacity c_p of NaCl-SWCNT.

energy stored in the LHS unit, an appropriate metal foam porosity is highly recommended in the practical engineering applications to balance the charging and discharging rates and energy storage capacity.

To achieve a deep understanding for solid–liquid phase change conjugate heat transfer in the LHS unit, a 3D pore-scale lattice Boltzmann modeling is necessary. Recently, Ren et al. investigated the charging process of woven metal fiber-PCM composite at pore-scale by numerically reconstructing the various morphologies of metal fibers with respect to fiber porosity, fiber diameter, fiber curvature, and the anisotropic degree of fibers [75]. As displayed in Figure 5.6, the energy storage rate in a LHS unit could be accelerated by 40% through consolidating the anisotropic degree of woven metal fibers from $\sigma = 0.5$ to $\sigma = 0$ because of the enhanced heat transfer capability in a desired vertical direction. Similarly, woven metal fibers with appropriate porosity need to be used in order to balance the heat conduction rate inside fiber skeleton and the energy storage capacity of LHS unit. In addition, the heat conduction inside metal fiber-PCM composite is observed to be consistently dominant over natural convection even though the role of natural convection is consolidated with the increment of fiber porosity or characteristic temperature difference.

The aforementioned researches demonstrate the effectiveness of investigating solid–liquid phase change conjugate heat transfer in the LHS unit using pore-scale lattice Boltzmann modeling, which contributes to improving heat transfer enhancement technologies for optimizing the energy storage rate and energy storage capacity of LHS system.

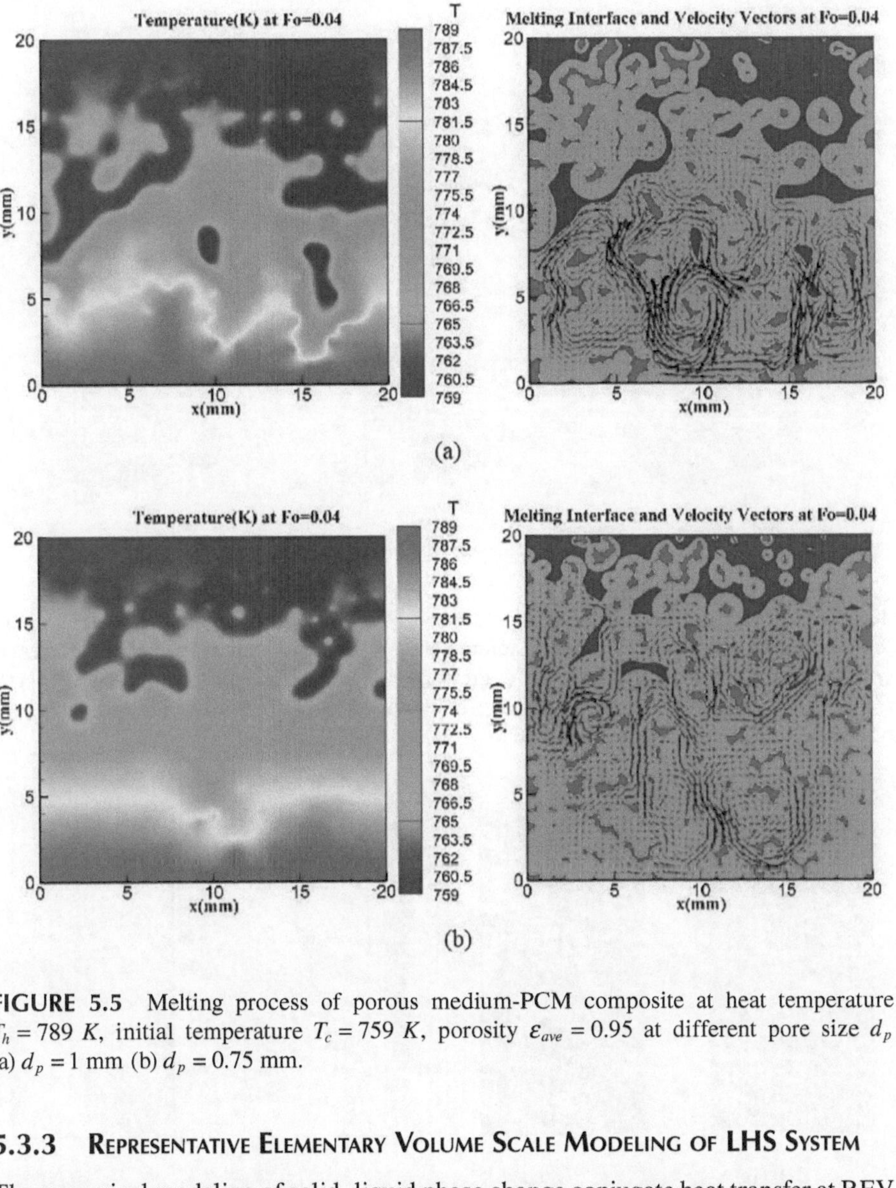

FIGURE 5.5 Melting process of porous medium-PCM composite at heat temperature $T_h = 789\ K$, initial temperature $T_c = 759\ K$, porosity $\varepsilon_{ave} = 0.95$ at different pore size d_p (a) $d_p = 1$ mm (b) $d_p = 0.75$ mm.

5.3.3 Representative Elementary Volume Scale Modeling of LHS System

The numerical modeling of solid–liquid phase change conjugate heat transfer at REV scale is significant for optimizing the thermal behavior of integrated LHS systems in engineering applications due to the relatively high computational efficiency compared with mesoscopic lattice Boltzmann modeling. Li et al. numerically investigated the open-cell metallic foams filled with paraffin at REV scale using FVM with consideration of the non-Darcy effect, local natural convection, and thermal non-equilibrium [55]. As shown in Figure 5.7, the numerical simulation using FVM at REV scale for solid–liquid phase change heat transfer in metal foam-PCM composite is consistent with the experimental results, which validates the accuracy of numerical

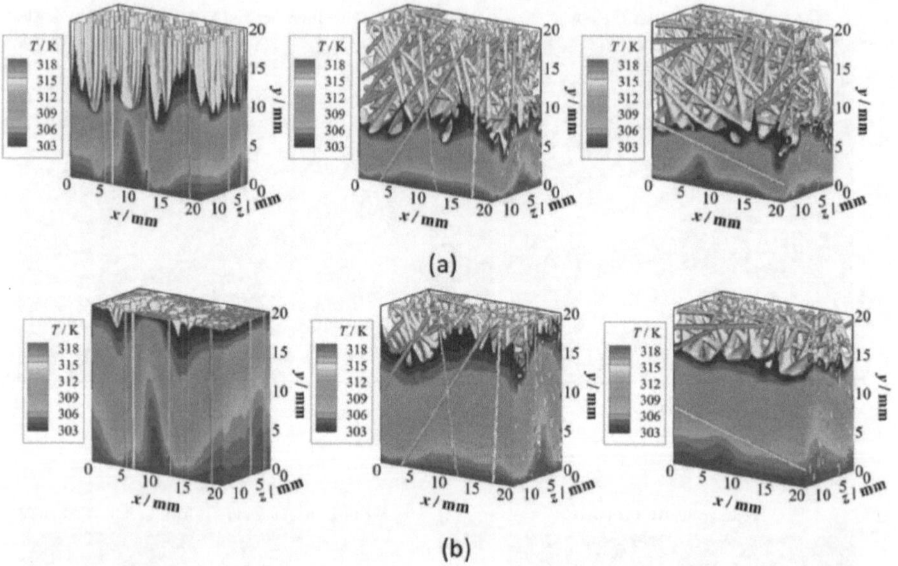

FIGURE 5.6 Solid–liquid phase change process of anisotropic straight woven metal fiber-PCM composite at porosity $\varepsilon = 0.90$, diameter $D = 0.4$ mm, and temperature difference $\Delta T = 15\,\mathrm{K}$ for different fiber anisotropic degree (a) From left to right: anisotropic degree $\sigma = 0$, $\sigma = 0.25$, and $\sigma = 0.50$ at time $t = 124.2$ s (b) From left to right: anisotropic degree $\sigma = 0$, $\sigma = 0.25$, and $\sigma = 0.50$ at time $t = 372.6$ s.

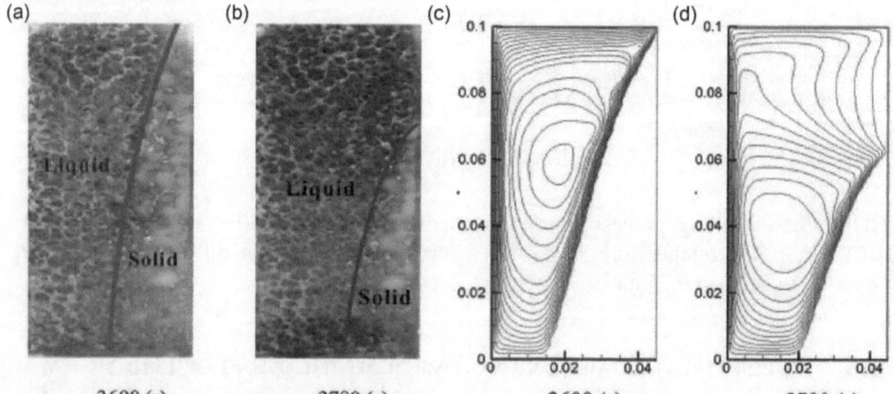

FIGURE 5.7 Comparison of the solid–liquid interface between the experimental photos and REV-scale numerical results at different times for metal foam at $\varepsilon = 0.90$, $\omega = 10$ PPI (a) Experimental photo at time $t = 3600$ s (b) Experimental photo at time $t = 3780$ s (c) Numerical results at time $t = 3600$ s (d) Numerical results at time $t = 3780$ s.

modeling in designing LHS system. Their results also indicate that the melting heat transfer is enhanced by the high thermal conductivity matrix foams even though the local natural convection is suppressed to some extent.

Furthermore, the commercial software ANSYS FLUENT based on FVM with SIMPLE algorithm is popular used by several researchers to investigate the solid–liquid phase change in LHS system with different heat transfer enhancement technologies at REV scale. For instance, Xu et al. applied ANSYS FLUENT to numerically investigate the melting performance of triplex-layer PCMs in a horizontal shell and tube LHS unit [86]. As shown in Figure 5.8, the transient melting process of LHS unit is analyzed with respect to different fin arrangements. The results show that the uniform radial fin arrangement in each PCM layers presents the maximum value of comprehensive storage density evaluation (CSDE), which is a significant criterion to evaluate the performance of LHS system. A large amount of research results for the thermal behavior of LHS system using FVM-based ANSYS FLUENT can be found in review Refs. [53,87].

5.4 CONCLUSIONS

Multi-scale numerical modeling is an effective tool for investigating solid–liquid phase change conjugate heat transfer in the latent heat TES applications. In this chapter, the multi-scale numerical methods including molecular dynamics simulation, LBM, and FVM for simulating thermal behavior of latent heat unit are systematically discussed with classical examples. Specifically, the molecular dynamics simulation is powerful

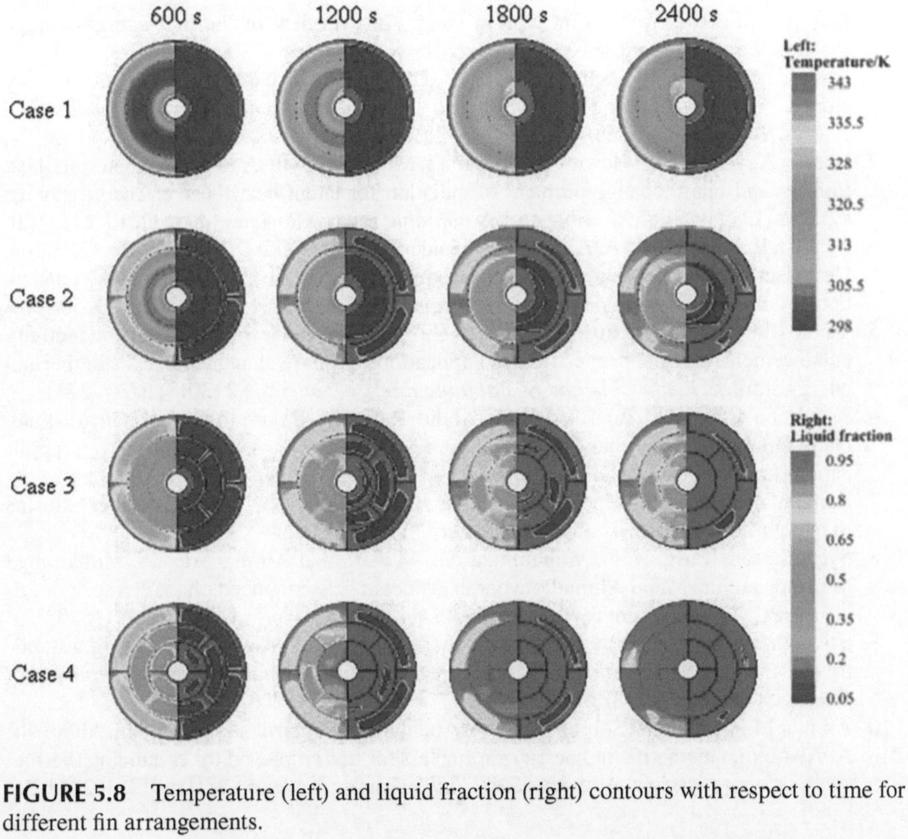

FIGURE 5.8 Temperature (left) and liquid fraction (right) contours with respect to time for different fin arrangements.

for computing the thermophysical properties of PCMs, laying foundations for mesoscopic and macroscopic numerical simulations. In addition, due to its kinetic nature, LBM is efficient for clarifying the heat transfer and fluid flow in porous medium-PCM composite at pore-scale due to its outstanding ability for handling complicated boundary conditions. Besides, as a traditional numerical method developed for more than 50 years, FVM is robust on simulating the macroscopic thermal performance of LHS system at REV scale. Based on the advantages of various numerical methods at different scales with their coupling scheme, multi-scale numerical modeling contributes to several research achievements in designing fantastic latent heat TES systems. With the rapid progress of numerical methods and computational hardware such as GPUs, the multi-scale modeling will definitely play an increasing role in optimizing practical latent heat TES applications in recent future.

ACKNOWLEDGMENTS

This work is supported by the project of the National Natural Science Foundation of China (No. 51806168), and the Foundation for Innovative Research Groups of the National Natural Science Foundation of China (No. 51721004).

REFERENCES

1. Guruprasad Alva, Yaxue Lin, Guiyin Fang, An overview of thermal energy storage systems, *Energy* 144 (2018) 341–378.
2. Atul Sharma, V. Veer Tyagi, Carl R. Chen, Dharam Buddhi, Review on thermal energy storage with phase change materials and applications, *Renewable and Sustainable Energy Reviews* 13 (2009) 318–345.
3. Francis Agyenim, Neil Hewitt, Philip Eames, Mervyn Smyth, A review of materials, heat transfer and phase change problem formulation for latent heat thermal energy storage systems (LHTESS), *Renewable and Sustainable Energy Reviews* 14 (2) (2010) 615–628.
4. Nasiru I. Ibrahim, Fahad A. Al-Sulaiman, Saidur Rahman, Bekir S. Yilbas, Ahmet Z. Sahin, Heat transfer enhancement of phase change materials for thermal energy storage applications: A critical review, *Renewable and Sustainable Energy Reviews* 74 (2017) 26–50.
5. Yaxue Lin, Yuting Jia, Guruprasad Alva, Guiyin Fang, Review on thermal conductivity enhancement, thermal properties and applications of phase change materials in thermal energy storage, *Renewable and Sustainable Energy Reviews.* 82 (2018) 2730–2742.
6. Kin Yuen Leong, Mohd Rosdzimin Abdul Rahman, Balamurugan A. Gurunathan, Nano-enhanced phase change materials: A review of thermo-physical properties, applications, and challenges, *Journal of Energy Storage* 21 (2019) 18–31.
7. Jianjian Wang, Ruiting Zheng, Jinwei Gao, Gang Chen, Heat conduction mechanisms in nanofluids and suspensions, *Nano Today* 7 (2) (2012) 124–136.
8. Syeda Laraib Tariq, Hafiz Muhammad Ali, Muhammad Ammar Akram, Muhammad Mansoor Janjua, Majid Ahmadlouydarab, Nanoparticle enhanced phase change materials (NePCMs) – A recent review, *Applied Thermal Engineering* 176 (2020) 115305.
9. Srikanth Salyan, S. Suresh, Study of the thermos-physical properties and cycling stability of D-Mannitol-copper oxide nanocomposites as phase change materials, *Journal of Energy Storage* 15 (2018) 245–255.
10. Ammar M. Abdulateef, Sohif Mat, Jasim Abdulateef, Kamaruzzaman Sopian, Abduljalil A. Al-Abidi, Geometric and design parameters for fins employed for enhancing thermal energy storage systems: A review, *Sustainable Energy Reviews* 82 (1) (2018) 1620–1635.

11. J.P. Holman, *Heat Transfer*, 10th ed. New York: MCGraw-Hill Companies, 2010.
12. Shuai Zhang, Daili Feng, Lei Shi, Li Wang, Yingai Jin, Limei Tian, Ziyuan Li, Guoyong Wang, Lei Zhao, Yuying Yan, A review of phase change heat transfer in shape-stabilized phase change materials (ss-PCMs) based on porous supports for thermal energy storage, *Renewable and Sustainable Energy Reviews* 135 (2021) 110127.
13. Hafiz Muhammad Ali, Muhammad Mansoor Janjua, Uzair Sajjad, Wei-Mon Yan, A critical review on heat transfer augmentation of phase change materials embedded with porous materials/foams, *International Journal of Heat and Mass Transfer* 135 (2019) 649–673.
14. Qinlong Ren, Ya-Ling He, Kai-Zhi Su, Cho Lik Chan, Investigation of the effect of metal foam characteristics on the PCM melting performance in a latent heat thermal energy storage unit by pore-scale lattice Boltzmann modeling, *Numerical Heat Transfer, Part A: Applications* 72 (10) (2017) 745–764.
15. Xinyi Li, Cong Niu, Xiangxuan Li, Ting Ma, Lin Lu, Qiuwang Wang, Pore-scale investigation on effects of void cavity distribution on melting of composite phase change materials, *Applied Energy* 275 (2020) 115302.
16. Ya-Ling He, Wen-Quan Tao, Multiscale simulations of heat transfer and fluid flow problems, *Journal of Heat Transfer* 134 (3) (2012) 031018.
17. Li Chen, Yong-Liang Feng, Chen-Xi Song, Lei Chen, Ya-Ling He, Wen-Quan Tao, Multi-scale modeling of proton exchange membrane fuel cell by coupling finite volume method and lattice Boltzmann method, *International Journal of Heat and Mass Transfer* 63 (2013) 268–283.
18. Ruiyuan Zhang, Ting Min, Li Chen, Qinjun Kang, Ya-Ling He, Wen-Quan Tao, Pore-scale and multiscale study of effects of Pt degradation on reactive transport processes in proton exchange membrane fuel cells, *Applied Energy* 253 (2019) 113590.
19. Zi-Xiang Tong, Ming-Jia Li, Xi Chen, Ya-Ling He, Direct coupling between molecular dynamics and lattice Boltzmann method based on velocity distribution functions for steady-state isothermal flow, *International Journal of Heat and Mass Transfer* 115 (2017) 544–555.
20. Wen-Jing Zhou, Zhi-Qiang Yu, Zhong-Zhen Li, Ya-Ling He, Wen-Quan Tao, Molecular dynamics-continuum hybrid simulation for the impingement of droplet on a liquid film, *Numerical Heat Transfer, Part A: Applications* 68 (2015) 512–525.
21. Li Chen, Ya-Ling He, Qinjun Kang, Wen-Quan Tao, Coupled numerical approach combining finite volume and lattice Boltzmann methods for multi-scale multi-physicochemical processes, *Journal of Computational Physics* 255 (2013) 83–105.
22. Zi-Xiang Tong, Ya-Ling He, A unified coupling scheme between lattice Boltzmann method and finite volume method for unsteady fluid flow and heat transfer, *International Journal of Heat and Mass transfer* 80 (2015) 812–824.
23. Daili Feng, Yanhui Feng, Lin Qiu, Pei Li, Yuyang Zang, Hanying Zou, Zepei Yu, Xinxin Zhang, Review on nanoporous composite phase change materials: Fabrication, characterization, enhancement and molecular simulation, *Renewable and Sustainable Energy Reviews* 109 (2019) 578–605.
24. Yong Zhang, Edward J. Maginn, A comparison of methods for melting point calculation using molecular dynamics simulations, *The Journal of Chemical Physics* 136 (2012) 144116.
25. YangChun Zou, ShiKai Xiang, ChengDa Dai, Investigation on the efficiency and accuracy of methods for calculating melting temperature by molecular dynamics simulation, *Computational Materials Science* 171 (2020) 109156.
26. Sirlok Srinivasan, Mouhamad S. Diallo, Sandip K. Saha, Opeyemi A. Abass, Aayush Sharma, Ganesh Balasubramanian, Effect of temperature and graphite particles fillers on thermal conductivity and viscosity of phase change material *n*-eicosane, *International Journal of Heat and Mass Transfer* 114 (2017) 318–323.

27. Changpeng Lin, Zhonghao Rao, Thermal conductivity enhancement of paraffin by adding boron nitride nanostructures: A molecular dynamics study, *Applied Thermal Engineering* 110 (2017) 1411–1419.
28. Yinsheng Yu, Yubing Tao, Ya-Ling He, Molecular dynamics simulation of thermophysical properties of NaCl-SiO_2 based molten salt composite phase change materials, *Applied Thermal Engineering* 166 (2020) 114628.
29. Yinsheng Yu, Chenyang Zhao, Yubing Tao, Xi Chen, Ya-Ling He, Superior thermal energy storage performance of NaCl-SWCNT composite phase change materials: A molecular dynamics approach, *Applied Energy* 290 (2021) 116799.
30. Shiyi Chen, Gary D. Doolen, Lattice Boltzmann method for fluid flows, *Annual Review of Fluid Mechanics* 30 (1998) 329–364.
31. Li Chen, Qinjun Kang, Yutong Mu, Ya-Ling He, Wen-Quan Tao, A critical review of the pseudopotential multiphase lattice Boltzmann model: Methods and applications, *International Journal of Heat and Mass Transfer* 76 (2014) 210–236.
32. W. Miller, The lattice Boltzmann method: A new tool for numerical simulation of the interaction of growth kinetics and melt flow, *Journal of Crystal Growth* 230 (1–2) (2001) 263–269.
33. I. Rasin, W. Miller, S. Succi, Phase-field lattice kinetic scheme for the numerical simulation of dendritic growth, *Physical Review E* 72 (6) (2005) 066705.
34. Yaling He, Qing Liu, Qing Li, Wenquan Tao, Lattice Boltzmann methods for single-phase and solid-liquid phase change heat transfer in porous media: A review, *International Journal of Heat and Mass Transfer* 129 (2019) 160–197.
35. Rongzong Huang, Huiying Wu, An immersed boundary-thermal lattice Boltzmann method for solid-liquid phase change, *Journal of Computational Physics* 227 (2014) 305–319.
36. Wen-Shu Jiaung, Jeng-Rong Ho, Chun-Pao Kuo, Lattice Boltzmann method for the heat conduction problem with phase change, *Numerical Heat Transfer, Part B: Fundamentals* 39 (2001) 167–187.
37. Mohsen Eshraghi, Sergio D. Felicelli, An implicit lattice Boltzmann model for heat conduction with phase change, *International Journal of Heat and Mass Transfer* 55 (9–10) (2012) 2420–2428.
38. Rongzong Huang, Huiying Wu, Ping Cheng, A new lattice Boltzmann model for solid-liquid phase change, *International Journal of Heat and Mass Transfer* 59 (2013) 295–301.
39. Rongzong Huang, Huiying Wu, Phase interface effects in the total enthalpy-based lattice Boltzmann model for solid-liquid phase change, *Journal of Computational Physics* 294 (2015) 346–362.
40. Dong Li, Qinlong Ren, Zi-Xiang Tong, Ya-Ling He, Lattice Boltzmann models for axisymmetric solid-liquid phase change, *International Journal of Heat and Mass Transfer* 112 (2017) 795–804.
41. Yutao Huo, Zhonghao Rao, Investigation of solid-liquid phase change in the spherical capsule using axisymmetric lattice Boltzmann method, *International Journal of Heat and Mass transfer* 119 (2018) 1–9.
42. Dong Li, Zi-Xiang Tong, Qinlong Ren, Ya-Ling He, Wen-Quan Tao, Three-dimensional lattice Boltzmann models for solid-liquid phase change, *International Journal of Heat and Mass Transfer* 115 (2017) 1334–1347.
43. Qing Liu, Xiang-Bo Feng, Ya-Ling He, Cai-Wu Lu, Qing-Hua Gu, Three-dimensional multiple-relaxation-time lattice Boltzmann models for single-phase and solid-liquid phase-change heat transfer in porous media at the REV scale, *Applied Thermal Engineering* 152 (2019) 319–337.
44. Wen-Zhen Fang, Yu-Qing Tang, Chun Yang, Wen-Quan Tao, Pore scale investigations on melting of phase change materials considering the interfacial thermal resistance, *International Communications in Heat and Mass Transfer* 115 (2020) 104631.

45. Xinyi Li, Ting Ma, Jun Liu, Hao Zhang, Qiuwang Wang, Pore-scale investigation of gravity effects on phase change heat transfer characteristics using lattice Boltzmann method, *Applied Energy* 222 (2018) 92–103.
46. Moran Wang, Ning Pan, Modeling and prediction of the effective thermal conductivity of random open-cell porous foams, *International Journal of Heat and Mass Transfer* 51 (2008) 1325–1331.
47. Yan Su, Tiniao Ng, Yinping Zhang, Jane H. Davidson, Three dimensional thermal diffusion in anisotropic heterogeneous structures simulated by a non-dimensional lattice Boltzmann method with a controllable structure generation scheme based on discrete Gaussian quadrature space and velocity, *International Journal of Heat and Mass Transfer* 108 (2017) 386–401.
48. S. Abishek, A.J.C. King, R. Mead-Hunter, V. Golkarfard, W. Heikamp, B.J. Mullins, Generation and validation of virtual nonwoven, foam and knitted filter (separator/coalesce) geometries for CFD simulations, *Separation and Purification Technology* 188 (2017) 493–507.
49. Qinlong Ren, Zexiao Wang,Jianjun Zhu, Zhiguo Qu, Pore-scale heat transfer of heat sink filled with stacked 2D metal fiber-PCM composite, *International Journal of Thermal Sciences* 161 (2021) 106739.
50. Qinlong Ren, Cho Lik Chan, GPU accelerated numerical study of PCM melting process in an enclosure with internal fins using lattice Boltzmann method, *International Journal of Heat and Mass Transfer* 100 (2016) 522–535.
51. Qinlong Ren, Penghua Guo, Jianjun Zhu, Thermal management of electronic devices using pin-fin based cascade microencapsulated PCM/expand graphite composite, *International Journal of Heat and Mass Transfer* 149 (2020) 119199.
52. Wenquan Tao, *Numerical Heat Transfer*, 2nd ed., Xi'an Jiaotong University Press, Xi'an 2001.
53. Hossein Asgharian, Ehsan Baniasadi, A review on modeling and simulation of solar energy storage systems based on phase change materials, *Journal of Energy Storage* 21 (2019) 186–201.
54. D.L Sun, Zhiguo Qu, Yaling He, Wenquan Tao, An efficient segregated algorithm for incompressible fluid flow and heat transfer problems-IDEAL (inner doubly iterative efficient algorithm for linked equations) part I: Mathematical formulation and solution procedure, *Numerical Heat Transfer, Part B: Fundamentals* 53 (2008) 1–17.
55. Wenqiang Li, Zhiguo Qu, YaLing He, WenQuan Tao, Experimental and numerical studies on melting phase change heat transfer in open-cell metallic foams filled with paraffin, *Applied Thermal Engineering* 37 (2012) 1–9.
56. Kong Ling, Wen-Quan Tao, A sharp-interface model coupling VOSET and IBM for simulations on melting and solidification, *Computers and Fluids* 178 (2019) 113–131.
57. Jasim M. Mahdi, Emmanuel C. Nsofor, Melting enhancement in triplex-tube latent heat thermal energy storage system using nanoparticles-metal foam combination, *Applied Energy* 191 (2017) 22–34.
58. Yang Xu, Qinlong Ren, Zhang-Jing Zheng, Ya-Ling He, Evaluation and optimization of melting performance for a latent heat thermal energy storage unit partially filled with porous media, *Applied Energy* 193 (2017) 84–95.
59. Chenzhen Ji, Zhen Qin, Swapnil Dubey, Fook Hoong Choo, Fei Duan, Simulation on PCM melting enhancement with double-fin length arrangements in a rectangular enclosure induced by natural convection, *International Journal of Heat and Mass Transfer* 127 (2018) 255–265.
60. Fan Yuan, Ming-Jia Li, Yu Qiu, Zhao Ma, Meng-Jie Li, Specific heat capacity improvement of molten salt for solar energy applications using charged single-walled carbon nanotubes, *Applied Energy* 250 (2019) 1481–1490.

61. C.Y. Zhao, Y.B. Tao, Y.S. Yu, Molecular dynamics simulation of nanoparticle effect on melting enthalpy of paraffin phase change material, *International Journal of Heat and Mass Transfer* 150 (2020) 119382.
62. C.Y. Zhao, Y.B. Tao, Y.S. Yu, Molecular dynamics simulation of thermal and phonon transport characteristics of nanocomposite phase change material, *Journal of Molecular Liquids* 329 (2021) 115448.
63. Accelrys Inc., *Materials Studio.* San Diego, CA: Accelrys Software Inc., 2010.
64. William Humphrey, Andrew Dalke, Klaus Schulten, VMD: Visual molecular dynamics, *Journal of Molecular Graphics* 14 (1) (1996) 33–38.
65. Qinlong Ren, Enhancement of nanoparticle-phase change material melting performance using a sinusoidal heat pipe, *Energy Conversion and Management* 180 (2019) 784–795.
66. Qinlong Ren, Fanlong Meng, Penghua Guo, A comparative study of PCM melting process in a heat pipe-assisted LHTES unit enhanced with nanoparticles and metal foams by immersed boundary-lattice Boltzmann method at pore-scale, *International Journal of Heat and Mass Transfer* 121 (2018) 1214–1228.
67. Zhao-Li Guo, Chu-Guang Zheng, Bao-Chang Shi, Non-equilibrium extrapolation method for velocity and pressure boundary conditions in the lattice Boltzmann method, *Chinese Physics* 11 (2002) 366.
68. Zhaoli Guo, Chuguang Zheng, An extrapolation method for boundary conditions in lattice Boltzmann method, *Physics of Fluids* 14(6) (2002) 2007–2010.
69. Chih-Hao Liu, Kuen-Hau Lin, Hao-Chueh Mai, Chao-An Lin, Thermal boundary conditions for thermal lattice Boltzmann simulations, *Computers & Mathematics with Applications* 59 (7) (2010) 2178–2193.
70. Rui Du, Baochang Shi, Xingwang Chen, Multi-relaxation-time lattice Boltzmann model for incompressible flow, *Physics Letters A* 359 (2006) 564–572.
71. Di Tian, Zhiguo Qu, Jianfei Zhang, Qinlong Ren, Enhancement of solar pond stability performance using an external magnetic field, *Energy Conversion and Management* 243 (2021) 114427.
72. Moran Wang, Jinku Wang, Ning Pan, Shiyi Chen, Mesoscopic predictions of the effective thermal conductivity for microscale random porous media, *Physical Review E* 75 (2007) 036702.
73. Moran Wang, Jihuan He, Jianyong Yu, Ning Pan, Lattice Boltzmann modelling of the effective thermal conductivity for fibrous materials, *International Journal of Thermal Sciences* 46 (2007) 848–855.
74. Moran Wang, Qinjun Kang, Ning Pan, Thermal conductivity enhancement of carbon fiber composites, *Applied Thermal Engineering* 29 (2009) 418–421.
75. Qinlong Ren, Zexiao Wang, Tao Lai, JianFei Zhang, Zhiguo Qu, Conjugate heat transfer in anisotropic woven metal fiber-phase change material composite, *Applied Thermal Engineering* 189 (2021) 116618.
76. Qinlong Ren, Cho Lik Chan, Numerical study of double-diffusive convection in a vertical cavity with Soret and Dufour effects by lattice Boltzmann method on GPU, *International Journal of Heat and Mass Transfer* 93 (2016) 538–553.
77. Qinlong Ren, Chenxing Liang, Zexiao Wang, Zhiguo Qu, Continuous trapping of bacteria in non-Newtonian blood flow using negative dielectrophoresis with quadrupole electrodes, *Journal of Physics D: Applied Physics* 54 (2020) 015401.
78. Qinlong Ren, Chenxing Liang, Insulator-based dielectrophoretic antifouling of nanoporous membrane for high conductive water desalination, *Desalination* 482 (2020) 114410.
79. Qinlong Ren, Cho Lik Chan, Natural convection with an array of solid obstacles in an enclosure by lattice Boltzmann method on a CUDA computation platform, *International Journal of Heat and Mass Transfer* 93 (2016) 273–285.

80. Qinlong Ren, Yichao Wang, Xixiang Lin, Cho Lik Chan, AC electrokinetic induced non-Newtonian electrothermal blood flow in 3D microfluidic biosensor with ring electrodes for point-of-care diagnostics, *Journal of Applied Physics* 126 (2019) 084501.
81. Yao Zhao, Changying Zhao, Zhiguo Xu, Numerical study of solid-liquid phase change by phase field method, *Computers and Fluids* 164 (2018) 94–101.
82. Li Chen, Huibao Luan, Yongliang Feng, Chenxi Song, Ya-Ling He, Wen-Quan Tao, Coupling between finite volume method and lattice Boltzmann method and its application to fluid flow and mass transport in proton exchange membrane fuel cell, *International Journal of Heat and Mass Transfer* 55 (13–14) (2012) 3834–3848.
83. H.B. Luan, H. Xu, L. Chen, D.L. Sun, W.Q. Tao, Numerical illustrations of the coupling between the lattice Boltzmann method and finite-type macro-numerical methods, *Numerical Heat Transfer, Part B: Fundamentals* 57 (2010) 147–171.
84. Hui Xu, Huibao Luan, Yaling He, Wenquan Tao, A lifting relation from macroscopic variables to mesoscopic variables in lattice Boltzmann method: Derivation, numerical assessments and coupling computations validation, *Computers & Fluids* 54 (2012) 92–104.
85. Zhonghao Rao, Shuangfeng Wang, Feifei Peng, Molecular dynamics simulations of nano-encapsulated and nanoparticle-enhanced thermal energy storage phase change materials, *International Journal of Heat and Mass Transfer* 66 (2013) 575–584.
86. Hongtao Xu, Ning Wang, Chenyu Zhang, Zhiguo Qu, Meng Cao, Optimization on the melting performance of triplex-layer PCMs in a horizontal finned shell and tube thermal energy storage unit, *Applied Thermal Engineering* 176 (2020) 115409.
87. Ben Xu, Peiwen Li, Cho Lik Chan, Application of phase change materials for thermal energy storage in concentrated solar thermal power plants: A review to recent developments, *Applied Energy* 160 (2015) 286–307.

6 Latent Heat of Fusion and Applications of Silicon-Metal Alloys

Ming Liu, Peter Majewski, Frank Bruno, Nikki Stanford, Rhys Jacob, Shane Sheoran
University of South Australia

Serge Bondarenko
Climate Change Technologies Pty Ltd

CONTENTS

6.1 INTRODUCTION

Renewable energy sources like wind and solar are well known to be intermittent, and efforts are being made to create energy storage systems to harvest excess energy from renewable energy systems for times of higher demand and lower energy generation (Liu, Belusko, and Bruno 2016). Apart from conventional batteries like lithium-ion batteries, zinc-bromide batteries, or more conventional lead acid batteries, the storage of thermal energy has attracted significant interest, as it generally provides

DOI: 10.1201/9781003213260-6

much higher energy density and efficiency at lower cost (Liu, Belusko, and Bruno 2016; Liu et al. 2016). Thermal energy storage (TES) applications may use different material properties to achieve energy storage. According to the thermal mechanism used to store energy, TES can be classified into three types: sensible (e.g. water and rock), latent (e.g. water/ice, salt hydrates, and salts), and thermo-chemical reactions (e.g. chemical reactions and sorption processes). Sensible storage happens when the temperature of a material is raised or lowered, whereas latent storage occurs when the phase of a material is changed (solid to liquid or liquid to vapour) without a change in temperature. Both mechanisms can occur in the same material. The third mechanism, a chemical reaction or a sorption process, takes place on the surface of a material. In all cases, heat can be either absorbed or released from the material.

Thermal energy can be stored in phase change materials (PCMs) which store the thermal energy in the molten stage of the material and release thermal energy during crystallisation (Liu, Belusko, and Bruno 2016; Liu et al. 2016; Liu, Sun, and Bruno 2020). At potential storage temperatures of up to 1000°C, a large number of salt mixtures and metal alloys have been identified, studied, and reported (Liu, Saman, and Bruno 2012; Liu et al. 2015; Liu et al. 2017; Liu, Fernández, and Segarra 2018; Liu et al. 2021; Kenisarin 2010). Salt-based PCMs typically have very low thermal conductivities of approximately 0.5–1.5 W/(m·K) (Kenisarin 2010), leading to poor heat transfer into and out of the PCM. Extensive research work has been carried out to reduce the thermal resistance between the heat transfer fluid (HTF) and the PCM by addition of fins and heat pipes, enhancement of the effective thermal conductivity of the PCM by combining it with a highly conductive graphite/metal porous matrix or doping with nanomaterials, and employing a PCM cascade configuration to improve the overall energy and exergy efficiencies (Gasia, Miró, and Cabeza 2016; Li et al. 2019).

Metal and metal alloy PCMs, overcoming the common disadvantage of low thermal conductivity of salt-based PCMs, has recently attracted significant attention. Other advantages of metallic PCMs include high volumetric densities, good cycling stability, and low volume expansion (Zhu, Nguyen, and Yonezawa 2021). In general, much higher storage temperatures provide more efficient energy storage. Silicon has emerged as a powerful storage media for thermal energy at very high temperatures, as it has the highest latent heat of fusion value of 1800 J/g among metals and a melting temperature of 1414°C (Datas et al. 2016). While the high latent heat of fusion of silicon is of benefit, the very high melting temperature can provide significant operational challenges like appropriate insulation, stability of heating elements, and compatible container material. To overcome these issues, the use of silicon alloys is considered to be an option to reduce the melting temperature (Wang et al. 2015). Four binary Al-Si composites and two ternary composites (Al-Si-Fe and Al-Si-Ni) were characterised in Wang et al.'s work (Wang et al. 2015) and they were found to be promising candidates as TES materials. Kenisarin (Kenisarin 2010) has reviewed high-temperature metal alloy PCMs with melting temperatures in a range of 300°C–950°C, which usually contains Al, Si, Cu, Mg, and Fe.

The melting temperatures and latent heats of fusion of silicon-based PCMs from (Kenisarin 2010; Wang et al. 2015; Sun et al. 2007) are presented in Figure 6.1, where the composition is in weight percentage. The cost estimation of those metal alloys was performed and plotted in Figure 6.2, using the cost of pure metal from Granta

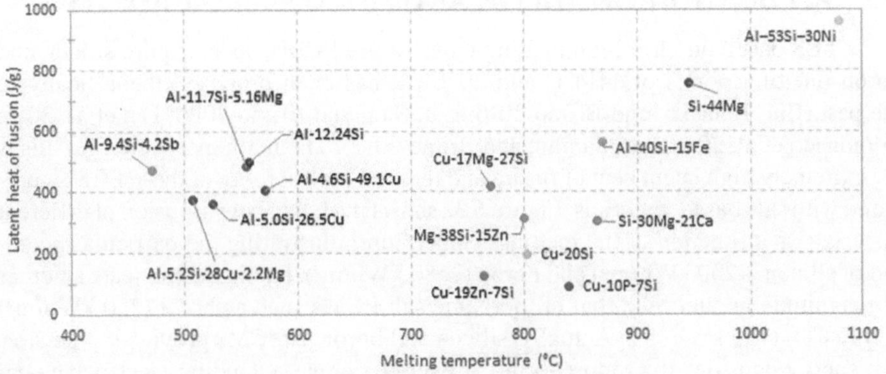

FIGURE 6.1 Melting temperatures and latent heats of fusion of silicon-based PCMs. (Based upon values from Kenisarin 2010 and Wang et al. 2015.)

EduPack 2021 (ANSYS Inc 2021). However, only the cost of raw material and its weight composition were considered when calculating the cost of the metal alloy. The processing and heat treatment required to prepare different metal alloys could be significantly distinct, therefore, it is impossible to estimate the cost without knowing the production procedures. As can be seen in Figure 6.2, the cost of material was found to be more expensive when the material's melting temperature is higher.

To the best of the authors' knowledge, only scarce experimental results of silicon alloys are available in the literature and most studied alloys have melting temperatures below 1000°C. This chapter presents a number of silicon alloys studied with respect to their melting temperatures (850°C–1400°C) and latent heats of fusion to provide new knowledge for the use of silicon alloys for TES. The material costs of the studied silicon alloys were estimated as well.

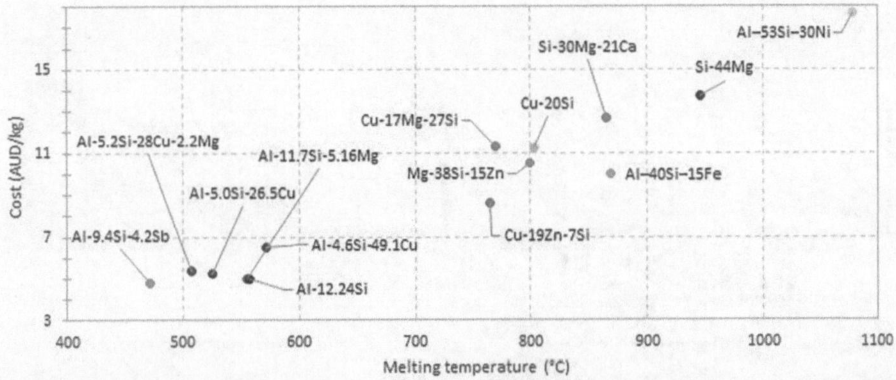

FIGURE 6.2 Melting temperatures and cost of silicon-based PCMs in Figure 6.1 (1 AUD = 0.73 USD).

6.2 APPLICATION OF SILICON AND SILICON-METAL ALLOYS

Solar TES based on ultra-high melting temperature PCMs, such as pure silicon and boron (melting points of 1414°C and 2076°C), has been proposed theoretically in the past (Liu, Belusko, and Bruno 2016; Liu, Sun, and Bruno 2020; Liu et al. 2016; Lomonaco et al. 2016; Liu, Saman, and Bruno 2012). The main motivation for this is the extremely high latent heat of fusion and thermal conductivity of these PCMs compared with salt-based materials. Figure 6.3a shows the latent heat of fusion of different materials as a function of the melting temperature, illustrating the particular potential of silicon (1230 kWh/m^3) and boron (2680 kWh/m^3), having latent heats an order of magnitude greater than that of inorganic salt PCMs such as NaCl (250 kWh/m^3) and K_2CO_3 (120 kWh/m^3). Actually, silicon and boron as PCMs provide higher storage energy densities than most forms of energy storage, including electrochemical batteries (Figure 6.3b).

The high storage temperature of silicon and silicon boron alloy makes them of particular interest for energy storage in thermophotovoltaic (TPV) systems. A TPV system converts thermal radiation into electrical energy using PV cells. The thermal radiation can be generated from various heat energy sources such as concentrated solar, waste heat, and chemical fuel. In solar TPV systems, TES enables the continuing production of electricity by storing excess heat during the day and withdrawing it after sunset. Many researchers have proposed various conceptual designs of integration of silicon or silicon boron PCM for TES in solar TPV systems and investigated the effect of the design on the performance of the TPV system. In Emziane and Alhosani's design, PCM is encapsulated in a cylinder and heat is transferred into the PCM from the receiver cavity through conduction (Emziane and Alhosani 2014). A vertically tapered configuration was studied by Datas, Chubb, and Veeraragavan (2013) and Veeraragavan, Montgomery, and Datas (2014). The proposed TPV system integrated with PCM storage is illustrated in Figure 6.4. As shown in Figure 6.5, by placing a mobile TPV generator in the middle of a PCM storage system, the power output is adjustable to meet the demand (Datas et al. 2016). In power-heat-power

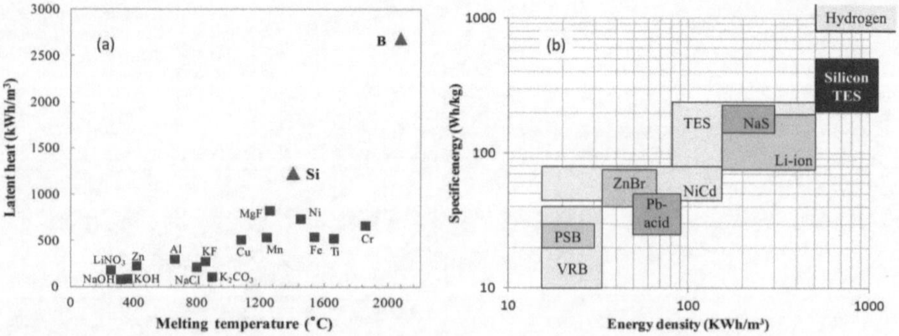

FIGURE 6.3 Latent heat of fusion of different materials as a function of the melting temperature (a) and specific energy of different energy storage systems as a function of energy density (b). (Datas et al. 2016.)

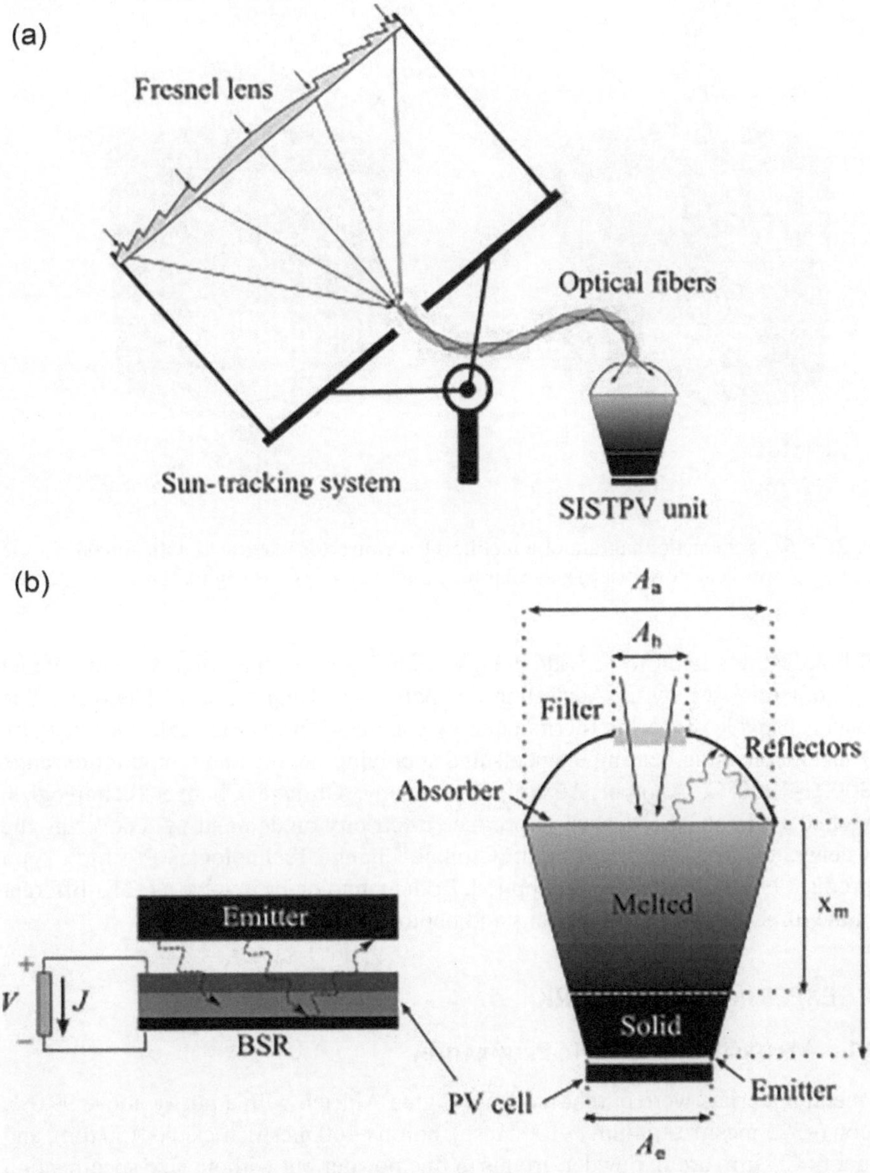

FIGURE 6.4 Schematic diagram of fibre optic concentrator and TPV system integrated with silicon storage (a) and vertical oriented TPV-silicon storage system (b). (Datas, Chubb, and Veeraragavan 2013.)

application (left-hand side of Figure 6.5), excess electricity in the grid can be stored by generating heat using an electric heating system and melting the PCM. Indicated by simulation results, the electric energy density of the proposed configuration can achieve 200–450 kWh_e/m^3, comparable to the best performing state-of-the-art lithium-ion batteries (Datas et al. 2016).

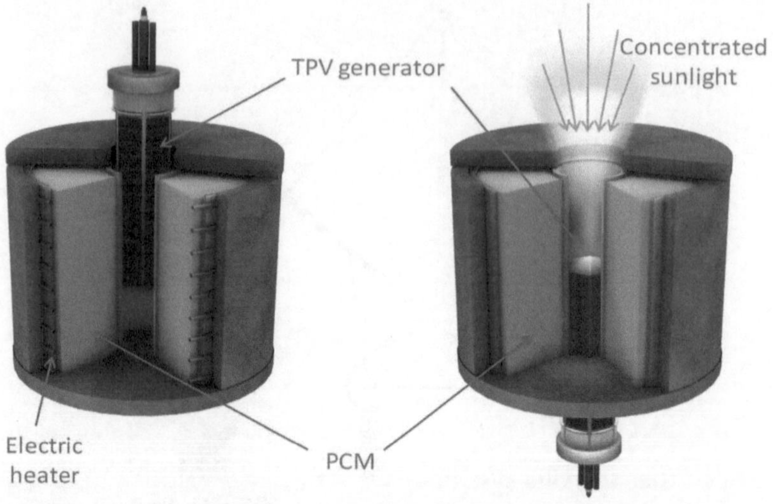

FIGURE 6.5 Schematic diagram of a mobile TPV convertor integrated with silicon storage system for power-heat-power (left) and solar-heat-power conversions (right). (Datas et al. 2016.)

SiBox, using silicon to provide energy storage and regeneration, was developed and commercialised by an Australian company 1414 Degrees (1414 Degrees). The product is capable of storing thermal energy generated from renewable electricity by melting silicon using heating elements and supplying hot air in a temperature range of 800°C–1000°C. Another Australian company, Climate Change Technologies, developed a silicon TES system to produce electricity on demand by recovering the heat generated from electrical input (Climate Change Technologies Pty Ltd). As a by-product, hot water of 78°C is supplied. Both technologies involve a HTF, different from the direct emission of electrons and photons in TPV technology.

6.3 EXPERIMENTAL WORK

6.3.1 Materials and Sample Preparation

All metal materials were purchased from Sigma Aldrich with a purity above 99.0%. Silicon (–325 mesh), titanium (–100 mesh), boron (–60 mesh), nickel (<150 μm), and copper (<425 μm) are in powder. Iron is in fine powder but without size specification provided. Chromium and cobalt are granular, and they were crushed into powder prior to sample preparation. Silicon alloy samples of 2 g were prepared by mixing the metals in the required weight ratio as presented in Table 6.1. The mixture was placed inside an alumina crucible and heated to 20°C above its reported melting temperature for 6 hours in a high-temperature tube furnace. Prior to the heating process, high-purity argon gas (>99.99%) was purged through the furnace for at least 2 hours in order to eliminate air and avoid metal oxidation during the sample preparation. The furnace was then cooled down to room temperature and the crystallised mixture was crushed and grinded into fine granules.

TABLE 6.1
Silicon and Silicon Alloys Identified from the Literature

Sample	Composition (mol.%)	Composition (wt.%)	Reaction Type	Reported Melting Temperature (°C)	Reported Latent Heat of Fusion (J/g)	References
Si	–	–	–	1414	1800	Datas et al. (2016)
Si92B8	Si-B: 92–8	96.8–3.2	Eutectic	1385	–	Zaitsev and Kodentsov (2001)
Si86B14	Si-B: 86–14	94.1–5.9	Eutectic	1362	–	Brosset and Magnusson (1960)
Si84Ti16	Si-Ti: 84–16	75.5–24.5	Eutectic	1330	–	Cahn (1991)
Si87Cr13	Si-Cr: 87–13	78.3–21.7	Eutectic	1328	–	Du and Schuster (2000)
Fe_2Si	Si-Fe: 34–66	20.6–79.4	Congruent melting compound	1212	–	von Goldbeck (1982)
Fe_3Si_7	Si-Fe: 71–29	55.2–44.8	Congruent melting compound	1220	–	von Goldbeck (1982)
Si77Co23	Si-Co: 77.5–22.5	62.1–37.9	Eutectic	1259	–	Ishida, Nishizawa, and Schlesinger (1991)
Si56Ni44	Si-Ni: 56–44	37.9–62.1	Eutectic	966	–	Nash and Nash (1987)
Si50Cu50	Si-Cu: 50–50	30.7–69.3	–	859	200	Olesinski and Abbaschian (1986), Birchenall et al. (1980)

6.3.2 Measurement Methods

The phase change temperature and enthalpy of all the samples were measured by using a simultaneous thermal analyser (Netzsch STA 449 F3 Jupiter®). The temperature and sensitivity calibration was conducted using indium, tin, zinc, aluminium, silver, and gold as reference materials. The sample of 10–20 mg, weighed by using Mettler Todedo balance XS105DU, was loaded in an 85 μL graphite crucible and covered with lid. The sample was subjected to one or two heating–cooling thermal cycle(s) with a heating/cooling rate of 10°C/min. The maximum temperature during heating was 50°C above the reported melting temperature. The experiment was conducted under constant argon flow at a rate of 20 ml/min. After the test was complete, the phase change temperature and enthalpy were calculated using the software Proteus® from Netzsch.

TABLE 6.2
Summary of the Measured Melting Temperatures and Latent Heat of Fusion for the Various Systems and the Estimated Material Costs

Sample	Measured Melting Temperature (°C)	Measured Latent Heat of Fusion (J/g)	Material cost (AUD/kg) (1 AUD = 0.73 USD)	Cost of Latent Heat Storage (AUD/kWh)	Comment
Si	1411.5	1850	21.4	41.6	
Si92B8	1402.6	1025	51.1	179.5	
Si86B14	1405.6	1287	75.5	211.2	
Si84Ti16	1333.1	419.4	22.0	188.7	
Si87Cr13	1330.6	730.2	20.5	100.9	
Fe_2Si	1212.4	167.5	5.3	115.0	Reaction with graphite crucible after melting
Fe_3Si_7	1220.1	646.6	12.3	68.7	
Si77Co23	1264.4	988.9	34.2	124.6	
Si56Ni44	949.5	187.8	28.4	544.8	
Si50Cu50	799.3 and 1125.4	146.6 and 261	12.6	309.6 and 173.9	Reaction with graphite crucible after melting

6.3.3 Results

The experimental results for each sample are discussed in the following subsections. The results are summarised in Table 6.2 at the end of this section, along with the material cost estimated by using the cost of pure metal from Granta EduPack 2021.

6.3.3.1 Silicon

As a reference sample, pure silicon was analysed. The heat flux curve of the sample shows a clear and sharp melting peak at 1411.5°C (Figure 6.6), which is slightly lower than the reported melting temperature of silicon at 1414°C. The melting event has a latent heat of fusion (ΔH) value of 1850 J/g which is slightly higher than reported ΔH values of pure silicon at about 1800 J/g (Datas et al. 2016).

6.3.3.2 Silicon Boron

The silicon boron system is characterised by a eutectic point at 92 mol.% silicon and 1385°C as well as a number of Si-B alloys in the boron rich region of the systems (Zaitsev and Kodentsov 2001). These alloys have melting temperatures in excess of 1800°C. Another study of the Si-B phase diagram shows the eutectic point at 86 mol.% silicon and a eutectic temperature of 1362°C (Brosset and Magnusson 1960). Therefore, two samples with the silicon composition of 92 and 86 mol.% were prepared, respectively. The differential scanning calorimetry (DSC) graph of the sample with 92 mol.% silicon in Figure 6.7 shows a sharp melting peak with the onset temperature of 1402.6°C which is higher than the reported melting temperature. The latent

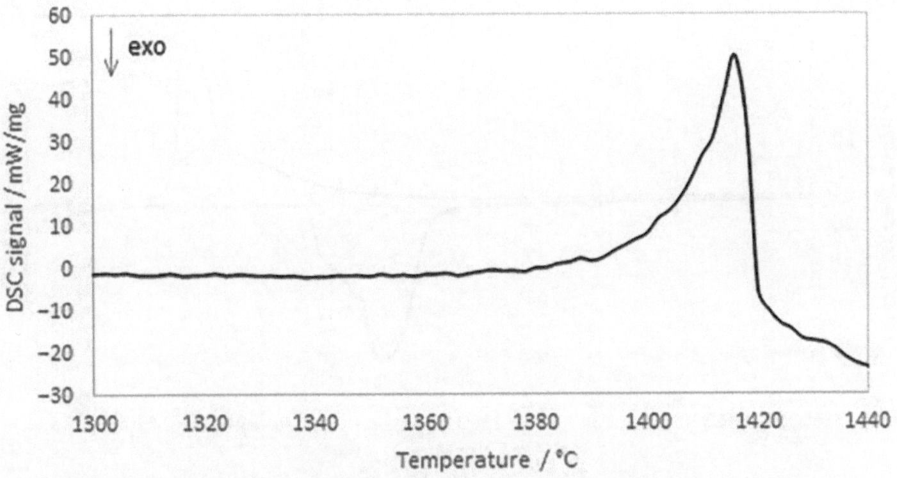

FIGURE 6.6 DSC curve of pure silicon sample.

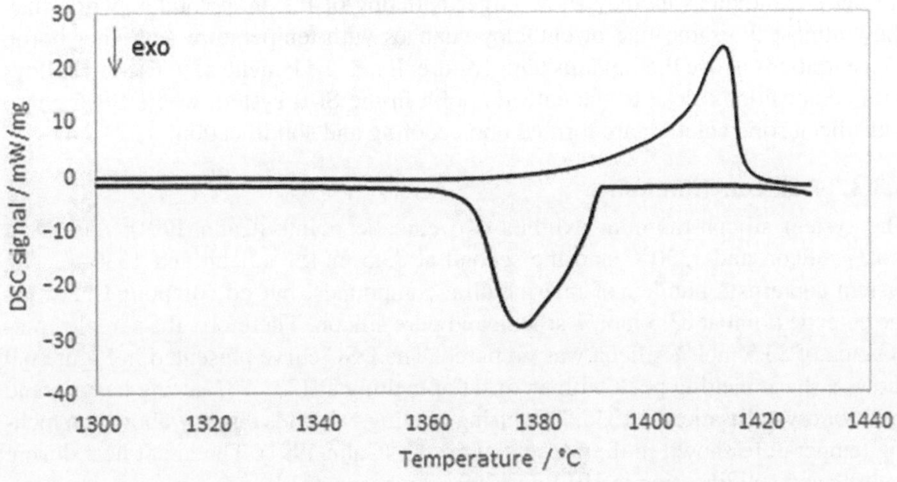

FIGURE 6.7 DSC curve of the Si-B sample with the composition of 92 mol.% silicon.

heat value of the melting and solidification event is 1025 and 1150 J/g, respectively. The DSC curve of the sample with 86 mol.% silicon shows that the sample appears not to be at the eutectic composition, as the peak during cooling exhibits a pronounced shoulder in Figure 6.8. The onset temperature of the main melting event is 1405.6°C, nearly identical to that of the sample with 92 mol.% silicon. However, the ΔH value is 1287 J/g (melting) and 1440 J/g (solidification), higher than that of the sample with 92 mol.% silicon. The results suggest that the eutectic point is at about 92 mol.% silicon and the eutectic temperature is about 1400°C.

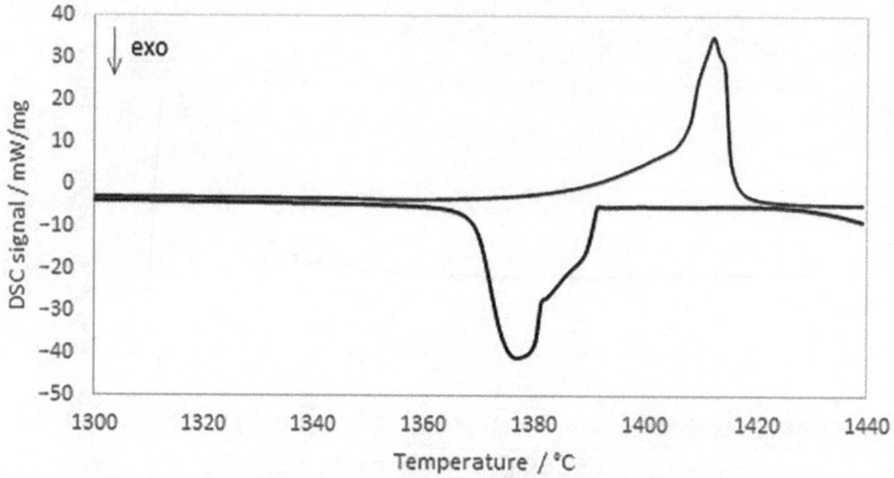

FIGURE 6.8 DSC curve of the Si-B sample with the composition of 86 mol.% silicon.

It may be concluded that the alloys closer to the eutectic point of the Si-B system are better candidates as they show larger enthalpy of fusion at solidus, while they show almost the same rate of enthalpy changes with temperature as higher boron concentrations above the liquidus temperature. It may be beneficial to use Si-B alloys with compositions close to the eutectic point in the Si-B system where SiB6 phase and silicon solid solution are formed upon cooling and solidification.

6.3.3.3 Silicon-Titanium

The system silicon-titanium exhibits two eutectic points (Cahn 1991): one at 14 mol.% silicon and 1330°C and the second at 83.5 mol.% silicon and 1330°C. The system contains a number of intermetallic compounds, but no compound between the eutectic point at 83.5 mol.% silicon and pure silicon. Therefore, the sample composition of 83.5 mol.% silicon was prepared. The DSC curve presented in Figure 6.9 shows a sharp melting peak with an onset of melting at 1333.1°C during heating and onset of crystallisation at 1336.5°C during cooling, which is slightly above the melting temperature shown in the phase diagram in (Cahn 1991). The latent heat during melting and solidification is 419.4 and 399.6 J/g, respectively.

6.3.3.4 Silicon-Chromium

The silicon-chromium system contains three eutectic points and a number of intermetallic compounds (Du and Schuster 2000). One eutectic point at 16 mol.% silicon and 1700°C, the second one at 58 mol.% and 1408°C, and the third eutectic point at 87 mol.% silicon and 1328°C. No intermetallic compound exists between the third eutectic point and pure silicon and, therefore, the sample composition having 87 mol.% of silicon was chosen. The DSC curve of this sample exhibits a sharp and clear peak at temperatures of 1330.6°C during heating and 1332.5°C during crystallisation of the sample which is slightly higher than the reported melting temperature (Figure 6.10). The ΔH value of the phase change event is 695.6 and 730.2 J/g during solidification and melting, respectively.

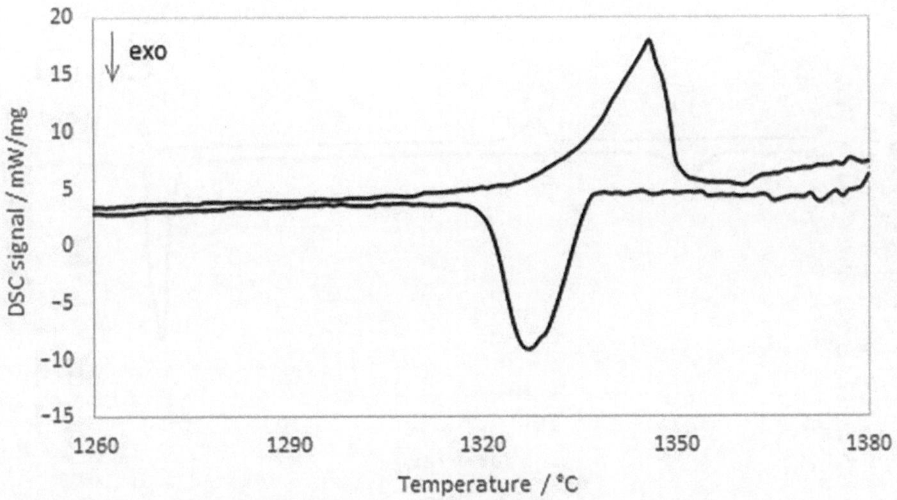

FIGURE 6.9 DSC curve of the Si-Ti sample with the composition of 83.5 mol.% silicon.

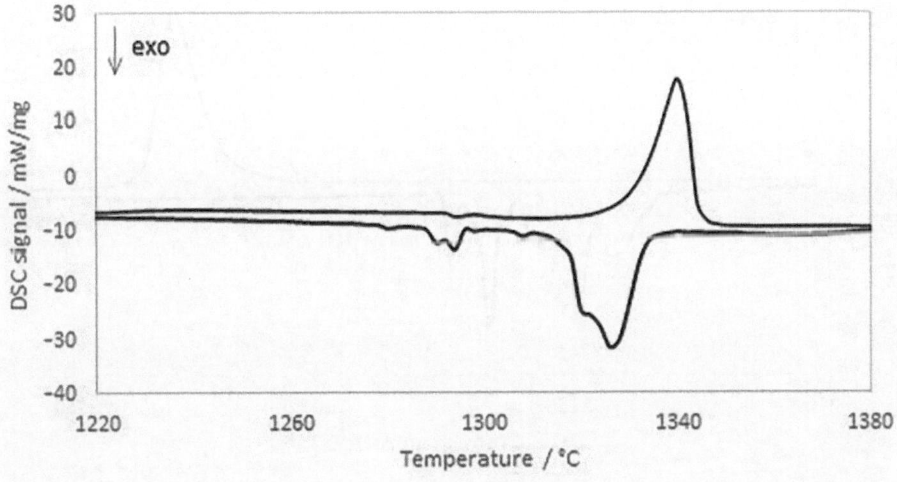

FIGURE 6.10 DSC curve of the Si-Cr sample with the composition of 87 mol.% silicon.

6.3.3.5 Silicon Iron

The system silicon iron has a number of eutectic points and intermetallic compounds (von Goldbeck 1982). At temperatures above 1100°C the compounds Fe_2Si, FeSi, and Fe_3Si_7 are stable and melt congruently at 1212°C, 1410°C, and 1220°C, respectively. Between these compounds eutectic points exist. Two samples with the compositions Fe_2Si and Fe_3Si_7 were prepared for the analysis. As presented in Figure 6.11, the DSC curve of the sample with the composition Fe_2Si shows a sharp peak at temperatures of 1209.8°C during cooling and 1212.4°C during heating, which is

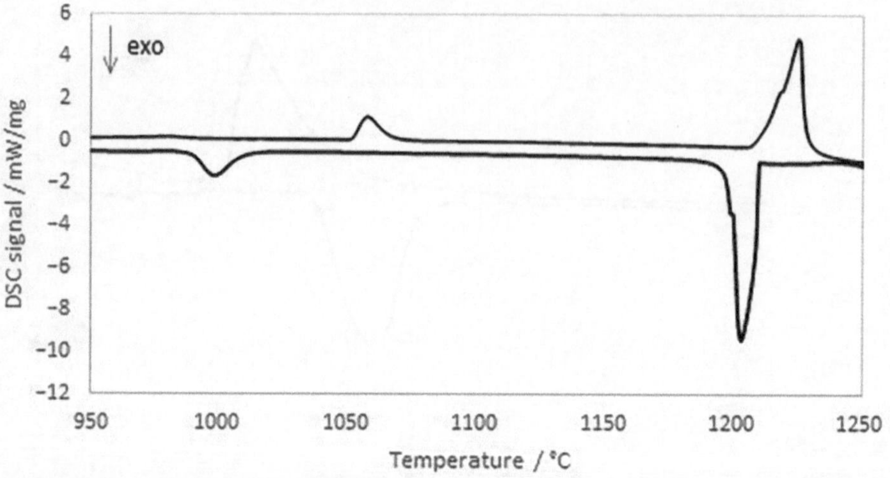

FIGURE 6.11 DSC curve of Fe_2Si sample.

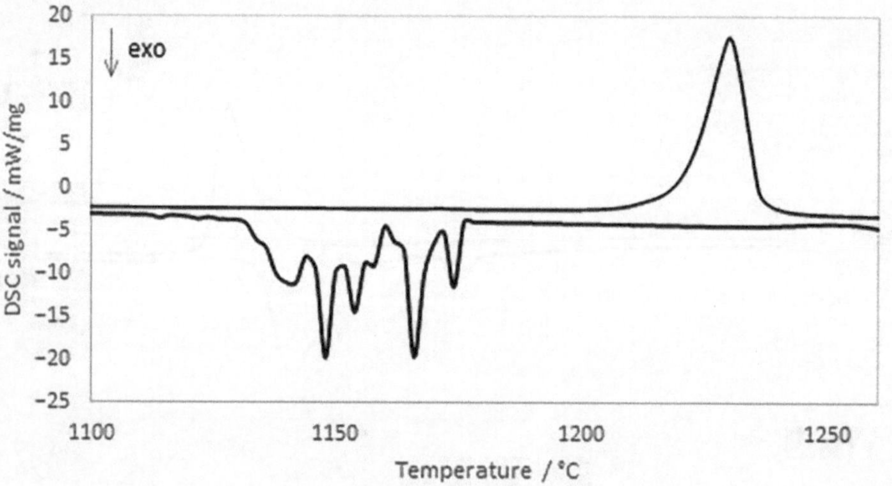

FIGURE 6.12 DSC curve of F_3Si_7 sample.

in reasonable agreement with the melting temperature of the Fe_2Si compound in the phase diagram in (von Goldbeck 1982). The ΔH value of the melting and solidification event is 167.5 J/g and 205.5 J/g, respectively. The DSC curve of the F_3Si_7 sample as shown in Figure 6.12 shows a sharp melting peak at 1220.1°C which correlates with the reported melting temperature for this compound. The ΔH value of the melting event is 646.6 J/g. However, during cooling the DSC curve shows various not well-pronounced peaks. This effect may be due to the observed reaction of this sample with the crucible material after the first melting event.

6.3.3.6 Silicon Cobalt

The silicon cobalt system exhibits a number of eutectic points and intermetallic compounds (Ishida, Nishizawa, and Schlesinger 1991). A eutectic point exists at 77.5 mol.% silicon and 1259°C. No intermetallic compound exists between this eutectic point and pure silicon and a sample with this composition was prepared for the DSC analysis. The DSC curve of the sample at the eutectic point of 77.5 mol.% silicon shows a slightly broadened melting peak for both heating and cooling as shown in Figure 6.13. The onset temperature is 1264.4°C during heating and 1265.4°C during crystallisation of the sample, which is higher than the reported temperature. The melting event has a ΔH value of 988.9 J/g in the second melting process.

6.3.3.7 Silicon-Nickel

The system silicon-nickel contains a number of eutectic points and intermetallic compounds (Nash and Nash 1987). Between the compounds Si-Ni and Si_2Ni a eutectic point exists at 56 mol.% silicon and 966°C. The sample with the composition at the eutectic point exhibits a melting event with the onset temperature of 964.5°C during heating of the sample and an onset temperature of 969.2°C during crystallisation, which is in good agreement with the phase diagram (Figure 6.14). The melting event has a ΔH value between 145.7 and 187.8 J/g. However, the DSC curve clearly indicates two additional thermal events at about 899°C and the second just before melting at about 949°C. This event only occurs during heating of the sample. There are no indications in the phase diagram that could explain these thermal events. It is well known that the compounds NiSi and $NiSi_2$ form polyforms and the observed thermal event may indicate crystallographic phase transitions of one of the compounds or both (Asayama et al. 2008; Lord et al. 2015). However, this could not be clarified within the scope of this study.

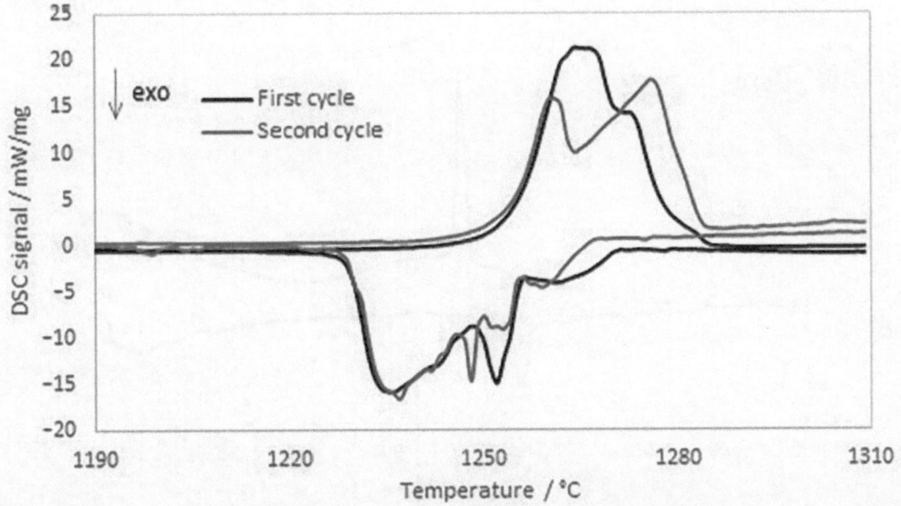

FIGURE 6.13 DSC curve of the Si-Co sample with the composition of 77.5 mol.% silicon.

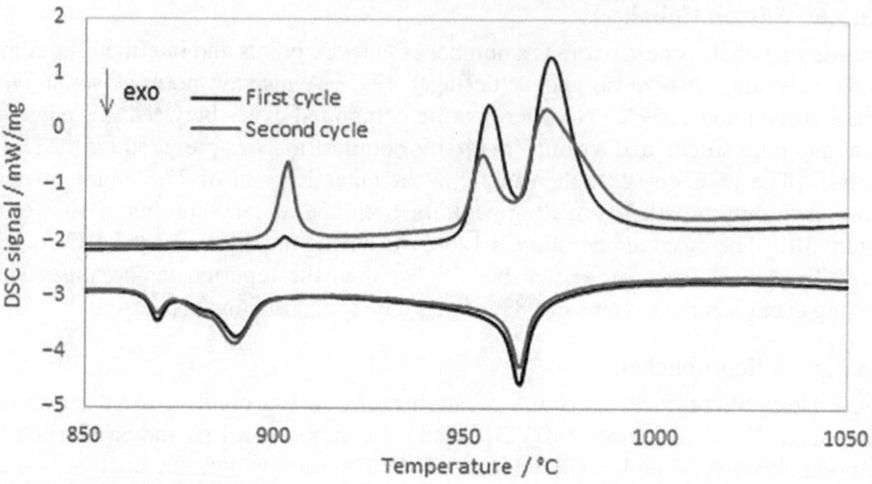

FIGURE 6.14 DSC curve of the Si-Ni sample with the composition of 56 mol.% silicon.

6.3.3.8 Silicon-Copper

The silicon-copper system contains a number of intermetallic compounds in the copper-rich region and below 859°C (Olesinski and Abbaschian 1986). Between the compound Cu_3Si, which melts at 859°C, and pure silicon, a eutectic point exists at 30 mol.% silicon and 802°C. No intermetallic compound exists between this eutectic point and silicon. A sample composition of 50 mol.% silicon was chosen to obtain ΔH values of the melting at the eutectic point and complete melting of the sample. As expected, the DSC curve exhibits two melting events with an onset temperature of 799.3°C and latent heat of fusion of 146.6 J/g for the first event, which is lower than the reported values of about 200 J/g (Birchenall et al. 1980) (Figure 6.15). The reason

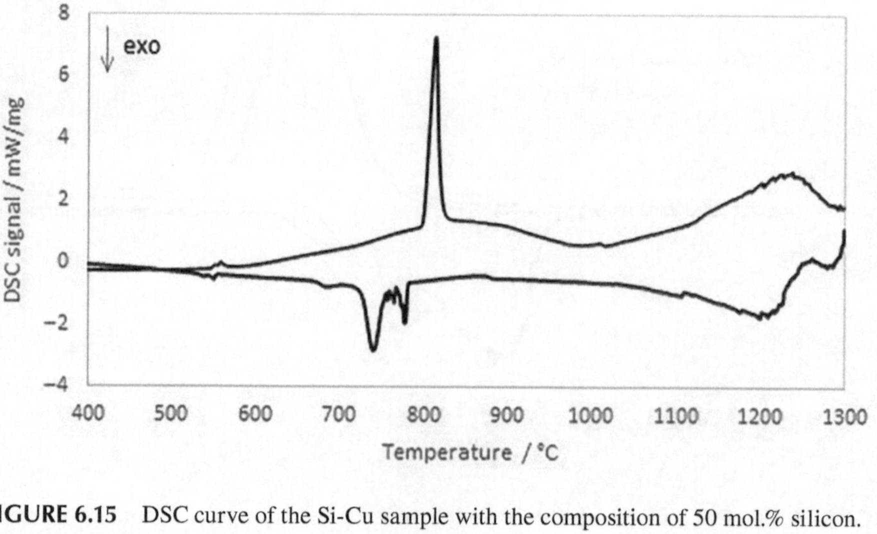

FIGURE 6.15 DSC curve of the Si-Cu sample with the composition of 50 mol.% silicon.

for this may be due to the slightly elevated baseline of the DSC curve between about 600°C and 1000°C during heating. The second event has an onset temperature of 1125.4°C and a latent heat of fusion of 261 J/g and can be attributed to the melting of the system at the composition of the sample, which is in good agreement with the phase diagram. Unfortunately, during melting some reaction with the crucible material occurred and, therefore, the DSC curve during cooling is distorted.

6.4 CONCLUSION

The study shows that several silicon-based alloys can be used for TES over a temperature range between 800°C and 1414°C. However, none of the investigated silicon alloys have a latent heat of fusion that is comparable with that of pure silicon. While the measurements in the Si-Fe system produced almost identical melting temperatures for the analysed compounds, in most studied samples, the measured melting temperatures slightly deviate from that reported in the literature, being either higher or lower. This effect can't be explained by a standard error of the used equipment and reviews of the phase diagrams appear to be appropriate. Among the studied alloys, silicon boron alloy shows the highest latent heat of fusion. However, the melting temperature is only slightly below that of silicon which is not a significant gain. All other alloys have latent heat of fusion values below about 1000 J/g. Among them, silicon cobalt alloy with a melting temperature of about 1265°C has the highest latent heat of fusion. The current high price of cobalt, however, may be prohibitive for the use of this metal in any large volume application. An interesting candidate can be silicon iron alloy due to a significantly lower melting temperature of about 1220°C compared to pure silicon, a reasonable latent heat of fusion value of about 646.6 J/g, and a moderate price for iron.

ACKNOWLEDGEMENT

The authors would like to thank the Government of South Australia for financial support through the Future Industries Accelerator scheme, project number FIA043. This research was performed as part of the Australian Solar Thermal Research Institute (ASTRI), which is funded by the Australian Government, through the Australian Renewable Energy Agency (ARENA). This work is also funded by the Department of Industry, Science, Energy and Resources (Australia) through Australia India Strategic Research Fund (AIRXII000124).

REFERENCES

1414 Degrees. https://1414degrees.com.au/.

Ansys Granta EduPack software, ANSYS, Inc., Cambridge, UK, 2020, (www.ansys.com/materials).

Asayama, K., N. Hashikawa, M. Kawakami, and H. Mori. 2008. High accuracy and resolution for the separation of nickel silicide polymorphs by improved analyses of EELS spectra. A. G. Cullis, P. A. Midgley, Editors. *Microscopy of Semiconducting Materials 2007. Springer Proceedings in Physics*, pp. 329–332, vol. 120. Springer, Dordrecht.

Birchenall, C. E., S. I. Gueceri, D. Farkas, M. B. Labdon, N. Nagaswami, and B. Pregger. 1981. *Heat Storage in Alloy Transformations*. University of Delaware, Newark. Report No. DOE/NASA/3184-2 (NASA CR-165355), https://www.osti.gov/servlets/purl/6213383.

Brosset, C., and B. Magnusson. 1960. The silicon-boron system. *Nature* 187 (4731):54–55.

Cahn, R. W. 1991. Binary alloy phase diagrams–Second edition. T. B. Massalski, Editor-in-Chief; H. Okamoto, P. R. Subramanian, L. Kacprzak, Editors. ASM International, Materials Park, OH. December 1990. xxii, 3589 pp., 3 vol., hardback. $995.00 the set. *Advanced Materials* 3 (12):628–629.

Climate Change Technologies Pty Ltd. https://www.cctenergy.com.au/.

Datas, A., D. L. Chubb, and A. Veeraragavan. 2013. Steady state analysis of a storage integrated solar thermophotovoltaic (SISTPV) system. *Solar Energy* 96:33–45.

Datas, A., A. Ramos, A. Martí, C. del Cañizo, and A. Luque. 2016. Ultra high temperature latent heat energy storage and thermophotovoltaic energy conversion. *Energy* 107:542–549.

Du, Y., and J. C. Schuster. 2000. Experimental reinvestigation of the CrSi-Si partial system and update of the thermodynamic description of the entire Cr-Si system. *Journal of Phase Equilibria* 21 (3):281–286.

Emziane, M., and M. Alhosani. 2014. Sensitivity analysis of a solar thermophotovoltaic system with silicon thermal storage. Paper read at *3rd International Symposium on Environmental Friendly Energies and Applications (EFEA)*, 19–21 November 2014, Paris, France.

Gasia, J., L. Miró, and L. F. Cabeza. 2016. Materials and system requirements of high temperature thermal energy storage systems: A review. Part 2: Thermal conductivity enhancement techniques. *Renewable and Sustainable Energy Reviews* 60:1584–1601.

Ishida, K., T. Nishizawa, and M. E. Schlesinger. 1991. The Co-Si (cobalt-silicon) system. *Journal of Phase Equilibria* 12 (5):578–586.

Kenisarin, M. M. 2010. High-temperature phase change materials for thermal energy storage. *Renewable and Sustainable Energy Reviews* 14 (3):955–970.

Li, Qi, C. Li, Z. Du, F. Jiang, and Y. Ding. 2019. A review of performance investigation and enhancement of shell and tube thermal energy storage device containing molten salt based phase change materials for medium and high temperature applications. *Applied Energy* 255:113806.

Liu, M., S. Bell, M. Segarra, et al. 2017. A eutectic salt high temperature phase change material: Thermal stability and corrosion of SS316 with respect to thermal cycling. *Solar Energy Materials and Solar Cells* 170:1–7.

Liu, M., M. Belusko, and F. Bruno. 2016. *High Temperature Phase Change Thermal Energy Storage for Concentrating Solar Power Plants*, Vol. 34. New South Wales: Australian Institute of Energy.

Liu, M., A. I. Fernández, and M. Segarra. 2018. Chapter 8 – Materials for phase change material at high temperature. In *High Temperature Thermal Storage Systems Using Phase Change Materials*, edited by L. F. Cabeza, and N. H. S. Tay. Academic Press, 195–230.

Liu, M., J. C. Gomez, C. S. Turchi, N. H. S. Tay, W. Saman, and F. Bruno. 2015. Determination of thermo-physical properties and stability testing of high-temperature phase-change materials for CSP applications. *Solar Energy Materials and Solar Cells* 139:81–87.

Liu, M., E. S. Omaraa, J. Qi, et al. 2021. Review and characterisation of high-temperature phase change material candidates between 500°C and 700°C. *Renewable and Sustainable Energy Reviews* 150:111528.

Liu, M., W. Saman, and F. Bruno. 2012. Review on storage materials and thermal performance enhancement techniques for high temperature phase change thermal storage systems. *Renewable and Sustainable Energy Reviews* 16 (4):2118–2132.

Liu, M., Y. Sun, and F. Bruno. 2020. A review of numerical modelling of high-temperature phase change material composites for solar thermal energy storage. *Journal of Energy Storage* 29:101378.

Liu, M., N.H.S. Tay, S. Bell, et al. 2016. Review on concentrating solar power plants and new developments in high temperature thermal energy storage technologies. *Renewable and Sustainable Energy Reviews* 53:1411–1432.

Lomonaco, A., D. Haillot, E. Pernot, E. Franquet, and J.-P. Bédécarrats. 2016. Sodium nitrate thermal behavior in latent heat thermal energy storage: A study of the impact of sodium nitrite on melting temperature and enthalpy. *Solar Energy Materials and Solar Cells* 149:81–87.

Lord, O. T., A. R. Thomson, Elizabeth T. H. Wann, Ian G. Wood, David P. Dobson, and Lidunka Vocadlo. 2015. The equation of state of the Pmmn phase of NiSi. *Journal of Applied Crystallography* 48 (6):1914–1920.

Nash, P., and A. Nash. 1987. The Ni–Si (nickel-silicon) system. *Bulletin of Alloy Phase Diagrams* 8 (1):6–14.

Olesinski, R. W., and G. J. Abbaschian. 1986. The Cu–Si (copper-silicon) system. *Bulletin of Alloy Phase Diagrams* 7 (2):170–178.

Sun, J. Q., R. Y. Zhang, Z. P. Liu, and G. H. Lu. 2007. Thermal reliability test of Al-34%Mg-6%Zn alloy as latent heat storage material and corrosion of metal with respect to thermal cycling. *Energy Conversion and Management* 48 (2):619–624.

Veeraragavan, A., L. Montgomery, and A. Datas. 2014. Night time performance of a storage integrated solar thermophotovoltaic (SISTPV) system. *Solar Energy* 108:377–389.

von Goldbeck, O. K. 1982. Fe–Si iron–silicon. In *IRON–Binary Phase Diagrams*. Berlin, Heidelberg: Springer, 136–139.

Wang, Z., H. Wang, X. Li, et al. 2015. Aluminum and silicon based phase change materials for high capacity thermal energy storage. *Applied Thermal Engineering* 89:204–208.

Zaitsev, A. I., and A. A. Kodentsov. 2001. Thermodynamic properties and phase equilibria in the Si-B system. *Journal of Phase Equilibria* 22 (2):126.

Zhu, S., M. T. Nguyen, and T. Yonezawa. 2021. Micro- and nano-encapsulated metal and alloy-based phase-change materials for thermal energy storage. *Nanoscale Advances* 3 (16):4626–4645.

7 Heat Transfer Augmentation of Latent Heat Thermal Storage Systems Employing Extended Surfaces and Heat Pipes

B. Kamkari and L. Darvishvand
Islamic Azad University

CONTENTS

7.1 INTRODUCTION

Energy crisis and global warming are major challenges in today's world. The effective use of renewable resources of energy such as solar energy plays a major role in overcoming these serious global problems. Despite the abundance and being environmentally friendly and freely available, solar energy is typically transient and intermittent in nature [1,2]. Consequently, efficient techniques are required to balance the unconformity between supply and demand of energy. Thermal energy storage techniques including sensible, latent, and chemical methods of heat storage provide efficient and sustainable solution to store and release a large amount of thermal energy [3,4]. Among these methods, latent heat thermal storage (LHTS) systems containing phase change materials (PCMs) have attracted a great attention and are

DOI: 10.1201/9781003213260-7

developing rapidly due to their significant advantages such as high energy storage density, uniform energy storage/supply, excellent chemical stability and controllability, and low price and compactness [5–9].

Although the LHTS systems have a high thermal energy density and are used as promising energy storage devices, the PCMs loaded in these systems possess relatively low thermal conductivity which is the most considerable drawback of LHTS units and drastically limits the thermal energy charging and discharging rates [6]. As a result, it is crucial to implement the thermal performance enhancement methods to tackle this disadvantage of the PCMs and pave the way for the deployment of LHTS in practical applications. Several techniques are utilized to improve the thermal performance of LHTS systems which can be categorized into three main groups as follows:

1. **Heat transfer enhancement methods** include employing extended surfaces, heat pipes, encapsulation, and multiple PCMs with different melting points.
2. **Thermal conductivity enhancement methods** include impregnation of porous materials with PCMs, dispersion of nanoparticles, and implementing low-density high thermal conductivity materials such as carbon fibers.
3. **Combination of different enhancement methods** include utilizing two or more enhancement approaches in a system simultaneously to further improve the heat transfer rate.

Figure 7.1 summarizes the thermal performance enhancement methods. The various heat transfer enhancement techniques in LHTS systems were reviewed by Ibrahim et al. [10]. Tao and He [3] presented a comparative review on different kinds of PCMs and performance enhancement methods for latent heat storage systems. Furthermore, they discussed the research gaps in the present performance enhancement methods and proposed some recommendations for further researches.

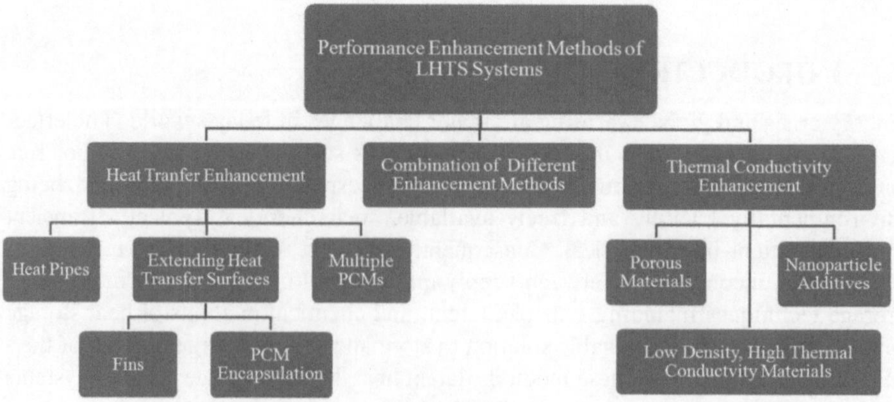

FIGURE 7.1 Classification of performance enhancement methods of LHTS systems.

Figure 7.2 shows some photos and schematic views regarding the enhancement techniques. In the following, brief descriptions are given for each of the enhancement techniques.

Increasing the heat transfer area between the heat transfer fluid (HTF) and PCM by employing the fins or encapsulating the PCMs are practical and effective methods to

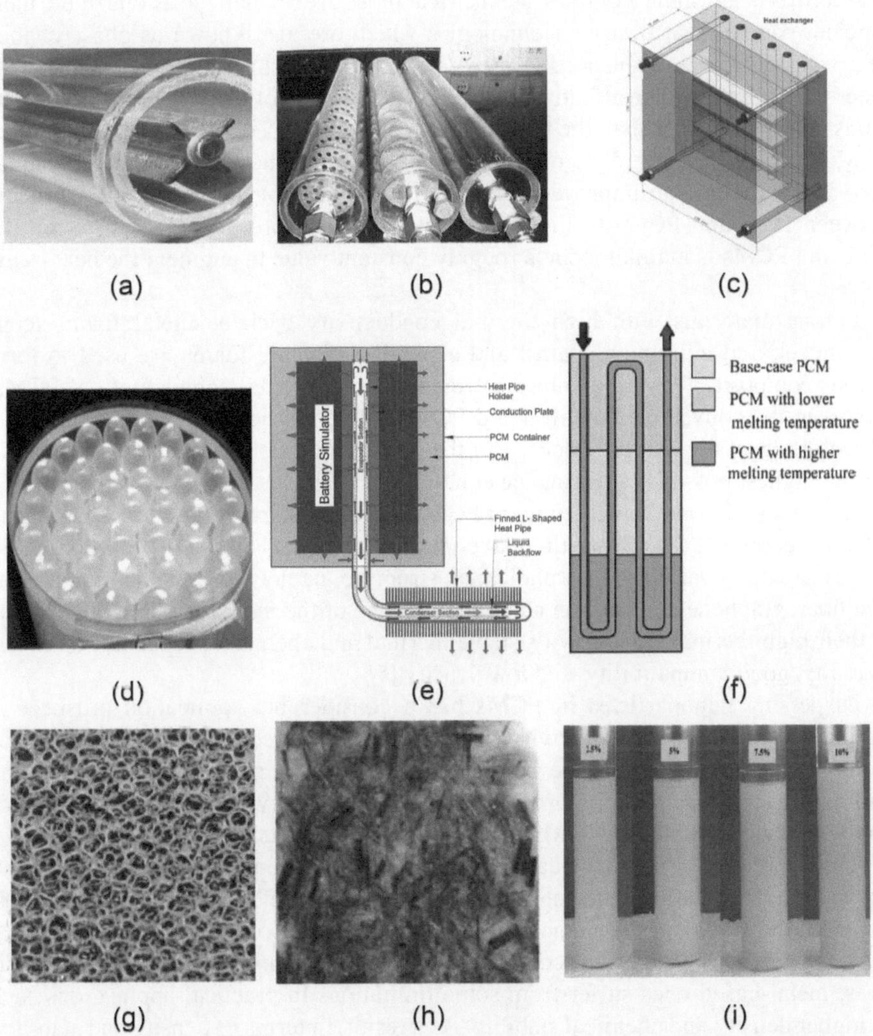

FIGURE 7.2 Photos and schematic views of thermal performance enhancement methods in LHTSs. (a) Longitudinal fin in cylindrical container [11]. (b) Angular fin in cylindrical container [12]. (c) Plate fin in rectangular enclosure [13]. (d) PCM encapsulated in spheres [14]. (e) PCM and heat pipe integration as passive battery cooling system [15]. (f) Multiple PCM with festoon design tube arrangement [16]. (g) Copper metal foam employed as porous materials [17]. (h) PCM + 4% carbon fiber 3 mm as low-density high thermal conductivity materials [18]. (i) Nanocomposite PCMs [19].

enhance the thermal performance of the LHTS systems. Adding fins to the heat transfer surface augments the heat transfer rate by increasing the surface area exposed to the PCM. In PCM encapsulation method, increase in effective heat transfer area is obtained by encapsulating the PCM in capsules with macro, micro, and nano scales namely macro-encapsulation, micro-encapsulation, and nano-encapsulation, respectively [3].

Integration of heat pipes with LHTS systems is an effective approach to enhance the thermal performance of the systems. Heat pipes are recognized as one of the most efficient passive heat transfer technologies which are also known as phase change heat transfer devices transporting large amount of heat by using latent heat of an inner working fluid. The effective thermal conductivities of these devices are several thousand times higher than the best metals [20].

Multiple PCMs with different melting points along the fluid flow direction are used in LHTS units to improve the rate and uniformity of the heat transfer process between HTF and PCM. In this method, the temperature difference between the HTF and PCMs is maintained at a roughly constant value to augment the heat transfer rate [4].

Porous materials with high thermal conductivity such as metal foams (copper, nickel, and aluminum foams) and expanded graphite foams are used to form porous composite PCM and enhance effectively the thermal conductivity of PCMs. However, the convective flow in liquid PCM will be restricted by porous materials. Therefore, porosity and pore density of the porous media should be chosen properly for the highest possible performance enhancement [3].

Enhancing thermal conductivity of PCMs by adding carbon-based materials is an effective approach to improve thermal conductivity of LHTS systems. Carbon-based materials with a variety of morphological structure, namely expanded graphite, carbon fiber, graphene, and carbon nanotube, are one of the most popular additives due to their high thermal conductivity, stable thermal and chemical properties, extensive usability, good compatibility, and low density [5].

Dispersing nanoparticles in PCMs has a considerable application prospect in enhancing the thermal conductivity of PCM due to the recent development of nanotechnology [5]. However, some practical problems such as particle sedimentation and aggregation under high particle loadings as well as viscosity augmentation of nano-enhanced PCMs (NePCMs) should be taken into account. Therefore, trade-off studies are required to evaluate the effect of thermal conductivity enhancement, convection degradation, and latent heat reduction in NePCMs [21]. Lin et al. [5] reviewed the methods for enhancing thermal conductivity of PCMs including adding substances with high thermal conductivity. Compared with the carbon-based additives, metal-based ones suffer from some limitations in practical applications such as higher density and chemical stability. As a result, in terms of density and stability, carbon-based additives possess a promising application prospects.

Recently, many researchers combined different enhancement methods to obtain further improvement in thermal performance of the LHTS systems. These include combining fins and nanoparticles [22,23], fins and heat pipe [15], heat pipe and metal foam [24]. Employing various augmentation techniques gives the opportunity to increase the heat transfer area and the thermal conductivity of PCM simultaneously leading to higher performance enhancement [10].

In the following sections, thermal performance enhancement techniques using extended surfaces and heat pipes will be discussed in details.

7.2 HEAT TRANSFER ENHANCEMENT USING EXTENDED SURFACES

Providing additional heat transfer surface between PCM and HTF by adding fins is one of the most efficient, cost-effective, and reliable methods for heat transfer augmentation in LHTS systems. On the other hand, embedding the fins in LHTS systems reduces the PCM volume, and consequently, thermal storage capacity of LHTS system decreases. Furthermore, the presence of the fins may hinder the natural convection flow as a dominant heat transfer mechanism in melting process [3]. Increase in the cost and weight of the LHTS system is another concern related to the employment of fins which should be taken into account. Hence, trade-off studies have been conducted extensively on fin numbers, fin patterns/shapes, geometries, and orientations of the system and fins to find the optimum performance of the system. The most typical geometries used for containers of LHTS systems can be classified into cylindrical, rectangular, and spherical. In the following sections, applications of fins in these common geometries are addressed.

7.2.1 Cylindrical Finned LHTS Containers

Many researchers have installed longitudinal [25,26], annular [27,28], and helical [29–31] fins in different cylindrical containers to intensify the melting performance of PCM by extending the heat transfer surfaces between PCM and HTF. A comparative study of the different fin types and shell-and-tube orientations was conducted by Zhang et al. [31]. They numerically simulated and compared melting process of PCM (RT-35) in vertical and horizontal shell-and-tube LHTS systems with different fin configurations (annular fin, double helical fin, quadruple helical fin, and longitudinal fin) as shown in Figure 7.3. They came to conclusion that fin configuration has a great influence on the charging process. Moreover, the results indicated that the vertically orientation LHTS systems with annular or double helical fins have superior heat transfer performance compared to the other cases. On the contrary, quadruple helical fin and longitudinal fin exhibited a better thermal behavior in the horizontal orientation.

Yang et al. [28] numerically investigated the effects of number, height, and thickness of annular fins on the melting process of commercial-grade paraffin (RT35) in a vertical shell-and-tube LHTS container. They selected fin number, height, and thickness as design variables and recommended an optimal group fin parameter for maximizing thermal performance. Paria et al. [32] experimentally investigated charging and discharging processes of paraffin as PCM in a horizontal shell-and-tube LHTS system equipped with annular fins. They evaluated the effects of fin density and Raynolds number on the latent thermal energy storage. It was demonstrated that increasing the Reynolds number and fin density reduce the melting and solidification times and increase the amount of stored energy. Agyenim et al. [33] experimentally compared the impact of installing annular and longitudinal fins in a

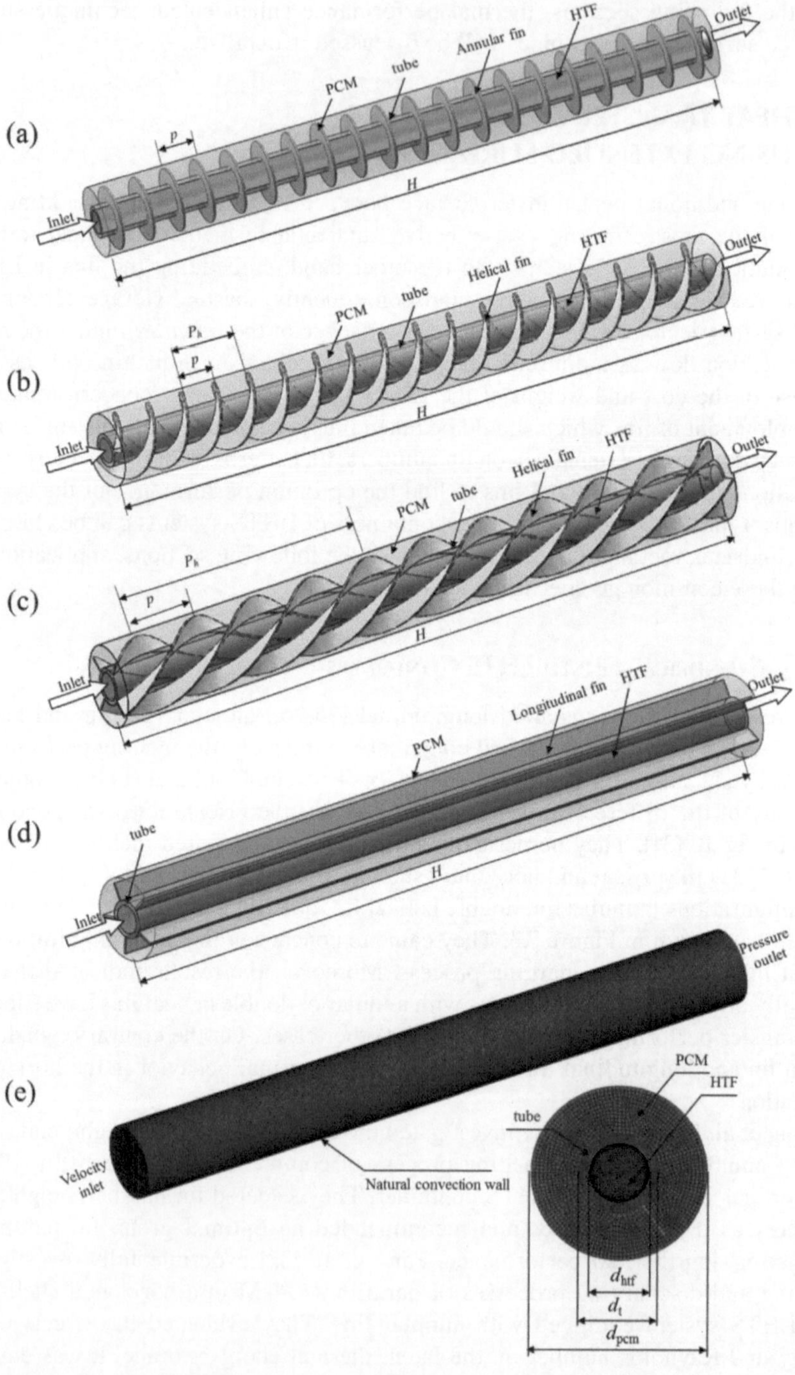

FIGURE 7.3 The three-dimensional geometry of shell-and-tube LHTS with different fin configurations investigated by Zhang et al. [31]: (a) annular fin; (b) double helical fin; (c) quadruple helical fin; (d) longitudinal fin; (e) unfinned tube.

horizontal shell-and-tube heat exchanger containing Erythritol with a melting point of 117.7°C. It was observed that the best performance was achieved by the system with longitudinal fins during charging and discharging processes. The heat transfer augmentation for a vertical shell-and-tube LHTS unit with three longitudinal fins was experimentally investigated by Rathod and Banerjee [26]. They performed a comparative analysis for the charging and discharging processes with and without fin and reported a reduction in melting and solidification times due to the longitudinal fin installation. Mehta et al. [29] carried out a comprehensive experimental analysis and numerical simulation to study the heat transfer enhancement in a vertical shell-and-tube-type LHTS system with incorporation of the helical fins. They indicated that employing the helical fins reduced the charging and discharging time by 41.48% and 22.16%, respectively.

Using branched fins is a beneficial technique to increase and manage the available surface area in order to improve the thermal performance. The impacts of employing bifurcated fin configurations and fin arrangement on melting behavior of paraffin inside the horizontal shell-and-tube heat exchangers were both experimentally and numerically studied by Safari et al. [34]. They also analyzed the effects of fin length and thickness for different fin arrangements shown in Figure 7.4. To gain a better insight into the effect of fin configuration and arrangement on the melting process,

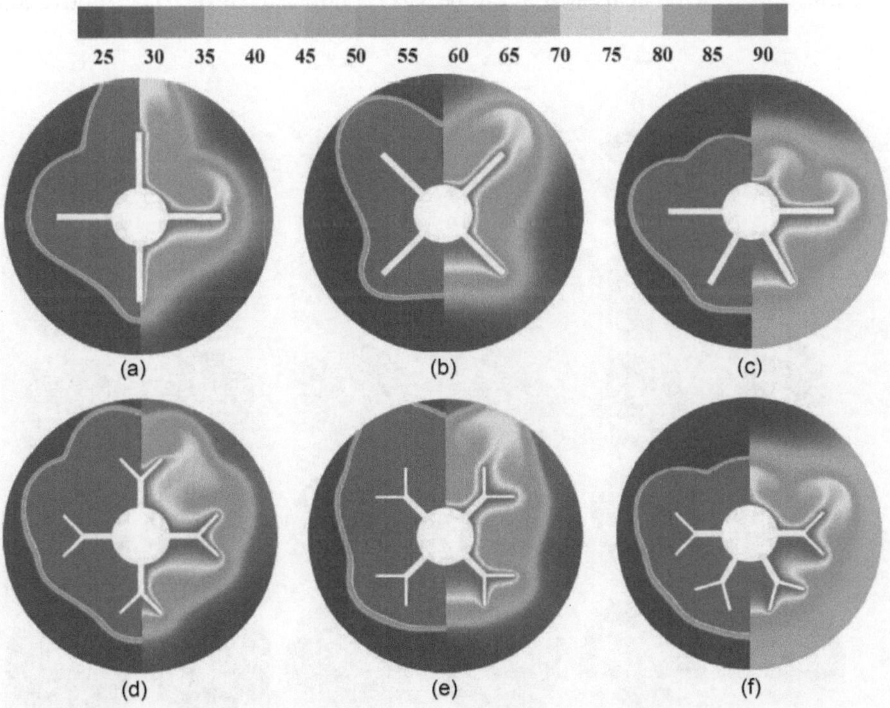

FIGURE 7.4 Liquid fraction and temperature distribution for the straight and bifurcated fins heat exchangers at time = 30 minutes obtained by Safari et al. [34]. (a) Straight-(a). (b) Straight-(a). (c) Straight-(a). (d) Bifurcated-(a). (e) Bifurcated-(b). (f) Bifurcated-(c).

liquid fractions and temperature distributions during the melting of PCM are represented in Figure 7.4 at equal melting times. It was observed that performance of bifurcated fin configuration is better than the straight fin configuration. Moreover, it was concluded that although the fin arrangement plays an important role in enhancing the thermal performance of LHTS system, in order to obtain the optimum fin arrangement, influence of fin length should be taken into account as well.

The eccentricity of the inner tube in a horizontal double pipe LHTS unit is a practically simple and beneficial technique to enhance thermal performance of the system studied by several researchers [35,36]. The combined effects of eccentricity of inner tube and fin configuration on melting behavior of paraffin inside horizontal shell-and-tube heat exchangers were analyzed experimentally and numerically by Safari et al. [11]. Heat exchangers with different fin configurations of straight and bifurcated were fabricated with the same total fin mass (Figure 7.5). It was found that the eccentricity of the inner tube had a significant impact on the melting rate acceleration due to the development of natural convection flows to a larger area of the annulus. Furthermore, they demonstrated that eccentric tube heat exchanger with a long upper bifurcated fin and a short lower straight fin exhibited the maximum melting time reduction compared to the other cases.

For the first time, Karami and Kamkari [12] experimentally investigated the effect of applying perforated fins on the thermal performance and energy storage of lauric acid in a vertical shell-and-tube LHTS heat exchanger. The motivation

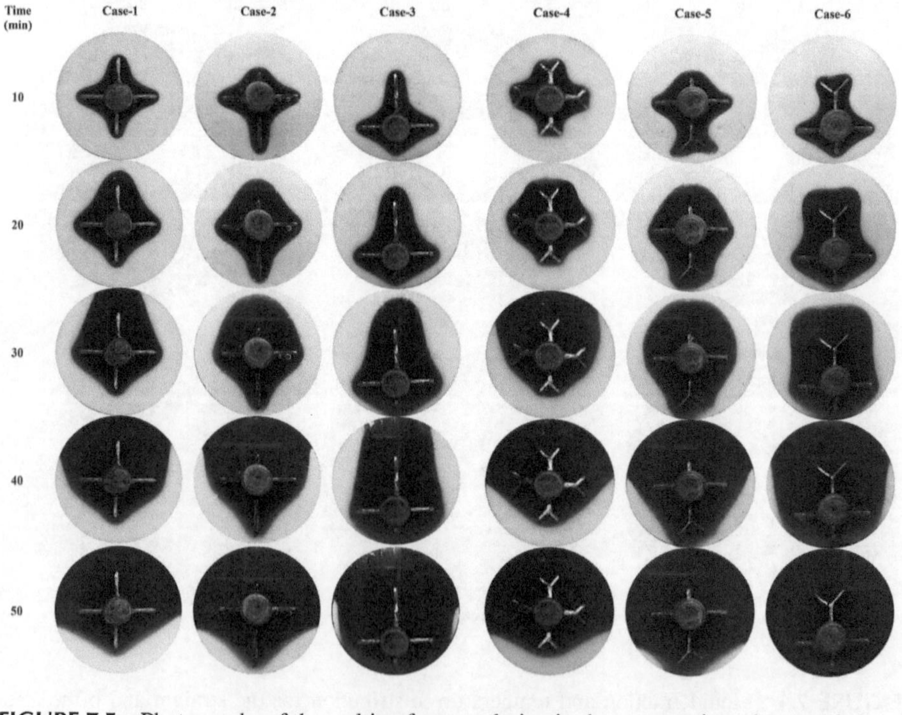

FIGURE 7.5 Photographs of the melting front evolution in the concentric and eccentric heat exchangers with straight and bifurcated fins proposed by Safari et al. [11].

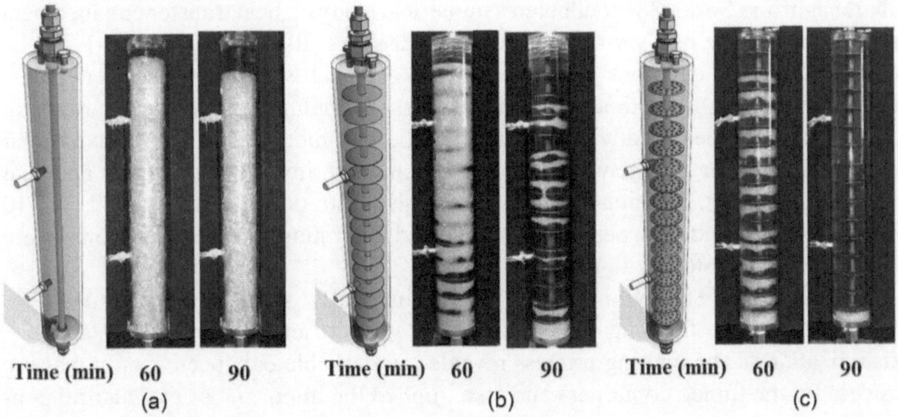

FIGURE 7.6 Photographs of the melting process investigated by Karami and Kamkari [12]. (a) Unfinned heat exchanger. (b) Solid finned heat exchanger. (c) Perforated finned heat exchanger.

of this research was to intensify the buoyancy-driven convection flows during the melting process. They compared the results with those of the unfinned and solid finned heat exchangers. Figure 7.6 shows the photographs of the melting process in unfinned, solid finned, and perforated finned heat exchangers. It was concluded that the time-averaged Nusselt number of the perforated finned heat exchanger is about 30% higher than that of the solid finned heat exchanger due to the minor hindering effect of the perforated fins on the development of the convection flows. Moreover, the total melting time of the perforated finned heat exchanger is about 7% lower than that of the solid finned heat exchanger while the total weight of the perforated fins is 16% less than that of the solid fins. Therefore, using perforated fins leads to a lighter and more cost-effective heat exchanger, and at the same time, improves the thermal performance of the system.

Mahdi et al. [37] numerically compared the melting behavior of PCM (RT82) in the presence of fins, nanoparticles (Al_2O_3), and combination of fins and nanoparticles in triplex-tube heat exchanger. Different fin arrangements were studied and concluded that embedding more long fins at the lower region maximizes the efficiency of the finned system. They compared the optimal fin configuration with two other systems containing the same amount of PCM with fins or nanoparticles. The volume of immersed fins and dispersed nanoparticles were equal. The proposed fin arrangement achieved significantly higher melting rate compared to the use of fin-nanoparticles combination or nano-enhanced PCM in the same volume of the heat exchanger.

7.2.2 Rectangular Finned LHTS Containers

Rectangular LHTS containers are basic heat storage units which are simple in geometry and boundary conditions. Plate fin and pin fin are the most popular high thermal conductive fins used in rectangular LHTS containers to enhance the system performance.

Gharebaghi and Sezai [38] conducted a numerical study on heat transfer enhancement in the horizontally and vertically oriented heat sinks filled with RT27 as PCM by inserting fins with different spacings and thicknesses. It was observed that decreasing the fin spacing leads to a significant increase in melting rate for both horizontal and vertical modules. Reddy [39] performed thermal modeling of PCM-based solar integrated collector storage water heating system to optimize the solar gain and heat loss characteristics. Rectangular enclosures filled with paraffin having 4, 9, and 19 fins were considered. The best performance and maximum latent energy stored were obtained by the system with nine fins.

Kamkari and Shokouhmand [40] experimentally studied the melting process in vertical rectangular LHTS container with different number of plate fins. Visualization of the melting process revealed remarkable enhancements in melting process for the finned containers and also implied the important role of natural convection flows on the melting process. The total melting time for the 1-fin and 3-fin containers were, on average, 18% and 37% less than that of the unfinned container. It was also mentioned that the increasing rate of the heat transfer rate decreases with adding the fin number. This thermal behavior implied that the beneficial impact of increase in heat transfer area is being offset by hindering the convection current. Furthermore, they presented two correlation equations to predict the Nusselt number and melt fraction.

Gharbi et al. [41] conducted a comparative experimental study on the thermal behavior of the different configurations of PCM-based heat sinks for cooling electronic devices. Four configurations were taken into account to extend critical time of system below a critical temperature: pure PCM, PCM in a silicone matrix, PCM in a graphite matrix, and pure PCM in a system of fins. Results indicated that combination of PCM with long, well-spaced fins presented a better thermal control of the system compared to the other enhancement techniques.

Pin fins have been widely studied in PCM-based heat sinks for the efficient thermal management of the electronic devices [42]. Arshad et al. [43] experimentally investigated the effect of insertion of aluminum square pin fin as well as the pin thickness and PCM volume fraction on thermal performance of a heat sink filled by paraffin wax as a PCM. It was revealed that the amount of PCM and number of fins have a great impact on effectiveness of a PCM-based heat sink. Baby and Balaji [42] developed an optimization algorithm using artificial neural network and genetic algorithm to obtain the optimum configuration of the pin fin in the n-eicosane-based heat sink to maximize the operating time of the electronic device. Ali et al. [44] performed an experimental analysis of different cross-sections of pin fins (triangular, rectangular, and circular) in heat sinks containing various PCMs. It was found that the triangular fin shape is the most effective pin-fin configuration.

7.2.3 Spherical Finned LHTS Containers

Spherical LHTS containers are mostly used in the form of spherical capsules containing PCM in the packed bed LHTS systems [45]. Although spherical containers provide a higher heat transfer areas and thermal performance compared to the other configuration, it is still essential to employ heat transfer enhancement method to overcome the low thermal conductivity of the PCM [46]. The effect of embedding the conducting pins into the PCM encapsulated in sphere was investigated experimentally and numerically by Aziz et al. [46] to enhance the thermal performance of the LHTS tank. They compared three PCM encapsulation types including a plain plastic sphere, plastic sphere with copper pins, and copper sphere with copper pins. It was shown that copper sphere with copper pins has the best performance among the three.

The melting process of a PCM inside an externally heated spherical capsule can be categorized into constrained and unconstrained melting. In unconstrained melting, unmelted PCM is not fixed and sinks to the bottom of the sphere. As a result, heat is directly conducted from the shell to the solid PCM. This type of heat transfer is known as close contact melting. Whereas, in constrained melting, unmelted solid bulk is fixed and does not migrate in the capsule. In this case, convection is the main heat transfer mechanism between the unmelted solid PCM and shell. Constrained melting heat transfer of a PCM in a circumferentially finned spherical capsule was studied numerically and experimentally by Fan et al. [47] as shown in Figure 7.7. They investigated the effect of fin height on the melting performance of the LHTS system. The results revealed that the best performance was achieved by annular fin with the highest fin height due to the enhanced heat conduction and inducing local natural convection in the presence of the fin.

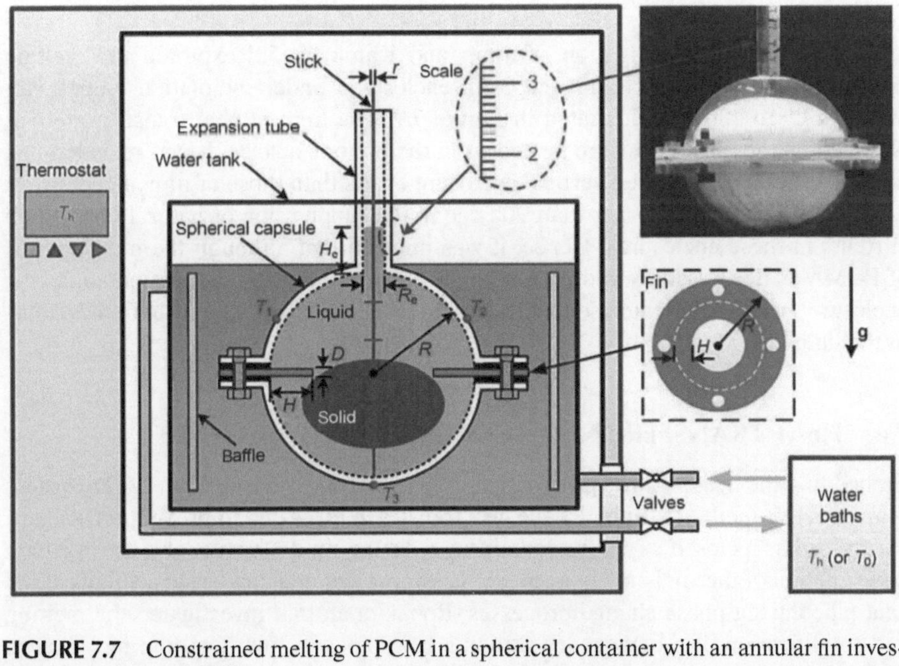

FIGURE 7.7 Constrained melting of PCM in a spherical container with an annular fin investigated by Fan et al. [47].

7.3 EFFECT OF CONTAINER ORIENTATION

Recent studies demonstrated the significant effect of the container orientation on the performance of the system [48]. It has been also proved that the influence of orientation may vary in the presence of the fins [25,31,49,50]. Kalapala and Devanuri [27] experimentally and numerically examined the melting behavior of PCM under various orientations of a cylindrical LHTS unit with annular fins attached to the tube. They evaluated the performance of the system in terms of total melting time, energy/exergy efficiency, energy/exergy stored, and energy/exergetic effectiveness efficiency. It was deduced that vertical configuration performs better in all aspects except energy efficiency. In another study [51] conducted by the same authors, the effect of inclination angle on melting and solidification processes was experimentally investigated. It was found that melting as a convection-dominated heat transfer process was remarkably affected by the inclination angle. However, no meaningful relation between solidification process and orientation was observed since the solidification is a conduction-dominated heat transfer process.

The effect of inclination angle on melting enhancement of lauric acid in finned rectangular enclosures has been studied experimentally by Kamkari and Groulx [13]. Melting process was visualized for the inclination angles of 90° (vertical), 45°, and 0° (horizontal) as shown in Figure 7.8. It was observed that melting rate was augmented by changing the inclination angle from vertical to horizontal for both finned and unfinned enclosures. Melting times of 0-fin horizontal and 3-fin vertical enclosures were reduced by 52% and 37%, respectively, compared to the 0-fin vertical enclosure. This reveals that an unfinned horizontal enclosure performs better than the 3-fin vertical enclosure.

In a numerical investigation, Karami and Kamkari [52] explored the melting behavior of PCM in the 1-fin and 3-fin enclosures under inclination angles varied from 0° to 180° as illustrated in Figure 7.9. The highest heat transfer rate was obtained by the 3-fin enclosure heated from the bottom surface. It was revealed that melting time in an unfinned vertical enclosure is less than those of finned enclosures with inclination angles larger than 90° due to the suppression of natural convection currents in these enclosures. Hence, it was inferred that although the melting time of PCM was decreased by adding fins to the enclosure at each inclination angle, the enclosure should be oriented so that it maximizes the intensity of convection currents in the liquid PCM.

7.4 HEAT TRANSFER ENHANCEMENT USING HEAT PIPES

Embedding the heat pipes as passive, two-phase thermal devices in an LHTS system is an effective method to enhance the heat transfer in PCM and improve its efficiency. Heat pipe has a closed structure containing working fluid which is able to transfer a large amount of thermal energy between the evaporator and condenser sections of the heat pipe during phase change processes. Robak et al. [53] investigated the melting and solidification of n-octadecane utilizing heat pipes or fins. The findings showed that the heat pipe-assisted systems reduce both melting and solidification times by 50% compared to the fin-assisted system. Figure 7.10 presents the Photographs of the

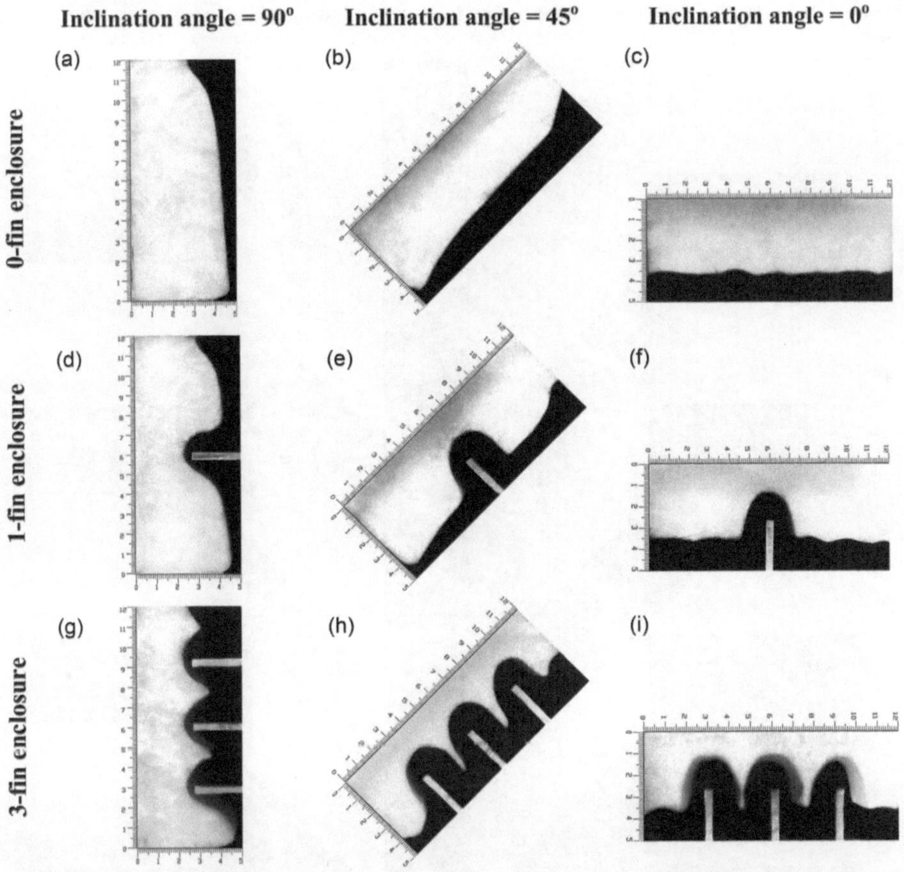

FIGURE 7.8 Comparison of melting photographs in 0-fin, 1-fin, and 3-fin enclosures with different inclination angles at $t = 45$ minutes obtained by Kamkari and Groulx [13].

melting and solidification processes in the benchmark, heat pipe-assisted, and fin-assisted enclosures. The remarkable impact of employing heat pipes on heat transfer enhancement is evident in both melting and solidification processes.

In addition to the decreasing thermal resistance of PCMs, incorporating the heat pipes into PCMs is a beneficial technique to separate the heat sink and source, control, and moderate the temperature in thermal systems [54]. Rashidi et al. [55] reviewed the real potentials of coupling PCM storage modules with the heat pipes as an important thermal system in terms of energy savings and thermal energy management.

Maldonado et al. [20] conducted a systematic review on numerical and experimental research papers which investigated hybrid devices combining LHTS technology and heat pipes. They performed a bibliometric analysis to show how employing of the heat pipes in LHTS has globally grown in popularity during time. Statistical survey showed that there is great attention on the development of PCM integrated heat pipe since 2014 due to the necessity of high-performance refrigeration in small devices like portable computers, and the integration of this hybrid technology with solar energy

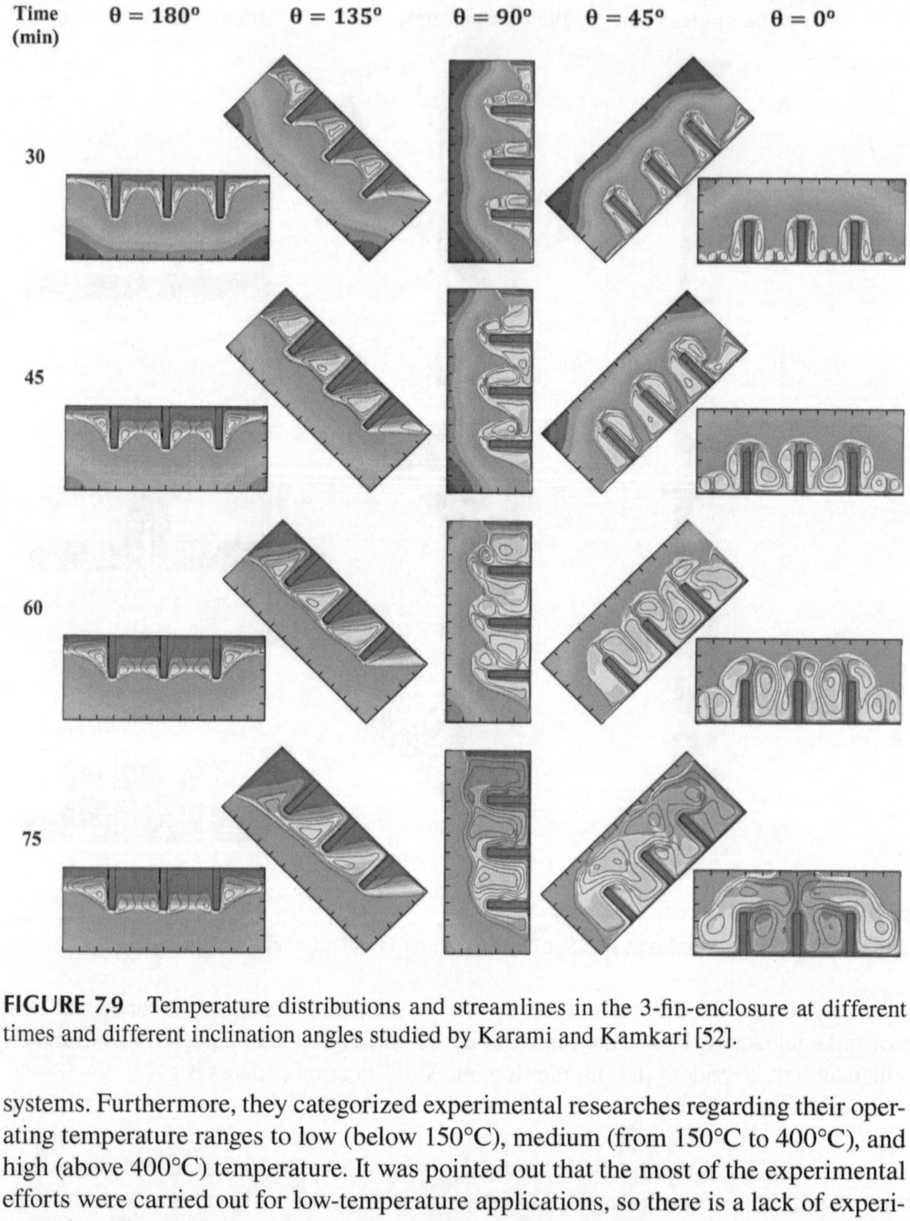

FIGURE 7.9 Temperature distributions and streamlines in the 3-fin-enclosure at different times and different inclination angles studied by Karami and Kamkari [52].

systems. Furthermore, they categorized experimental researches regarding their operating temperature ranges to low (below 150°C), medium (from 150°C to 400°C), and high (above 400°C) temperature. It was pointed out that the most of the experimental efforts were carried out for low-temperature applications, so there is a lack of experimental studies at high temperatures to validate the numerical simulations. Since the biggest potential of this hybrid technology is solar applications, it was suggested to perform more research at a high temperature range (>150°C). The numerical studies can be classified into three main methodologies (Figure 7.11) as follows:

I. Considering the heat pipe as a highly thermal conduction element using three different methods:
   a. Assigning a constant high conductivity to the heat pipe [56,57]

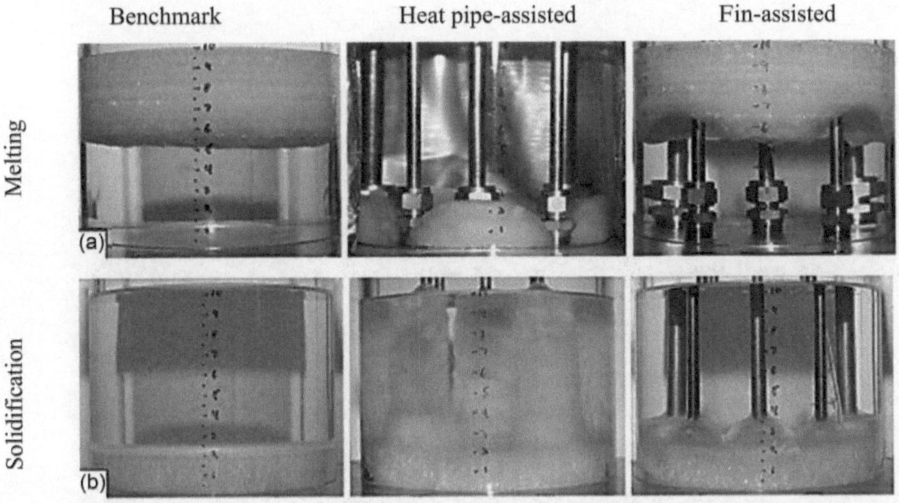

FIGURE 7.10 Photographs of phase change processes of n-octadecane in enclosures with heat pipe, fin and no heat transfer enhancement techniques (benchmark) studied by Robak et al. [53]: (a) melting and (b) solidification.

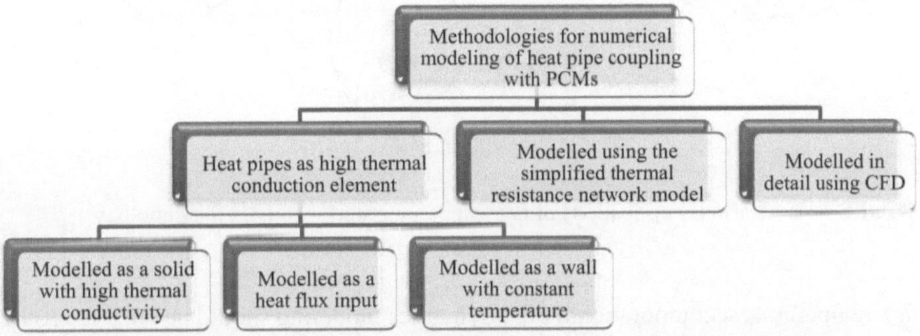

FIGURE 7.11 Methodologies for numerical modeling of heat pipe coupling with PCM.

b. Simulating the heat pipe as a constant heat flux input [58,59]
c. Considering a constant temperature along the heat pipe outer wall [60,61]

II. Modeling based on the simplified thermal resistance network model [62,63],
III. Modeling the heat pipe in detail using computational fluid dynamics and applying the continuity, momentum, and energy equations to the fluid inside the heat pipe [64,65].

They concluded that although modeling the heat pipe as a highly thermal conductive solid (for example, considering heat pipe as a copper pipe with thermal conductivity of 90 times higher) is a good estimation validated with experimental tests at low temperatures in many researches, in the case of designing a new equipment especially at

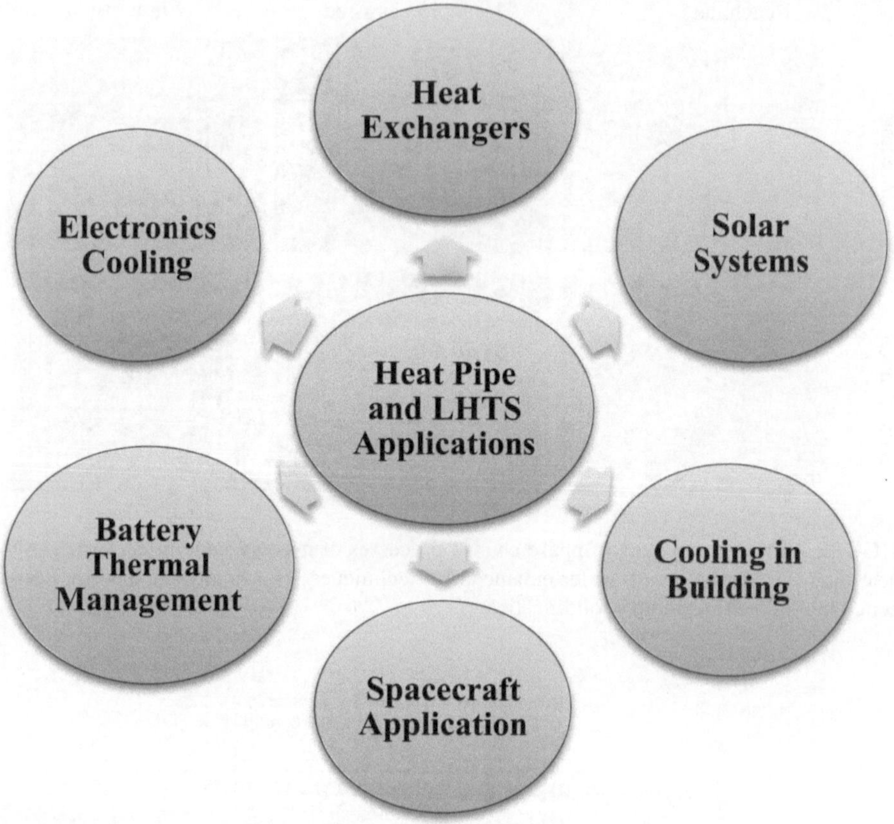

FIGURE 7.12 Various applications of heat pipe integrated with LHTS technology.

high temperatures, computational fluid dynamics modeling should be done to ensure a safer approximation.

Heat pipes integrated in LHTS technology can be used in various applications such as heat exchangers [66,67], solar systems [68,69], cooling in electronics [70,71] and building [72,73], thermal management of electric batteries [74,75], and spacecraft applications [76,77] as shown in Figure 7.12.

The PCM-based system incorporated with heat pipe is a recent innovative cooling method for thermal management applications of Lithium-ion batteries. Zhang et al. [78] designed and developed a battery thermal management system including heat pipe, porous metal foam, PCM, and fan to enhance heat dissipation performance and temperature uniformity in a LiFePO4 battery pack as shown in Figure 7.13. Metal foam was employed to improve the PCM thermal conductivity, store the PCM, and prevent liquid-phase PCM leakage. In case that the PCM was melted completely or the battery temperature was too high, fan started to work to increase the heat transfer into the environment. In addition, aluminum fins were connected to the copper foams to intensify the PCM latent heat release to the environment. The results

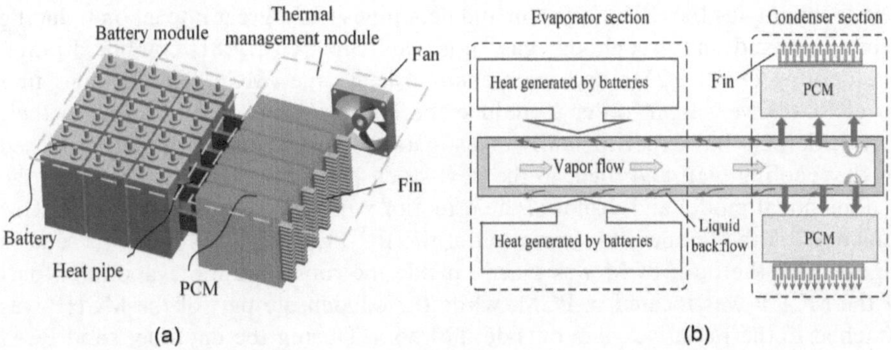

FIGURE 7.13 Schematic of the battery pack and the separation–type battery thermal management design presented by Zhang et al. [78]. (a) Heat pipe-assisted PCM-based thermal management system. (b) Schematic illustrations of heat transfer in battery thermal management system assembly.

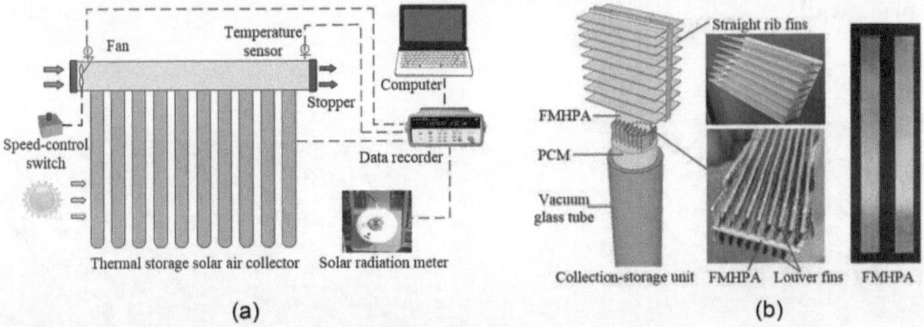

FIGURE 7.14 Thermal storage solar air heater proposed by Wang et al. [79]. (a) Test system of the thermal storage solar air heater. (b) Heat transfer principle of flat micro-heat pipe arrays.

pointed out that the proposed system performs better and reduces the temperature gradient within the battery pack more efficiently than other methods.

Wang et al. [79] carried out an experimental study as represented in Figure 7.14 to analyze the thermal performance of a thermal storage solar air heater (TSSAH) composed of a vacuum glass tube, flat micro-heat pipe arrays (FMHPA) as the core heat transfer element, and paraffin as a thermal storage material. In order to improve the heat transfer process, the aluminum louver and straight rib fins were considered for the FMHPA evaporator and condenser sections, respectively. They drew the conclusion that their proposed TSSAH can be an efficient solar air thermal-collection storage device which is able to collect and release a large amount of thermal energy during thermal discharge for space heating and crop drying applications.

Nowadays, 40% of global energy in developed countries is used in building sector which most of this energy consumption is related to building heating and cooling systems [80]. Hence, design of energy-efficient buildings by developing new technologies like LHTS system and heat pipes has a great impact on reducing energy demand and greenhouse gas emissions. Shen et al. [81] combined radiative cooling (RC), PCMs and microchannel heat pipe (MCHP) techniques in a novel RC-PCM wall in order to reduce the cooling loads in buildings. In their first work [82], they experimentally investigated the performance of their proposed all-day cooling wall and then, in the next study [81] in 2020, they developed the mathematical model and explored the effect of wind speed, emissivity, and PCM thickness on the temperature variation of the RC-PCM wall. As it can be seen in Figure 7.15, the bulk PCM was placed inside the room and the evaporation part of the MCHP was located in PCM, while the condensate part of the MCHP was attached to the radiative plate outside the room. During the daytime, solid PCM stores the thermal energy and gradually melts to reduce the indoor cooling load in comparison with the brick. In contrast, at night, the stored energy in PCM is transferred to the outdoor through the external wall, radiative plate, and gravity-type MCHP. The radiative plate accelerates the condensing process by radiative cooling which is a passive cooling mechanism. The results indicated that implementing the RC-PCM wall decreased the cooling load by 25% compared to the same thickness of brick wall.

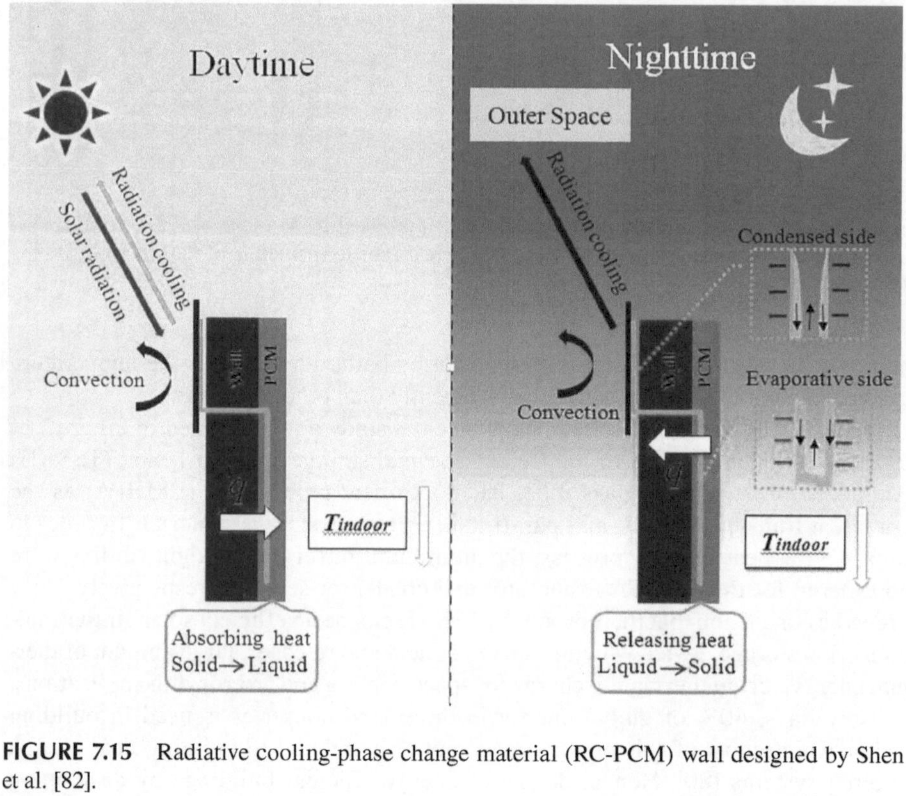

FIGURE 7.15 Radiative cooling-phase change material (RC-PCM) wall designed by Shen et al. [82].

7.5 CONCLUSION

In order to enhance the thermal performance of LHTS systems, augmentation methods were categorized and briefly described in this chapter including employing heat pipes, fins, encapsulated PCM, multiple PCMs, porous materials, nanoparticles, and low-density high thermal conductivity materials as well as the combination of these techniques. Afterward, two of the most effective enhancement methods: fins and heat pipes as thermal conduits between HTF and PCM in LHTS system were reviewed in more detail. Common geometries for LHTS systems are rectangular, cylindrical, and spherical. It was found that both fin configuration (annular, longitudinal, pin, branch) and arrangement have remarkable impact on the melting and solidification processes. Increasing the fin number increases the heat transfer area but reduces the convective heat transfer coefficient due to hindering the convection flows in the container which adversely affects the total heat transfer rate. Therefore, proper fin spacing is important to maximize the beneficial effect of the fins. Eccentricity of the heat transfer tube and orientation of the LHTS container are two other important factors that should be taken into account to further improve thermal performance of the systems. Hence, designers should evaluate all the mentioned influencing factors for the successful design of the LHTS systems. Furthermore, combining LHTS technology and heat pipes were introduced as an effective method to enhance the heat transfer in PCM with various applications in heat exchangers, solar systems, electronics devices and building cooling, thermal management of electric batteries, and spacecraft.

REFERENCES

1. K. Bhagat, S.K. Saha, Numerical analysis of latent heat thermal energy storage using encapsulated phase change material for solar thermal power plant, *Renew. Energy.* 95 (2016) 323–336. doi:10.1016/j.renene.2016.04.018.
2. R. Kumar, P. Verma, An experimental and numerical study on effect of longitudinal finned tube eccentric configuration on melting behaviour of lauric acid in a horizontal tube-in-shell storage unit, *J. Energy Storage.* 30 (2020) 101396. doi:10.1016/j.est.2020.101396.
3. Y.B. Tao, Y. He, A review of phase change material and performance enhancement method for latent heat storage system, *Renew. Sustain. Energy Rev.* 93 (2018) 245–259. doi:10.1016/j.rser.2018.05.028.
4. M. Liu, W. Saman, F. Bruno, Review on storage materials and thermal performance enhancement techniques for high temperature phase change thermal storage systems, *Renew. Sustain. Energy Rev.* 16 (2012) 2118–2132. doi:10.1016/j.rser.2012.01.020.
5. Y. Lin, Y. Jia, G. Alva, G. Fang, Review on thermal conductivity enhancement, thermal properties and applications of phase change materials in thermal energy storage, *Renew. Sustain. Energy Rev.* 82 (2018) 2730–2742. doi:10.1016/j.rser.2017.10.002.
6. S. Jegadheeswaran, S.D. Pohekar, Performance enhancement in latent heat thermal storage system: A review, *Renew. Sustain. Energy Rev.* 13 (2009) 2225–2244. doi:10.1016/j.rser.2009.06.024.
7. S. Jegadheeswaran, S.D. Pohekar, Exergy analysis of particle dispersed latent heat thermal storage system for solar water heaters, *J. Renew. Sustain. Energy.* 2 (2010) 023105. doi:10.1063/1.3427221.
8. W. Wang, L. Wang, Y. He, The energy efficiency ratio of heat storage in one shell-and-one tube phase change thermal energy storage unit, *Appl. Energy.* 138 (2015) 169–182. doi:10.1016/j.apenergy.2014.10.064.

9. P. Bose, V. Arasu, A review on thermal conductivity enhancement of paraf fi nwax as latent heat energy storage material, *Renew. Sustain. Energy Rev.* 65 (2016) 81–100. doi:10.1016/j.rser.2016.06.071.
10. N.I. Ibrahim, F.A. Al-Sulaiman, S. Rahman, B.S. Yilbas, A.Z. Sahin, Heat transfer enhancement of phase change materials for thermal energy storage applications: A critical review, *Renew. Sustain. Energy Rev.* 74 (2017) 26–50. doi:10.1016/j.rser.2017.01.169.
11. V., H. Abolghasemi, L. Darvishvand, B. Kamkari, Thermal performance investigation of concentric and eccentric shell and tube heat exchangers with different fin configurations containing phase change material, *J. Energy Storage.* 37 (2021) 102458. doi:10.1016/j.est.2021.102458.
12. R. Karami, B. Kamkari, Experimental investigation of the effect of perforated fins on thermal performance enhancement of vertical shell and tube latent heat energy storage systems, *Energy Convers. Manag.* 210 (2020) 112679. doi:10.1016/j.enconman.2020.112679.
13. B. Kamkari, D. Groulx, Experimental investigation of melting behaviour of phase change material in finned rectangular enclosures under di ff erent inclination angles, *Exp. Therm. Fluid Sci.* 97 (2018) 94–108. doi:10.1016/j.expthermflusci.2018.04.007.
14. C. Arkar, B. Vidrih, S. Medved, Efficiency of free cooling using latent heat storage integrated into the ventilation system of a low energy building, *Int. J. Refrig.* 30 (2007) 134–143. doi:10.1016/j.ijrefrig.2006.03.009.
15. N. Putra, A. Fahrizal, B. Ariantara, N. Abdullah, T. Meurah, I. Mahlia, Performance of beeswax phase change material (PCM) and heat pipe as passive battery cooling system for electric vehicles, *Case Stud. Therm. Eng.* 21 (2020) 100655.
16. J.C. Kurnia, A.P. Sasmito, S. V. Jangam, A.S. Mujumdar, Improved design for heat transfer performance of a novel phase change material (PCM) thermal energy storage (TES), *Appl. Therm. Eng.* 50 (2013) 896–907. doi:10.1016/j.applthermaleng.2012.08.015.
17. D. Zhou, C.Y. Zhao, Experimental investigations on heat transfer in phase change materials (PCMs) embedded in porous materials, *Appl. Therm. Eng.* 31 (2010) 970–977. doi:10.1016/j.applthermaleng.2010.11.022.
18. F. Frusteri, V. Leonardi, S. Vasta, G. Restuccia, Thermal conductivity measurement of a PCM based storage system containing carbon fibers, *Appl. Therm. Eng.* 25 (2005) 1623–1633. doi:10.1016/j.applthermaleng.2004.10.007.
19. M. Nourani, N. Hamdami, J. Keramat, A. Moheb, Thermal behavior of paraffin-nano-Al_2O_3 stabilized by sodium stearoyl lactylate as a stable phase change material with high thermal conductivity, *Renew. Energy.* 88 (2016) 474–482. doi:10.1016/j.renene.2015.11.043.
20. J.M. Maldonado, A. de Gracia, L.F. Cabeza, Systematic review on the use of heat pipes in latent heat thermal energy storage tanks, *J. Energy Storage.* 32 (2020). doi:10.1016/j.est.2020.101733.
21. T. Xiong, L. Zheng, K.W. Shah, Nano-enhanced phase change materials (NePCMs): A review of numerical simulations, *Appl. Therm. Eng.* 178 (2020) 115492. doi:10.1016/j.applthermaleng.2020.115492.
22. M.E. Nakhchi, M. Hatami, M. Rahmati, A numerical study on the effects of nanoparticles and stair fins on performance improvement of phase change thermal energy storages, *Energy.* 215 (2021) 119112. doi:10.1016/j.energy.2020.119112.
23. M. Arıcı, E. Tütüncü, Ç. Yıldız, D. Li, Enhancement of PCM melting rate via internal fin and nanoparticles, *Int. J. Heat Mass Transf.* 156 (2020). doi:10.1016/j.ijheatmasstransfer.2020.119845.
24. M.A. Hayat, H.M. Ali, M.M. Janjua, W. Pao, C. Li, M. Alizadeh, Phase change material/heat pipe and copper foam-based heat sinks for thermal management of electronic systems, *J. Energy Storage.* 32 (2020) 101971. doi:10.1016/j.est.2020.101971.

25. M.S. Mahdi, A.F. Hasan, H.B. Mahood, A.N. Campbell, A.A. Khadom, A.M.E.A. Karim, A.O. Sharif, Numerical study and experimental validation of the effects of orientation and configuration on melting in a latent heat thermal storage unit, *J. Energy Storage*. 23 (2019) 456–468. doi:10.1016/j.est.2019.04.013.
26. M.K. Rathod, J. Banerjee, Thermal performance enhancement of shell and tube Latent Heat Storage Unit using longitudinal fins, *Appl. Therm. Eng.* 75 (2015) 1084–1092. doi:10.1016/j.applthermaleng.2014.10.074.
27. L. Kalapala, J.K. Devanuri, Effect of orientation on thermal performance of a latent heat storage system equipped with annular fins – An experimental and numerical investigation, *Appl. Therm. Eng.* 183 (2021) 116244. doi:10.1016/j.applthermaleng.2020.116244.
28. X. Yang, Z. Lu, Q. Bai, Q. Zhang, L. Jin, J. Yan, Thermal performance of a shell-and-tube latent heat thermal energy storage unit: Role of annular fins, *Appl. Energy*. 202 (2017) 558–570. doi:10.1016/j.apenergy.2017.05.007.
29. D.S. Mehta, B. Vaghela, M.K. Rathod, J. Banerjee, Thermal performance augmentation in latent heat storage unit using spiral fin: An experimental analysis, *J. Energy Storage*. 31 (2020) 101776. doi:10.1016/j.est.2020.101776.
30. A. Rozenfeld, Y. Kozak, T. Rozenfeld, G. Ziskind, Experimental demonstration, modeling and analysis of a novel latent- heat thermal energy storage unit with a helical fin, *Int. J. Heat Mass Transf.* 110 (2017) 692–709. doi:10.1016/j.ijheatmasstransfer.2017.03.020.
31. S. Zhang, L. Pu, L. Xu, R. Liu, Y. Li, Melting performance analysis of phase change materials in different finned thermal energy storage, *Appl. Therm. Eng.* 176 (2020) 115425. doi:10.1016/j.applthermaleng.2020.115425.
32. S. Paria, S. Baradaran, A. Amiri, A.A.D. Sarhan, S.N. Kazi, Performance evaluation of latent heat energy storage in horizontal shell-and-finned tube for solar application, *J. Therm. Anal. Calorim.* 123 (2016) 1371–1381. doi:10.1007/s10973-015-5006-1.
33. F. Agyenim, P. Eames, M. Smyth, A comparison of heat transfer enhancement in a medium temperature thermal energy storage heat exchanger using fins, *Sol. Energy*. 83 (2009) 1509–1520. doi:10.1016/j.solener.2009.04.007.
34. V. Safari, H. Abolghasemi, B. Kamkari, Experimental and numerical investigations of thermal performance enhancement in a latent heat storage heat exchanger using bifurcated and straight fins, *Renew. Energy*. (2021). doi:10.1016/j.renene.2021.04.076.
35. X. Cao, Y. Yuan, B. Xiang, F. Haghighat, Effect of natural convection on melting performance of eccentric horizontal shell and tube latent heat storage unit, *Sustain. Cities Soc.* 38 (2018) 571–581. doi:10.1016/j.scs.2018.01.025.
36. M.A. Alnakeeb, M.A. Abdel Salam, M.A. Hassab, Eccentricity optimization of an inner flat-tube double-pipe latent-heat thermal energy storage unit, *Case Stud. Therm. Eng.* 25 (2021) 100969. doi:10.1016/j.csite.2021.100969.
37. J.M. Mahdi, S. Lohrasbi, D.D. Ganji, E.C. Nsofor, Accelerated melting of PCM in energy storage systems via novel configuration of fins in the triplex-tube heat exchanger, *Int. J. Heat Mass Transf.* 124 (2018) 663–676. doi:10.1016/j.ijheatmasstransfer.2018.03.095.
38. M. Gharebaghi, I. Sezai, Enhancement of heat transfer in latent heat storage modules with internal fins, *Numer. Heat Transf. Part A Appl.* 53 (2008) 749–765. doi:10.1080/10407780701715786.
39. K.S. Reddy, Thermal modeling of PCM-based solar integrated collector storage water heating system, *J. Sol. Energy Eng.* 129 (2007) 458. doi:10.1115/1.2770753.
40. B. Kamkari, H. Shokouhmand, Experimental investigation of phase change material melting in rectangular enclosures with horizontal partial fins, *Int. J. Heat Mass Transf.* 78 (2014) 839–851. doi:10.1016/j.ijheatmasstransfer.2014.07.056.
41. S. Gharbi, S. Harmand, S. Ben Jabrallah, Experimental comparison between different configurations of PCM based heat sinks for cooling electronic components, *Appl. Therm. Eng.* 87 (2015) 454–462. doi:10.1016/j.applthermaleng.2015.05.024.

42. R. Baby, C. Balaji, Thermal optimization of PCM based pin fin heat sinks: An experimental study, *Appl. Therm. Eng.* 54 (2013) 65–77. doi:10.1016/j.applthermaleng.2012.10.056.
43. A. Arshad, H.M. Ali, M. Ali, S. Manzoor, Thermal performance of phase change material (PCM) based pin-finned heat sinks for electronics devices: Effect of pin thickness and PCM volume fraction, *Appl. Therm. Eng.* 112 (2017) 143–155. doi:10.1016/j.applthermaleng.2016.10.090.
44. H. Muhammad Ali, M. Junaid Ashraf, A. Giovannelli, M. Irfan, T. Bin Irshad, H. Muhammad Hamid, F. Hassan, A. Arshad, Thermal management of electronics: An experimental analysis of triangular, rectangular and circular pin-fin heat sinks for various PCMs, *Int. J. Heat Mass Transf.* 123 (2018) 272–284. doi:10.1016/j.ijheatmasstransfer.2018.02.044.
45. M.M. Kenisarin, K. Mahkamov, S.C. Costa, I. Makhkamova, Melting and solidification of PCMs inside a spherical capsule: A critical review, *J. Energy Storage.* 27 (2020) 101082. doi:10.1016/j.est.2019.101082.
46. S. Aziz, N.A.M. Amin, M.S. Abdul Majid, M. Belusko, F. Bruno, CFD simulation of a TES tank comprising a PCM encapsulated in sphere with heat transfer enhancement, *Appl. Therm. Eng.* (2018). doi:10.1016/j.applthermaleng.2018.08.013.
47. L.W. Fan, Z.Q. Zhu, S.L. Xiao, M.J. Liu, H. Lu, Y. Zeng, Z.T. Yu, K.F. Cen, An experimental and numerical investigation of constrained melting heat transfer of a phase change material in a circumferentially finned spherical capsule for thermal energy storage, *Appl. Therm. Eng.* 100 (2016) 1063–1075. doi:10.1016/j.applthermaleng.2016.02.125.
48. D.S. Mehta, K. Solanki, M.K. Rathod, J. Banerjee, Influence of orientation on thermal performance of shell and tube latent heat storage unit, *Appl. Therm. Eng.* 157 (2019) 113719. doi:10.1016/j.applthermaleng.2019.113719.
49. B. Kamkari, H.J. Amlashi, Numerical simulation and experimental verification of constrained melting of phase change material in inclined rectangular enclosures, *Int. Commun. Heat Mass Transf.* 88 (2017). doi:10.1016/j.icheatmasstransfer.2017.07.023.
50. B. Kamkari, H. Shokouhmand, F. Bruno, Experimental investigation of the effect of inclination angle on convection-driven melting of phase change material in a rectangular enclosure, *J. Heat Mass Transf.* 72 (2014) 186–200. doi:10.1016/j.ijheatmasstransfer.2014.01.014.
51. L. Kalapala, J.K. Devanuri, Energy and exergy analyses of latent heat storage unit positioned at different orientations – An experimental study, *Energy.* 194 (2020) 116924. doi:10.1016/j.energy.2020.116924.
52. R. Karami, B. Kamkari, Investigation of the effect of inclination angle on the melting enhancement of phase change material in finned latent heat thermal storage units, *Appl. Therm. Eng.* 146 (2019) 45–60. doi:10.1016/j.applthermaleng.2018.09.105.
53. C.W. Robak, T.L. Bergman, A. Faghri, Enhancement of latent heat energy storage using embedded heat pipes, *Int. J. Heat Mass Transf.* 54 (2011) 3476–3484. doi:10.1016/j.ijheatmasstransfer.2011.03.038.
54. D.A. Reay, Thermal energy storage: The role of the heat pipe in performance enhancement, *Int. J. Low-Carbon Technol.* 10 (2015) 99–109. doi:10.1093/ijlct/ctv009.
55. S. Rashidi, H. Shamsabadi, J.A. Esfahani, S. Harmand, *A Review on Potentials of Coupling PCM Storage Modules to Heat Pipes and Heat Pumps*, Springer International Publishing, 2019. doi:10.1007/s10973-019-08930-1.
56. A. Faghri, *Heat Pipe Science and Technology*, first edition, Taylor & Francis, Washington, DC, 1995.
57. M. Mahdavi, S. Tiari, V. Pawar, Heat transfer analysis of a low-temperature heat pipe-assisted latent heat thermal energy storage system with nano-enhanced PCM, in: *Proceedings of the ASME's International Mechanical Engineering Congress & Exposition*, 2018. doi:10.1115/IMECE2018-86609.

58. S. Costa, D. Mullen, Solar salt latent heat thermal storage for a small solar organic rankine cycle plant, (2020). doi: 10.1115/1.4044557.
59. X. Gui, W. Qu, B. Lin, X. Yuan, Two-dimensional transient thermal analysis of a phase-change-material canister of a heat-pipe receiver under gravity, *J. Therm. Sci.* 19 (2010) 160–166. doi:10.1007/s11630-010-0160-z.
60. C. Pan, N. Vermaak, C. Romero, S. Neti, S. Hoenig, C. Chen, Efficient optimization of a longitudinal finned heat pipe structure for a latent thermal energy storage system, *Energy Convers. Manag.* 153 (2017) 93–105. doi:10.1016/j.enconman.2017.09.064.
61. C. Pan, N. Vermaak, C. Romero, S. Neti, S. Hoenig, C.H. Chen, R. Bonner, Cost estimation and sensitivity analysis of a latent thermal energy storage system for supplementary cooling of air cooled condensers, *Appl. Energy.* 224 (2018) 52–68. doi:10.1016/j.apenergy.2018.04.080.
62. E.G. Jung, J.H. Boo, Thermal analytical model of latent thermal storage with heat pipe heat exchanger for concentrated solar power, *Sol. Energy.* 102 (2014) 318–332. doi:10.1016/j.solener.2013.11.008.
63. K. Nithyanandam, R. Pitchumani, Computational studies on metal foam and heat pipe enhanced latent thermal energy storage, *J. Heat Transfer.* 136 (2014). doi:10.1115/1.4026040.
64. N. Sharifi, S. Wang, T.L. Bergman, A. Faghri, Heat pipe-assisted melting of a phase change material, *Int. J. Heat Mass Transf.* 55 (2012) 3458–3469. doi:10.1016/j.ijheatmasstransfer.2012.03.023.
65. H. Shabgard, A. Faghri, T.L. Bergman, C.E. Andraka, Numerical simulation of heat pipe-assisted latent heat thermal energy storage unit for Dish-Stirling systems, *J. Sol. Energy Eng.* 136 (2013) 021025. doi:10.1115/1.4025973.
66. Z. Liu, Z. Wang, C. Ma, An experimental study on heat transfer characteristics of heat pipe heat exchanger with latent heat storage. Part I: Charging only and discharging only modes, *Energy Convers. Manag.* 47 (2006) 944–966. doi:10.1016/j.enconman.2005.06.004.
67. Z. Liu, Z. Wang, C. Ma, An experimental study on the heat transfer characteristics of a heat pipe heat exchanger with latent heat storage. Part II : Simultaneous charging/discharging modes, 47 (2006) 967–991. doi:10.1016/j.enconman.2005.06.007.
68. T. Wang, Y. Diao, T. Zhu, Y. Zhao, J. Liu, X. Wei, Thermal performance of solar air collection-storage system with phase change material based on flat micro-heat pipe arrays, *Energy Convers. Manag.* 142 (2017) 230–243. doi:10.1016/j.enconman.2017.03.039.
69. Z. Wang, Y. Diao, Y. Zhao, C. Chen, L. Liang, T. Wang, Thermal performance investigation of an integrated collector–storage solar air heater on the basis of lap joint-type flat micro-heat pipe arrays: Simultaneous charging and discharging mode, *Energy.* 181 (2019) 882–896. doi:10.1016/j.energy.2019.05.197.
70. J. Krishna, P.S. Kishore, A.B. Solomon, Heat pipe with nano enhanced-PCM for electronic cooling application, *Exp. Therm. Fluid Sci.* 81 (2017) 84–92. doi:10.1016/j.expthermflusci.2016.10.014.
71. X.-H. Yang, S.-C. Tan, Z.-Z. He, J. Liu, Finned heat pipe assisted low melting point metal PCM heat sink against extremely high power thermal shock, *Energy Convers. Manag.* 160 (2018) 467–476. doi:10.1016/J.ENCONMAN.2018.01.056.
72. J.R. Turnpenny, D.W. Etheridge, D.A. Reay, Novel ventilation system for reducing air conditioning in buildings. Part II: Testing of prototype, *Appl. Therm. Eng.* 21 (2001) 1203–1217. doi:10.1016/S1359-4311(01)00003-5.
73. D. Etheridge, K. Murphy, D. Reay, A PCM/heat pipe cooling system for reducing air conditioning in buildings: Review of options and report on field tests, *Build. Serv. Eng. Res. Technol.* 27 (2006) 27–39. doi:10.1191/0143624406bt142oa.
74. J. Zhao, Z. Rao, C. Liu, Y. Li, Experiment study of oscillating heat pipe and phase change materials coupled for thermal energy storage and thermal management, *Int. J. Heat Mass Transf.* 99 (2016) 252–260. doi:10.1016/j.ijheatmasstransfer.2016.03.108.

75. J. Zhao, Z. Rao, C. Liu, Y. Li, Experimental investigation on thermal performance of phase change material coupled with closed-loop oscillating heat pipe (PCM/CLOHP) used in thermal management, *Appl. Therm. Eng.* 93 (2016) 90–100. doi:10.1016/j.applthermaleng.2015.09.018.
76. X. Song, Q. Song, X. Gui, S. Liang, D. Tang, Influence of void ratio on phase change in thermal storage canister of heat pipe receiver, *Heat Transf. Eng.* 36 (2015) 1154–1162. doi:10.1080/01457632.2015.987630.
77. X. Gui, X. Song, B. Nie, Thermal analysis of void cavity for heat pipe receiver under microgravity, *J. Therm. Sci.* 26 (2017) 138–143. doi:10.1007/s11630-017-0922-y.
78. W. Zhang, J. Qiu, X. Yin, D. Wang, A novel heat pipe assisted separation type battery thermal management system based on phase change material, *Appl. Therm. Eng.* 165 (2020). doi:10.1016/j.applthermaleng.2019.114571.
79. T. Wang, Y. Zhao, Y. Diao, C. Ma, Y. Zhang, X. Lu, Experimental investigation of a novel thermal storage solar air heater (TSSAH) based on flat micro-heat pipe arrays, *Renew. Energy.* 173 (2021) 639–651. doi:10.1016/j.renene.2021.04.027.
80. A. Kasaeian, L. bahrami, F. Pourfayaz, E. Khodabandeh, W.M. Yan, Experimental studies on the applications of PCMs and nano-PCMs in buildings: A critical review, *Energy Build.* 154 (2017) 96–112. doi:10.1016/j.enbuild.2017.08.037.
81. D. Shen, C. Yu, W. Wang, Investigation on the thermal performance of the novel phase change materials wall with radiative cooling, *Appl. Therm. Eng.* 176 (2020). doi:10.1016/j.applthermaleng.2020.115479.
82. W. He, C. Yu, J. Yang, B. Yu, Z. Hu, D. Shen, X. Liu, M. Qin, H. Chen, Experimental study on the performance of a novel RC-PCM-wall, *Energy Build.* 199 (2019) 297–310. doi:10.1016/j.enbuild.2019.07.001.

8 Fin-Metal Foam Hybrid Structure for Enhancing Solid–Liquid Phase Change

Xiaohu Yang and Ming-Jia Li
Xi'an Jiaotong University

Kamel Hooman
University of Queensland

CONTENTS

8.1 INTRODUCTION

8.1.1 Thermal Energy Storage for Solar Thermal Utilization

The convening of the Copenhagen World Climate Conference prompts countries around the world to actively respond to energy-saving and emission-reduction policies,

DOI: 10.1201/9781003213260-8

with the aim of reducing excessive consumption of fossil fuels and accelerating the utilization of renewable energy [1]. Abundant clean solar energy can be used to meet our ever-increasing demand for energy [2]. The utilization of solar energy is in its developing stage, which promises diversification of energy supply and consumption in the world thereby responding to clean and efficient energy initiatives [3]. The International Energy Agency (IEA) report gives a forecast of energy supply, pointing out that solar energy will become the main source of energy supply, especially in the power industry, by 2050 [4]. Thermal energy storage can be employed to address the randomness and intermittency of solar energy [5–8]. Latent heat thermal energy storage (LHTES) can play a vital role in solar energy utilization because of its reliability, stability, and high efficiency. Hence, a great deal of information on the very topic can be found in the literature [9,10].

Figure 8.1 demonstrates a household solar energy utilization system, a boosted photovoltaic thermal (PVT), with LHTES tank. The solar energy is utilized in two ways: the incoming energy is converted directly to electricity via solar photovoltaic which shows higher efficiency (and longer life) once they are cooled. Hence, a circulating heat transfer fluid (HTF) removes the heat from the photovoltaic (PV) back panels. The HTF is further heated using a booster heater to reach a higher temperature. The heat is then stored using solid–liquid phase change in a thermal energy storage tank. For a typical LHTES system, phase change materials (PCMs) filled in in thermal energy storage (TES) tanks are mainly used to absorb a large amount of heat while melting, so as to achieve the purpose of energy storage. However, the common PCMs have the disadvantage of low thermal conductivity leading to slow energy charging/discharging process [8]. Note that a fast charging (melting) is crucial given

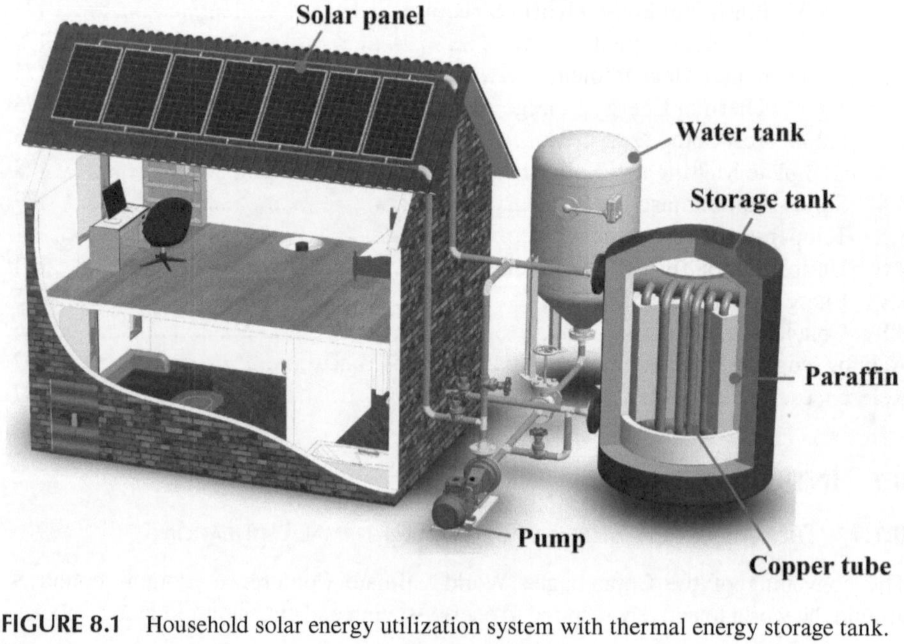

FIGURE 8.1 Household solar energy utilization system with thermal energy storage tank.

the short sun hours during colder months of the year when heat is needed most. Therefore, we aim at enhancing the thermal conductivity to attain an efficient and timely heat storage design as will be shown in this chapter.

8.1.2 Shell-and-Tube Latent Heat Thermal Energy Storage System

Shell-and-tube heat storage configuration in a TES tank is one of the most mature structures with low cost and a simple design to be adopted in the phase change heat storage market (see Figure 8.2). HTF goes through the tube side while the PCM fills the shell side. Typically, a single shell-and-tube unit is analyzed, similar to single-phase heat exchangers (see Figure 8.3a).

A great deal of experimental and numerical studies have been carried out to explore the thermal energy charging and discharging cycles [11], which have improved the efficiency of the shell-and-tube structure. Seddegh et al. [12] numerically studied the heat storage and release rate on the casing-type heat exchanger. By adjusting the temperature and flow rate of the HTF, a series of comparisons among various configurations for the horizontal and vertical TES tubes were made. It was found that while the thermal storage time was greatly affected by the HTF inlet temperature, its mass flow rate only weakly influenced the melting rate. Lorente et al. [13] established an experimental device to study the heat transfer characteristics in a vertical

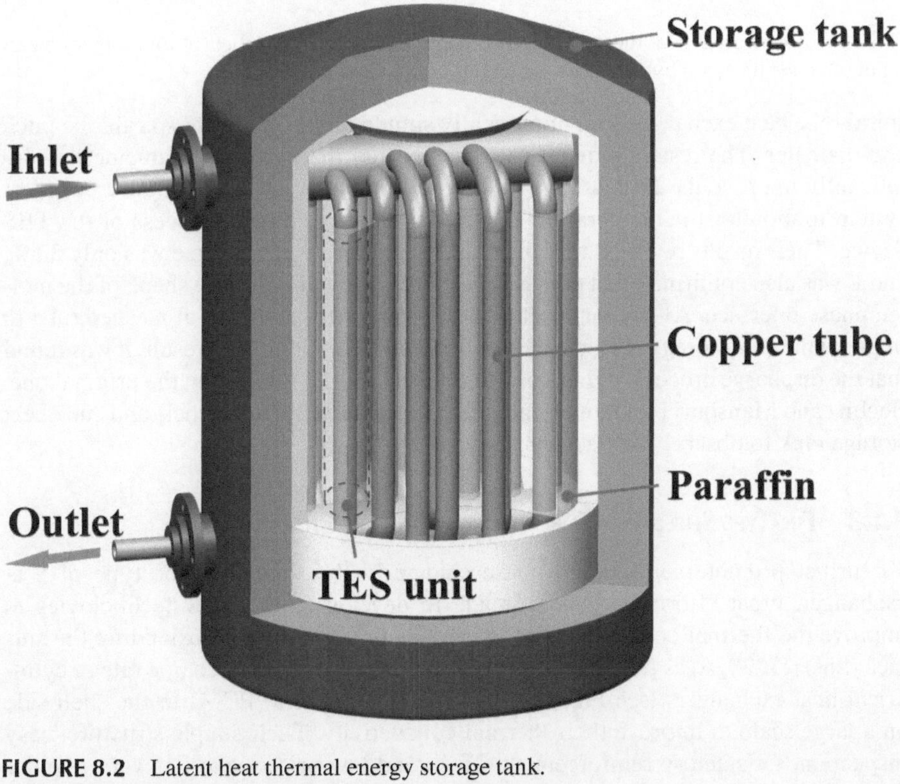

FIGURE 8.2 Latent heat thermal energy storage tank.

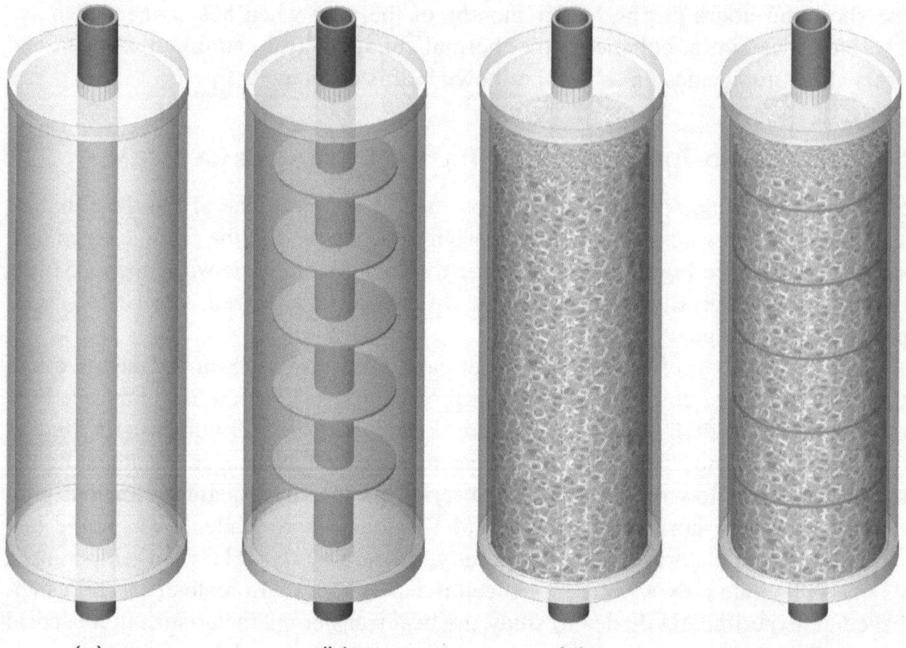

FIGURE 8.3 Latent heat thermal energy storage tubes: (a) bare tube; (b) fin tube; (c) metal foam tube; (d) fin-foam hybrid tube.

spiral tube heat exchanger, and numerically simulated the effects of spiral tube pitch and diameter. The results indicated that increasing the pitch and diameter can significantly improve the heat storage rate. Agyenim et al. [14] used a newly built test system to monitor the temperature change in the phase change process of the TES device. Their results revealed that the radial temperature change rate was only 3.5%, and it was also confirmed that convection significantly affected the shape of the molten phase interface. Al-Mudhafar et al. [15] proposed a new type of meshed tube to improve the heat storage performance of the LHTES system. As a result, it was found that the discharge process of this improved tube was 41% faster than the original one. Bechiri and Mansouri [16] numerically modeled a partially filled shell-and-tube heat storage tank to observe satisfactory results.

8.1.3 Fin-Type Shell-and-Tube Thermal Energy Storage Tube

To further promote solid–liquid phase change in the shell-and-tube type of heat exchanger, great efforts have been made to develop several new technologies to improve the thermal conductivity of PCMs, mainly focusing on extending the surface (fins) [17–20] so as to improve the thermal energy charge/discharge rate of cylindrical heat exchangers (see Figure 8.3b). Fins are added to PCMs in the shell side on a large scale to improve their thermal conductivity. Their simple structure, easy inspection, satisfactory reinforcement effect, and low price make fins very popular

[21]. Fins can be consolidated or disconnected. Righetti et al. [22] tested the use of a wire screen, as interconnected fins, for a storage tank. Opolot et al. [23] developed a numerical model, along with a theoretical one, to evaluate thermal contact resistance for such designs when different techniques are used to attach the wire screen to the outer surface of tubes in a shell-and-tube heat exchanger. Zhao et al. [24] conducted a parametric study to optimize the use of such periodic structures in thermal storage tanks. Favero et al. [25] explored the use of additive manufacturing for thermal storage tanks.

Erek et al. [26] studied the annular finned tube to observe that the Reynolds number (*Re*), the Stefan number (*Ste*), fin dimension, and spacing significantly influence the solidification process of PCM. Yang et al. [27] established a simulation model, verified by their experiments, of shell-and-tube heat storage tubes embedded with annular fins to study the charging characteristics. The optimum value of fin number and thickness were recommended for radial fins. The complete melting time was reduced by 65% for the optimized layout. Zhang and Faghri [28] studied the influence of the number, thickness, and height of the fins on the heat transfer performance of the shell-and-tube heat exchanger to report that these three can significantly accelerate the heat transfer process if designed properly. Tiari et al. [29] proved that natural convection played a key role in the melting process of PCMs. The longer the fin, the more uniform the temperature distribution, according to those authors. Relevant studies had shown that longitudinal fins [30], pinned fins [31], and triplex fins [32] can also improve phase change energy storage. Tay et al. [33] analyzed and compared the effects of different structures on thermal performance by numerical simulation of tubes with pins or fins. They concluded that the effect of pin fin on the tube was much better than that of pin on the tube. Mat et al. [34] numerically simulated the melting process of PCMs in a triplex tube as well as a smooth tube subject to three different boundary conditions. It was found that adding fins in the tube can effectively shorten the melting time, but both the volume of the equipment and the cost increased, compared with the bare pipe. The fin length cannot be unboundedly increased knowing that the fin effective length can be reached. Moreover, the fin spacing is a key parameter affecting the interaction between convection and conduction. Hence, while studies on fin optimization are still needed, other avenues should be investigated in parallel.

8.1.4 Metal Foam Type Shell-and-Tube Thermal Energy Storage Tube

Metal foam has an open porous structure with a series of advantages such as high thermal conductivity, high porosity, and large specific surface area (see Figure 8.4a). It has a periodic structure similar to that of a tetrakaidecahedron. PCMs fill the pores made by the interconnected ligaments as depicted in Figure 8.4b. The high thermal conductivity skeleton can efficiently transfer heat into the PCM and evenly spread it in a larger volume. While accelerating the heat transfer process, thanks to the high porosity (>90%) of metal foam which can be achieved without compromising the structural integrity the TES capacity will not be significantly reduced [35,36].

The use of metal foams in TES tubes has received enthusiastic attention from the academic community [37–40] similar to graphite foams for high-temperature

(a) (b)

FIGURE 8.4 Metal foam structures: (a) partly enlarged by scanning electron microscope (SEM); (b) topological reconstruction.

applications [41] (Figure 8.3c). Yang et al. [42] visually inspected the metal foam heat storage process to find that compared with pure PCM, the melting time was shortened by one-third. Chen et al. [43] tested a SEBS/paraffin/HDPE composite with paraffin wax as PCM and copper foam as supporting structure which resulted in a 7.9-fold higher conductivity compared with the PCM. Siahpush et al. [44] developed a composite PCM that was composed of 99% pure eicosane and copper foam. The phase change processes of two different PCMs under the same working conditions were compared. Results demonstrated that the phase change time of composite PCM was reduced by half. Wang et al. [45] conducted research on adding copper foam to PCM and suggested that adding copper foam with large pores can complete the phase change process faster, saving nearly 40% of time with a more uniform heat transfer in spatial. Li et al. [46] proposed that the propagation of interface was caused mainly by the natural convection under melting. Yang et al. [47] experimentally demonstrated that a cascaded gradient design can significantly accelerate the phase transition interface during solidification.

8.1.5 Fin-Metal Foam Hybrid Structure

Among the above measures to enhance heat transfer of PCM, adding disconnected fin is a traditional way to enhance heat transfer. The main principle is to increase the heat transfer area, where heat is transferred from the heating wall to the inside of the PCM through the fins. At the initial stage of phase change, only a small part of PCM near the fins can absorb the heat transferred from the fins, hence a slow melting rate. Only by increasing the fin number density the PCM melting rate can be accelerated, but there is the disadvantage of reducing the TES capacity. Instead, open-cell metal foam and composite designs can address this issue. Although the thin tortuous metallic ligaments are capable of spreading thermal energy uniformly, the heat conduction capacity is obviously insufficient for a metal foam compared with fins.

If the above two measures are combined, forming the fin-metal foam hybrid structure, the synergies in thermal conducting and heat spreading will be maximized (Figure 8.3d). Thermal energy is first transported to PCM around fins and most of thermal energy will be further distributed deeply into PCMs through the foam ligaments filling the tank volume. It is expected that the fin-metal foam hybrid structure will outperform the single component and the other competing structures for improving solid–liquid phase change.

8.1.6 Chapter Content

This chapter will introduce and test a new heat transfer enhancement method of fin-metal foam hybrid structure to pursue better heat transfer. For the sake of comparison, fin-foam hybrid tubes, bare tubes, finned tubes, and foam-covered tubes are experimentally investigated. The melting and solidification time, melting and solidification front propagation, temperature change at selected points, temperature field uniformity, and temperature response rate are compared for the aforementioned four designs. Finally, we systematically evaluate the thermal performance of these different designs.

8.2 EXPERIMENTAL MEASUREMENT

8.2.1 Thermal Energy Storage Tubes

Four different test specimens are designed and fabricated (see Figure 8.5). A 300-mm-concentric cylindrical TES tube is constructed and paraffin wax (see Table 8.1 for thermophysical properties) is filled in the concentric cylinder up to a height of 270 mm; a 30-mm-gap is left at the top to account for the PCM volume expansion. A total of 17 T-type thermocouples are arranged to allow for monitoring the temporal and spatial variation of the temperature during the experiment. Two of them are monitoring the HTF inlet and outlet temperatures; the rest are located equally spaced over five layers from bottom to top (see Figure 8.5). Copper foam with a porosity of 0.97 and a pore density of 10 pore per inch (PPI) is packed in the foamed test specimen. Five holes are drilled to hold the wooden stick on which three thermocouples are attached (see Figure 8.6). The annular copper foams are devised and filled with paraffin using the vacuum wax injection method. Thermally conductive glue with a high conductivity of 25 W/(m·K) is employed on the interface between copper foam and the tube outer surface.

8.2.2 Test Setup

An experimental test system is designed and built for studying the solid–liquid phase change process in shell-and-tube TES tubes (see Figure 8.7). Three main components of HTF circulation, phase change tests, and data collection were devised. Thermostatic bath (70°C) heats the HTF that flowed through the pipes. The micropump inside

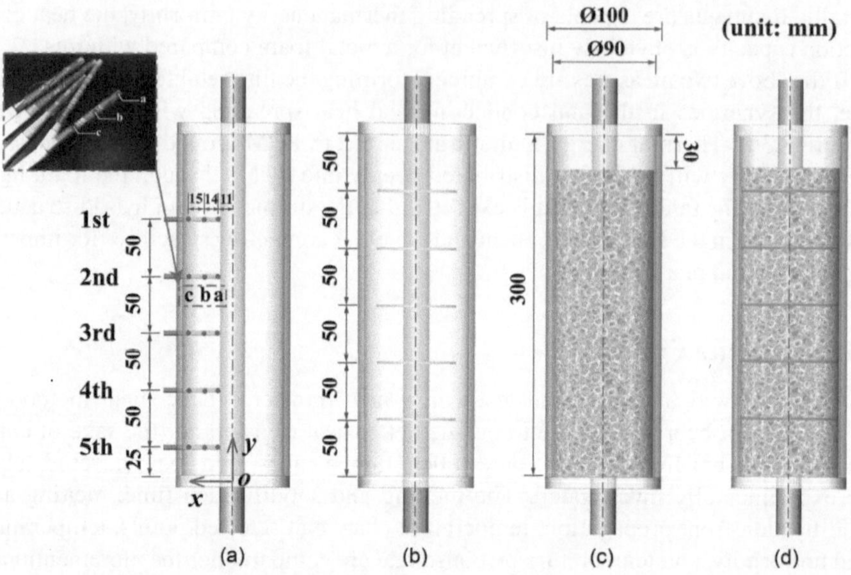

FIGURE 8.5 Four designed TES tubes filled with (a) pure PCM, (b) fins, (c) metal foam, and (d) fin-metal foam hybrid structure.

TABLE 8.1
Thermophysical Properties of PCM (Paraffin 52–56), Copper and Water

Material	Parameter	Value
Paraffin	Melting temperature (°C)	52–56
	Latent heat (kJ/kg)	180.5
	Density (kg/m^3) (solid–liquid)	785/707
	Special heat capacity (J/(kg·K))	3250
	Thermal conductivity (W/(m·K)) (solid–liquid)	0.2/0.1
	Kinematic viscosity (m^2/s)	3.65×10^{-3}
	Thermal expansion coefficient	0.001
Copper	Density (kg/m^3)	8920
	Special heat capacity (kJ/(kg·K))	380
	Thermal conductivity (W/(m·K))	400
Water	Density (kg/m^3)	998.2
	Special heat capacity (J/(kg·K))	4182
	Thermal conductivity (W/(m·K))	0.6
	Kinematic viscosity (m^2/s)	1×10^{-3}

the thermostatic bath and valves controlled the HTF flow with a constant velocity of 0.15 m/s. Data acquisition system including flow meter and temperature collector (Agilent 34970A) measured and collected data during the tests. The test section mainly consists of an annular tube composed of a transparent Perspex outer tube and

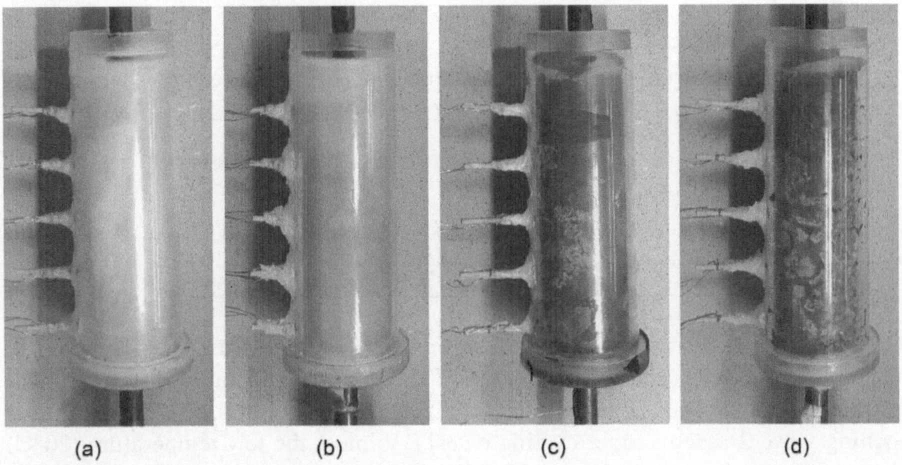

FIGURE 8.6 Preparation and assembly with thermocouples for four designed TES tubes: (a) pure PCM tube, (b) fin tube, (c) metal foam tube, and (d) fin-metal foam hybrid tube.

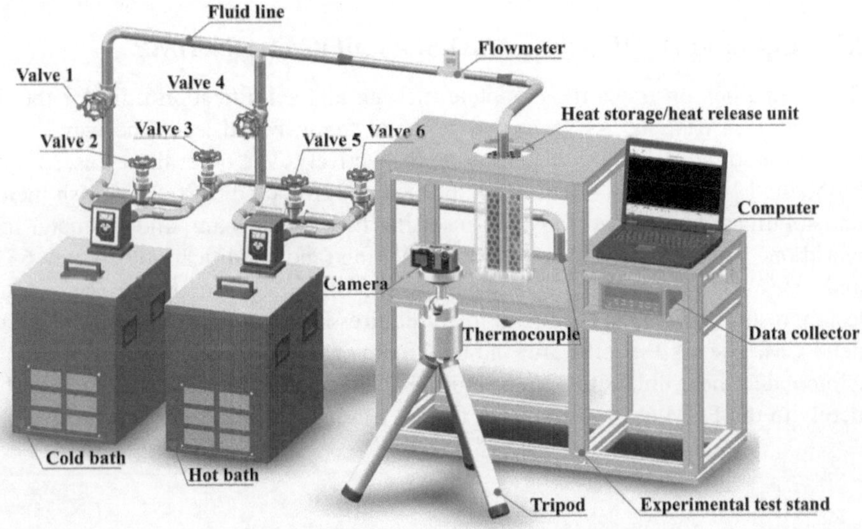

FIGURE 8.7 Experimental system for solid–liquid phase change tests.

a retractable inner copper tube allowing for testing of four different specimens. The outer wall of the TES tube (test section) and all pipes are covered by thermal insulation cottons which would effectively reduce heat dissipation from the outer wall and prevent temperature disturbance from the ambient environment. The critical radius of the insulation layer was tested to be 50 mm.

The room temperature of 20°C is maintained before experiments. Opening valve 5 but closing valve 4 and 6 provides an internal circulation for fluid, toward making

TABLE 8.2
Measurement Uncertainty for the Thermal Energy Storage Test

Parameters	Uncertainty
Measuring temperature (°C)	±0.3°C
Temperature of HTF by water bath (°C)	±0.1°C
Flow rate (L/min)	±1%

the fluid temperature reach the desired constant temperature. Then opening valves 4, 6 but closing valves 1 and 5, and pumping HTF into the TES tube, allows for heating the test section wall. Similarly for solidification tests, closing valves 1 and 3 but opening valve 2 assures inner cycling of HTF to meet the low temperature (20°C) requirement for discharge of energy. The high-definition camera captures images at 5 min intervals till the process is completed. Table 8.2 summarizes the measurement uncertainties of the components for the present experiments. Tests were repeated several times to ensure repeatability of the experiments.

8.3 COMPLETE MELTING AND SOLIDIFICATION TIME

Figure 8.8a demonstrates the complete melting and solidification time for the four different TES designs. As seen, the fin-metal foam hybrid leads the way in 6970 and 4630 s for charging and discharging, respectively. The other three cases require 47,450 (and 19,800 s), 16,490 (and 14,260 s), and 8210 s (and 6700 s) to finish melting (and solidification), respectively. By inserting fins, metal foam, and fin-metal foam hybrid one significantly reduces the total melting (and solidification) time by 65.2% (and 27.9%), 82.7% (and 66.2%), and 85.4% (and 76.7%), compared with the bare tube design, respectively. Besides, metal foam requires less than half of the finned tube to melt PCM, The high specific area of metal foam and high conducting metallic matrix help conduct heat uniformly deep into the PCM. Comparatively, fins conduct heat merely to the PCM at the vicinity of the fin leaving a stratified region farther away

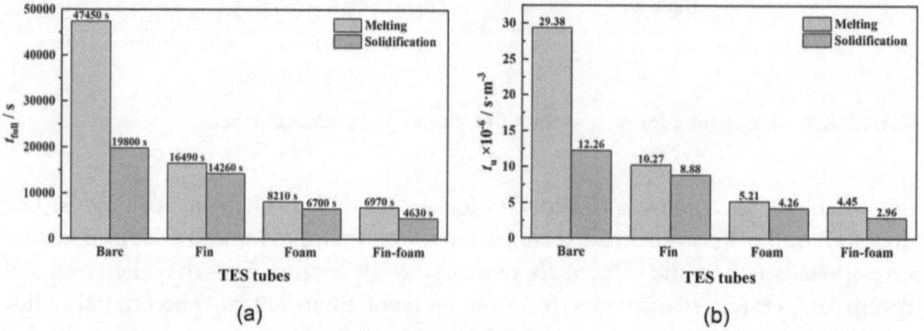

FIGURE 8.8 (a) Complete melting time and (b) melting time per unit mass of PCM in four TES tubes during heat storage and release.

from the fins, hence a nonuniform distribution of heat leading to overheating at the fin and a lag in transporting heat deep into the tank. Nonetheless, a fin-metal foam hybrid structure, leads to 15.1% and 30.9% reduction in melting and solidification time can further be achieved, compared to the fin case. Note that solidification time is shorter than that of melting regardless of the TES design. This can be attributed to a higher wall-solidification temperature difference (34°C) compared with that of wall melting (24°C).

Note in essence that four different designs are tested here associated with different PCM mass for each design. Setting the bare tube as the benchmark, the biggest deviation (7.1%) from the PCM mass occurs with the fin-metal foam hybrid tube with finned and foam mass differences being less than 3.8%. To account for the influence of PCM mass on the melting and solidification processes, the melting and solidification time per unit mass of PCM is depicted in Figure 8.8b. As seen, the fin-foam hybrid structure still demonstrates superior performance compared with the other three designs.

8.4 SOLID–LIQUID PHASE INTERFACE

The transient propagation of solid–liquid phase interface, melt or solidification front, provides an insight into the phase change processes encountered here. Figure 8.9 demonstrates the instantaneous locations for solid–liquid interface for four different TES designs. Note that the HTF flows downwards. A nearly horizontal level (marked in red dash line) is moving from top to bottom regardless of heat transfer enhancement methods for melting process.

There is a similar trend for interface movement for the bare tube and the finned one. The phase interface drops gradually with time. By comparison, the cases with filled metal foam or fin-metal foam hybrid demonstrate a period of color deepening process for PCM. For instance, before $t = 45$ min (30 min for fin-foam hybrid tube), it seems that no PCM is melted. However, this period is of great significance for melting. A large amount of thermal energy is transported from heat source deeply to PCM through the interconnected extended surface area of the ligaments constituting the foam. The temperature of solid PCM in the lower part of the TES tube raises faster to trigger melting. After that period, a fast melting speed is maintained for the two designs (Figure 8.9).

As for solidification, there is clear solidification front evolution at the initial stage but the interface vanishes with time. For bare and finned tubes, the inner copper tube and annular fins are fully covered by a solidified layer (in white). Bright white circles can clearly be seen for the finned case. On the contrary, no distinctive pigmentation can be noted with foams which can be attributed to the conduction-dominated phase change process leading to a more uniform solidification.

Furthermore, a nonlinear movement of the interface is noticed. At the initial and middle stages of melting, the large temperature difference between HTF (70°C) and PCM (20°C) serves as the driving force for melting PCM. Heat conduction and natural convection in the melted PCM both work well for transporting thermal energy to the solid–liquid phase front. With more and more PCMs melted, natural convection dominates the overall heat transfer process, leading to the imbalance of solid–liquid phase change in space. That is, the upper PCM melts much faster than the lower one.

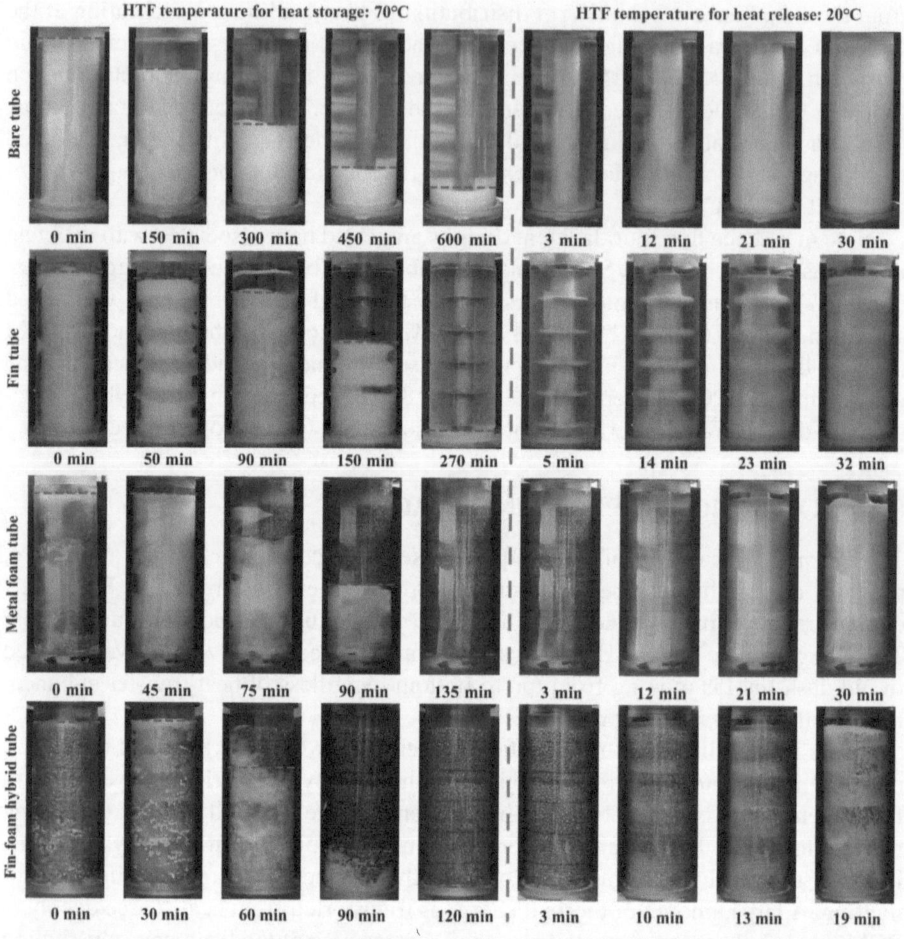

FIGURE 8.9 The solid–liquid interface of paraffin at different times for four TES tubes.

What's worse, in the late heat storage period, the bottom paraffin is quite difficult to melt, shrining stage [48], and consequently there is still a small amount of paraffin remaining solid until the end of the process as depicted in Figure 8.10. Although it separately takes 180 and 45 min to melt the bottom PCMs for bare and finned tube designs, there is little change in solid–liquid phase interface locations. The low thermal conductivity of the PCM and the imbalance of heat exchange due to the natural convection work together to slow down the melting during shrinkage process. The way around it could be to aim at a partial melting for PCM, say till 95%, rather than fully melting the PCM for a fin-type TES tube. Alternatively, the insertion of metal foam can address the issue. Although solidification front also evolves in a nonlinear way, i.e. the solidification front moves at a faster speed at the initial stages, no such kind localized (island) behavior is observed for solidification process as heat is mainly conducted by the solidified PCM and the fins.

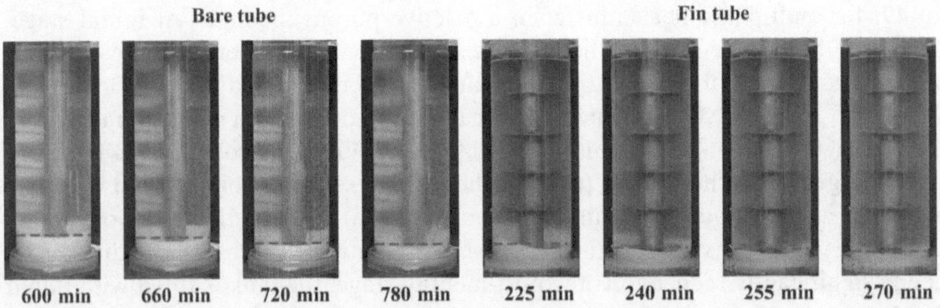

FIGURE 8.10 Solid–liquid interface diagram of bare tube and fin tube phase change materials in the late heat storage period.

8.5 TEMPERATURE RESPONSE

Figure 8.11 illustrates the temperature histories at the middle points of the TES tubes. As observed, a nearly similar temperature developing trend is noticed for the four types of TES tubes. A first rapid and then moderate increasing temperature is observed. The points on the third level, are employed for illustration; as referred to Figure 8.1. Since point 3a is located in the vicinity of the copper tube, its temperature and response rate are much higher than the others (3b and 3c). The temperature history for melting period can be divided into three stages: sensible heat (pre-melting), latent heat (melting), and sensible heat (post-melting); similar

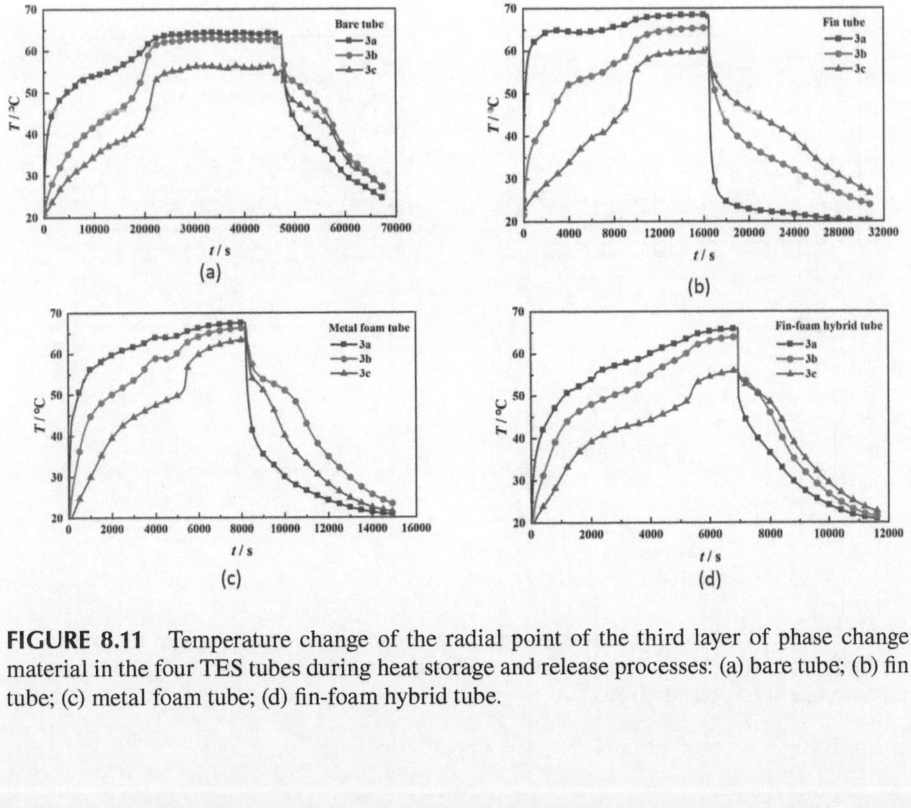

FIGURE 8.11 Temperature change of the radial point of the third layer of phase change material in the four TES tubes during heat storage and release processes: (a) bare tube; (b) fin tube; (c) metal foam tube; (d) fin-foam hybrid tube.

to [49] for multiphase heat transfer in a reactive porous medium. At initial stage, the closer the PCM gets to the heat source, the faster the temperature rises. A temperature gradient in the radial direction can be attributed to radial heat conduction. With time, more PCMs get melted and local natural convection can kick in to carry the heat away to the solid–liquid phase interface. When the solid–liquid interface is submerged in the horizontal plane of these three selected points, there will be a mass exchange of liquid paraffin (natural convection) in the radial direction, which also brings heat transfer. At this stage, the temperature difference in radial direction is significantly reduced. At the post-melting stage, the tank is filled with liquid PCM forming a number of natural convection cells leading to higher heat transfer rate and a more uniform PCM temperature. The time for temperature to plateau differs for each of the four cases; much longer with bare and finned tube compared with the foam and hybrid design. As for solidification, the temperature drops more rapidly. The temperature difference between the three points is less than that during the melting process.

Figure 8.12a and b compare the temperature history at three selected points (3b and 3c) among the four TES designs. Temperature rises (drops) the fastest for the case of fin-metal foam hybrid structure, followed by the metal foam tube, finned tube, and the bare tube in view of the whole melting (and solidification) process. As for the melting (solidification) stage, the temperature in the fin-metal foam hybrid tube is slightly higher (lower) than that in the metal foam tube. This indicates that the main contribution to heat transfer enhancement lies in the extending ligaments of the foam. The role of fins is to help transport thermal energy first to porous metals,

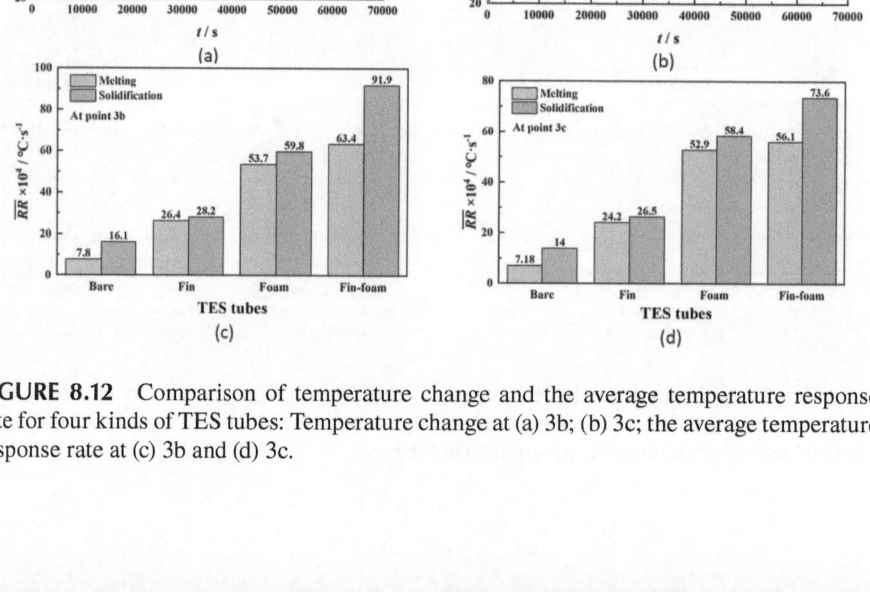

FIGURE 8.12 Comparison of temperature change and the average temperature response rate for four kinds of TES tubes: Temperature change at (a) 3b; (b) 3c; the average temperature response rate at (c) 3b and (d) 3c.

and the energy is further spread out by the ligaments which markedly reduces the thermal resistance of PCMs.

An integral mean index is introduced here, to quantitatively account for the temperature change, expressed as:

$$\overline{RR} = \int_0^{t_{\text{full}}} \frac{1}{t_{\text{full}}} \frac{T(t_i) - T(t_{i-1})}{(t_i - t_{i-1})} dt \tag{8.1}$$

where $\overline{RR}$ stands for the temperature response rate, $T(t_i)$ and $T(t_{i-1})$ denote the temperatures at time t_i and t_{i-1}, and t_{full} represents the complete melting time for each case. As can be seen in Figure 8.12c and d, fin-foam hybrid tube has the highest temperature response rate. Setting the bare tube as the benchmark, the maximal 8.13 and 5.71 times higher response rates for temperature at point 3b are found by the combination of metal foam and fin for melting and solidification, respectively. The segment achieves 6.88 and 3.71 times faster temperature response rate for metal foam tube; 3.38 and 1.75 times faster temperature response rate for finned tube. Similar phenomenon can be found at point 3c, and the temperature response rate at 3c is lower than the one at 3b.

Figure 8.13 presents the temperature development along the axial direction for the four TES designs. As seen, during the pre-melting stage, the temperature curves are highly converged for the four TES tubes. This is the transient response of the designs manifested in their overall heat capacity. Different heat transfer paths through fins, metal foam, and fin-metal foam hybrids lead to different onset and propagation of

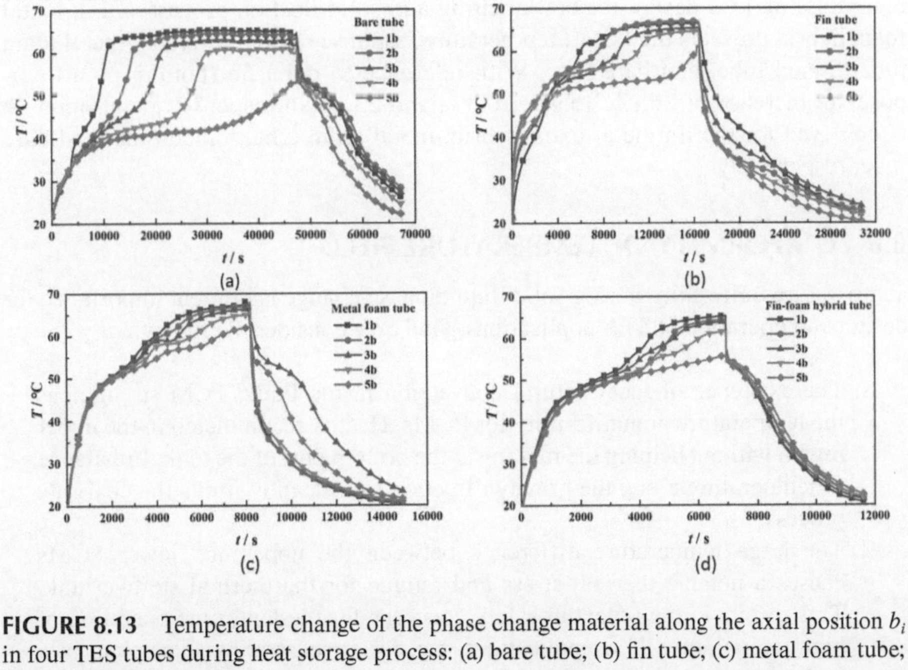

FIGURE 8.13 Temperature change of the phase change material along the axial position b_i in four TES tubes during heat storage process: (a) bare tube; (b) fin tube; (c) metal foam tube; (d) fin-foam hybrid tube.

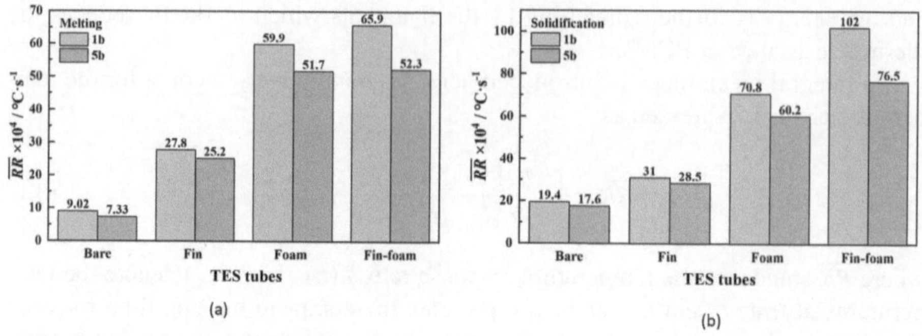

FIGURE 8.14 Comparison of temperature response of selected axial points in four TES tubes during (a) heat storage process and (b) heat release process.

melting. Depending on available flow area and local temperature difference within the liquid PCM, a unique natural convection pattern is anticipated for each design. With time convection gets stronger and then with loss in driving force, it loses its strength. With strong convective flow, the temperature gradient of the PCM along the axial direction progressively expands for all designs considered here. As for the solidification process, compared with melting, the temperate curves are more bundled in the absence of strong convective flows leading to low temperature difference among the five points.

Figure 8.14 compares the temperature response rate at selected points 1b and 5b for four TES designs. The response rate at 1b is much faster than that at 5b regardless of TES design for both melting and solidification processes. Fin-metal foam hybrid presents the fastest temperature response rate, followed by metal foam tube, finned tube, and bare tube. With reference to point 5b (bottom point) purpose, an increase of 26.1%, 15.9%, 10.3%, and 23.1% in response rate at point 1b is observed for the fin-metal foam hybrid, metal foam tube, finned tube, and bare tube, respectively.

8.6 UNIFORMITY OF TEMPERATURE FIELD

Temperature uniformity during solid–liquid phase change is of great importance for design and operation of TES applications. One may consider the following:

1. The existence of local natural convection in the liquid PCM strengthens the temperature nonuniformity for PCMs. Heat is accumulated in the upper region without helping the melting at the bottom part of the tank. Imbalance of temperature along the gravity direction significantly limits the charging process.
2. The large temperature difference between the upper and lower PCMs causes a notable thermal stress and fatigue for the thermal storage tank. This results in the potential safety concerns for the long-time operation of high-temperature TES applications, e.g. for concentrating solar plant in a high-temperature range of 600°C–800°C.

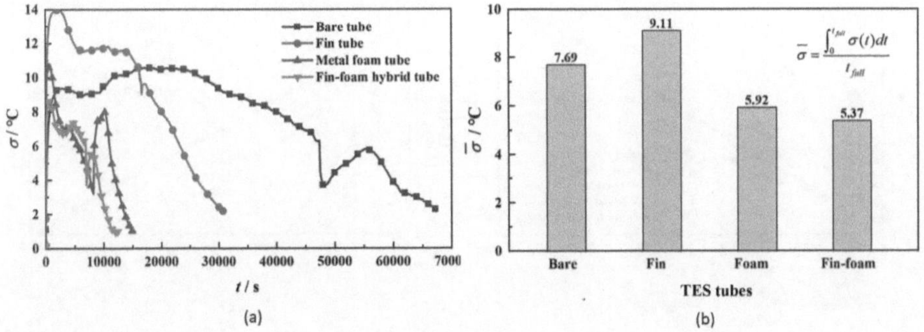

FIGURE 8.15 Uniformity of the temperature field in the four TES tubes during the heat storage process (a) radial and (b) axial temperature field uniformity.

3. The existing large temperature difference in a TES tank is not conducive to stable release of heat. Particularly for district heating, a stable temperature for the heat source plays a key role. A more uniform temperature inside a TES tank helps better operation of the equipment.

To assess temperature uniformity, an index is proposed as:

$$\sigma = \sqrt{\frac{1}{n}\sum_{i=1}^{n}\left[T(p_i,t) - T_{\text{ave}}(t)\right]^2} \tag{8.2}$$

where $T(p_i,t)$ denotes the PCM temperature at each point, n is the total number of measurement points during experiments, $T_{\text{ave}}(t)$ represents the arithmetic mean value for temperature at each time. Figure 8.15 illustrates the temperature uniformity indexes and the integral mean value during the entire melting and solidification process, respectively; the lower the index, the better the temperature uniformity. Overall, the fin-foam hybrid case has the lowest value for the transient temperature uniformity index while the difference between the foam tube and the hybrid one is negligible. Fin-foam hybrid structure has the integral mean uniformity index of 5.37°C while that of the foam design is 5.92°C. As expected, the uniformity for finned tubes and the bare ones is poor. However, while adding fins significantly improves the phase change rate, the temperature uniformity of PCM with finned tubes is the worst. The uniformity indexes for the finned and bare tube are 9.11°C and 7.69°C, respectively. Fin-foam hybrid structure achieves the best temperature uniformity for PCMs among the four structures though the fins, when not combined with foam, resulting in the lowest PCM temperature uniformity.

8.7 ENERGY STORAGE DENSITY

Energy balance analysis is conducted to estimate the thermal energy stored in the tank. To reveal the transient feature for energy storage, an index accounting for the stored thermal energy at a given time t_i to the one at time t_{full} is proposed:

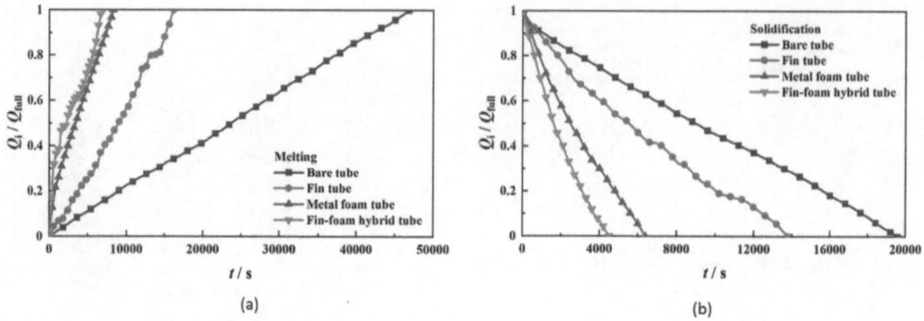

FIGURE 8.16 Thermal energy storage coefficient of four TES tubes during (a) melting process and (b) solidification process.

$$\frac{Q_i}{Q_{\text{full}}} = \frac{\int_0^{t_i} c_p V_{\text{HTF}} A_{\text{tube}} \rho_{\text{HTF}} (T_{\text{in}} - T_{\text{out}}) dt}{\int_0^{t_{\text{full}}} c_p V_{\text{HTF}} A_{\text{tube}} \rho_{\text{HTF}} (T_{\text{in}} - T_{\text{out}}) dt} \tag{8.3}$$

where Q_i and Q_{full} denote the stored energy at two selected times of t_i and t_{full}, t_{full} represents the complete melting time, c_p and ρ_{HTF} are the specific heat capacity and density of HTF (water in the present study), V_{HTF} represents the flow rate of HTF, A_{tube} is the surface area of inner copper tube, T_{in} and T_{out} indicate the HTF temperatures at inlet and outlet, respectively.

Figure 8.16 compares the energy storage index for the four TES tubes. The energy storage index is calculated according to the operation condition in the experiment: a high temperature of 70°C and a velocity of 0.15 m/s are selected for charging; a low temperature of 20°C and a velocity of 0.15 m/s are selected for discharging the energy. The thermophysical properties for the HTF and PCM are listed in Table 8.1. The energy storage index takes a value from 0 to 1 and presents a monotonic increasing trend regardless of TES design. It demonstrates a similar variation trend to the melting characteristics of PCM with special heat transfer enhancement. As expected, the energy storage rate for the fin-foam hybrid tube is the highest among the four types, followed by the metal foam tube, finned tube, and the original bare tube. As for solidification, a similar trend for energy storage index is observed. The fin-foam hybrid structure still leads the way of energy release. This further confirms the approach of fin-metal foam hybrid as heat transfer enhancement for TES applications.

8.8 CONCLUDING REMARKS

This chapter introduces a fin-metal foam hybrid structure as a potentially competitive heat transfer approach for improving solid–liquid phase change. To compare and highlight the fin-metal foam hybrid structure, other three TES tubes filled by pure phase change material (namely bare tube), fins, and metal foam are investigated with respect to full melting and solidification time, phase interface evolution, and temperature response.

It can be concluded as follows:

1. Fin-metal foam hybrid structure is capable of accelerating phase change process. An up to 85.4% (and 76.7%) reduction in full melting (and solidification) time is achieved by the hybrid structure, compared with the bare tube. The fin tube and metal foam tube can also reduce the complete melting (and solidification) time 65.2% (and 27.9%) and 82.7% (and 66.2%), respectively.
2. Fin-metal foam tube leads the way of increasing temperature response rate by maximally 8.13 and 5.71 times higher for melting and solidification, with compared to the bare tube. The segment achieves 6.88 and 3.71 times high temperature response rate for metal foam tube; 3.38 and 1.75 times high temperature response rate for fin tube.
3. Compared to the original case without any heat transfer enhancement structures, metal foam helps improve the temperature uniformity for PCMs by 23.1%, while fins deteriorate the temperature uniformity by 18.5%.

ACKNOWLEDGMENT

This work was supported by the National Natural Science Foundation of China (51976155). Xiaohu Yang gratefully acknowledges the support of K. C. Wong Education Foundation.

REFERENCES

1. S.D. Musa, Z.H. Tang, A.O. Ibrahim, M. Habib, China's energy status: A critical look at fossils and renewable options, *Renewable & Sustainable Energy Reviews* (81) (2018) 2281–90.
2. J. Qin, E. Hu, X. Li, Solar aided power generation: A review, *Energy and Built Environment* (1) (2020) 11–26.
3. E. Kabir, P. Kumar, S. Kumar, A.A. Adelodun, K.H. Kim, Solar energy: Potential and future prospects, *Renewable & Sustainable Energy Reviews* (82) (2018) 894–900.
4. F. Dincer, The analysis on photovoltaic electricity generation status, potential and policies of the leading countries in solar energy, *Renewable & Sustainable Energy Reviews* (15) (2011) 713–20.
5. G. Alva, L.K. Liu, X. Huang, G.Y. Fang, Thermal energy storage materials and systems for solar energy applications, *Renewable & Sustainable Energy Reviews* (68) (2017) 693–706.
6. Z. Yang, S.V. Garimella, Thermal analysis of solar thermal energy storage in a molten-salt thermocline, *Solar Energy* (84) (2010) 974–85.
7. B. Xie, C. Li, B. Zhang, L. Yang, G. Xiao, J. Chen, Evaluation of stearic acid/coconut shell charcoal composite phase change thermal energy storage materials for tankless solar water heater, *Energy and Built Environment* (1) (2020) 187–98.
8. N. Zhang, Y. Yuan, Synthesis and thermal properties of nanoencapsulation of paraffin as phase change material for latent heat thermal energy storage, *Energy and Built Environment* (1) (2020) 410–6.
9. A. Sharma, V.V. Tyagi, C.R. Chen, D. Buddhi, Review on thermal energy storage with phase change materials and applications, *Renewable & Sustainable Energy Reviews* (13) (2009) 318–45.

10. D. Li, Q.L. Ren, Z.X. Tong, Y.L. He, Lattice Boltzmann models for axisymmetric solid-liquid phase change, *International Journal of Heat and Mass Transfer* (112) (2017) 795–804.
11. F. Agyenim, N. Hewitt, P. Eames, M. Smyth, A review of materials, heat transfer and phase change problem formulation for latent heat thermal energy storage systems (LHTESS), *Renewable & Sustainable Energy Reviews* (14) (2010) 615–28.
12. S. Seddegh, X.L. Wang, A.D. Henderson, A comparative study of thermal behaviour of a horizontal and vertical shell-and-tube energy storage using phase change materials, *Applied Thermal Engineering* (93) (2016) 348–58.
13. S. Lorente, A. Bejan, J.L. Niu, Constructal design of latent thermal energy storage with vertical spiral heaters, *International Journal of Heat and Mass Transfer* (81) (2015) 283–8.
14. F. Agyenim, P. Eames, M. Smyth, Heat transfer enhancement in medium temperature thermal energy storage system using a multitube heat transfer array, *Renewable Energy* (35) (2010) 198–207.
15. A.H.N. Al-Mudhafar, A.F. Nowakowski, F.C.G.A. Nicolleau, Performance enhancement of PCM latent heat thermal energy storage system utilizing a modified webbed tube heat exchanger, *Energy Reports* (6) (2020) 76–85.
16. M. Bechiri, K. Mansouri, Study of heat and fluid flow during melting of PCM inside vertical cylindrical tube, *International Journal of Thermal Sciences* (135) (2019) 235–46.
17. M. Parsazadeh, X. Duan, Numerical study on the effects of fins and nanoparticles in a shell and tube phase change thermal energy storage unit, *Applied Energy* (216) (2018) 142–56.
18. A. Mostafavi, M. Parhizi, A. Jain, Semi-analytical thermal modeling of transverse and longitudinal fins in a cylindrical phase change energy storage system, *International Journal of Thermal Sciences* (153) (2020) 106352.
19. M. Alizadeh, K. Hosseinzadeh, M.H. Shahavi, D.D. Ganji, Solidification acceleration in a triplex-tube latent heat thermal energy storage system using V-shaped fin and nano-enhanced phase change material, *Applied Thermal Engineering* (163) (2019) 114436.
20. X. Yang, J. Guo, B. Yang, H. Cheng, P. Wei, Y.-L. He, Design of non-uniformly distributed annular fins for a shell-and-tube thermal energy storage unit, *Applied Energy* (279) (2020) 115772.
21. M. Lacroix, Study of the heat-transfer behavior of a latent-heat thermal-energy storage unit with a finned tube, *International Journal of Heat and Mass Transfer* (36) (1993) 2083–92.
22. G. Righetti, R. Lazzarin, M. Noro, S. Mancin, Phase change materials embedded in porous matrices for hybrid thermal energy storages: Experimental results and modeling, *International Journal of Refrigeration* (106) (2019) 266–77.
23. M. Opolot, C. Zhao, M. Liu, S. Mancin, F. Bruno, K. Hooman, Investigation of the effect of thermal resistance on the performance of phase change materials, *International Journal of Thermal Sciences* (164) (2021) 106852.
24. C. Zhao, M. Opolot, M. Liu, F. Bruno, S. Mancin, R. Flewell-Smith, et al., Simulations of melting performance enhancement for a PCM embedded in metal periodic structures, *International Journal of Heat and Mass Transfer* (168) (2021) 120853.
25. G. Favero, M. Bonesso, P. Rebesan, R. Dima, A. Pepato, S. Mancin, Additive manufacturing for thermal management applications: from experimental results to numerical modeling, *International Journal of Thermofluids* (10) (2021) 100091.
26. A. Erek, Z. Ilken, M.A. Acar, Experimental and numerical investigation of thermal energy storage with a finned tube, *International Journal of Energy Research* (29) (2005) 283–301.

27. X.H. Yang, Z. Lu, Q.S. Bai, Q.L. Zhang, L.W. Jin, J.Y. Yan, Thermal performance of a shell-and-tube latent heat thermal energy storage unit: Role of annular fins, *Applied Energy* (202) (2017) 558–70.
28. Y.W. Zhang, A. Faghri, Heat transfer enhancement in latent heat thermal energy storage system by using the internally finned tube, *International Journal of Heat and Mass Transfer* (39) (1996) 3165–73.
29. S. Tiari, S.G. Qiu, M. Mahdavi, Numerical study of finned heat pipe-assisted thermal energy storage system with high temperature phase change material, *Energy Conversion and Management* (89) (2015) 833–42.
30. M.J. Hosseini, A.A. Ranjbar, M. Rahimi, R. Bahrampoury, Experimental and numerical evaluation of longitudinally finned latent heat thermal storage systems, *Energy and Buildings* (99) (2015) 263–72.
31. A. Abdoli, G. Jimenez, G.S. Dulikravich, Thermo-fluid analysis of micro pin-fin array cooling configurations for high heat fluxes with a hot spot, *International Journal of Thermal Sciences* (90) (2015) 290–7.
32. H. Eslamnezhad, A.B. Rahimi, Enhance heat transfer for phase-change materials in triplex tube heat exchanger with selected arrangements of fins, *Applied Thermal Engineering* (113) (2017) 813–21.
33. N.H.S. Tay, F. Bruno, M. Belusko, Comparison of pinned and finned tubes in a phase change thermal energy storage system using CFD, *Applied Energy* (104) (2013) 79–86.
34. S. Mat, A.A. Al-Abidi, K. Sopian, M.Y. Sulaiman, A.T. Mohammad, Enhance heat transfer for PCM melting in triplex tube with internal-external fins, *Energy Conversion and Management* (74) (2013) 223–36.
35. E. Fleming, S.Y. Wen, L. Shi, A.K. da Silva, Experimental and theoretical analysis of an aluminum foam enhanced phase change thermal storage unit, *International Journal of Heat and Mass Transfer* (82) (2015) 273–81.
36. P. Chen, X.N. Gao, Y.Q. Wang, T. Xu, Y.T. Fang, Z.G. Zhang, Metal foam embedded in SEBS/paraffin/HDPE form-stable PCMs for thermal energy storage, *Solar Energy Materials and Solar Cells* (149) (2016) 60–5.
37. L. Liang, Y.H. Diao, Y.H. Zhao, Z.Y. Wang, F.W. Bai, Numerical and experimental investigations of latent thermal energy storage device based on a flat micro-heat pipe array–metal foam composite structure, *Renewable Energy* (161) (2020) 1195–208.
38. X. Yang, P. Wei, X. Wang, Y.-L. He, Gradient design of pore parameters on the melting process in a thermal energy storage unit filled with open-cell metal foam, *Applied Energy* (268) (2020) 115019.
39. K. Venkateshwar, S.H. Tasnim, H. Simha, S. Mahmud, Influence of metal foam morphology on phase change process under temporal thermal load, *Applied Thermal Engineering* (180) (2020) 115874.
40. G.K. Marri, C. Balaji, Experimental and numerical investigations on the effect of porosity and PPI gradients of metal foams on the thermal performance of a composite phase change material heat sink, *International Journal of Heat and Mass Transfer* (164) (2021) 120454.
41. C. Zhao, M. Opolot, M. Liu, F. Bruno, S. Mancin, K. Hooman, Phase change behaviour study of PCM tanks partially filled with graphite foam, *Applied Thermal Engineering* (196) (2021) 117313.
42. J.L. Yang, L.J. Yang, C. Xu, X.Z. Du, Experimental study on enhancement of thermal energy storage with phase-change material, *Applied Energy* (169) (2016) 164–76.
43. K. Chen, L.J. Guo, X.D. Xie, W.Z. Liu, Experimental investigation on enhanced thermal performance of staggered tube bundles wrapped with metallic foam, *International Journal of Heat and Mass Transfer* (122) (2018) 459–68.
44. A. Siahpush, J. O'Brien, J. Crepeau, Phase change heat transfer enhancement using copper porous foam, *Journal of Heat Transfer* (130) (2008) 082301.

45. C.H. Wang, T. Lin, N. Li, H.P. Zheng, Heat transfer enhancement of phase change composite material: Copper foam/paraffin, *Renewable Energy* (96) (2016) 960–5.
46. W.Q. Li, Z.G. Qu, Y.L. He, W.Q. Tao, Experimental and numerical studies on melting phase change heat transfer in open-cell metallic foams filled with paraffin, *Applied Thermal Engineering* (37) (2012) 1–9.
47. X.H. Yang, W.B. Wang, C. Yang, L.W. Jin, T.J. Lu, Solidification of fluid saturated in open-cell metallic foams with graded morphologies, *International Journal of Heat and Mass Transfer* (98) (2016) 60–9.
48. P. Jany, A. Bejan, Scaling theory of melting with natural convection in an enclosure, *International Journal of Heat and Mass Transfer* (31) (1988) 1221–35.
49. K. Hooman, U. Maas, Theoretical analysis of coal stockpile self-heating, *Fire Safety Journal* (67) (2014) 107–12.

9 Micro- and Nano-Encapsulated PCM Fluids

M. Mehrali, M. Shahi, and A. Mahmoudi
University of Twente

CONTENTS

9.1 INTRODUCTION: BACKGROUND AND DRIVING FORCES

Encapsulated phase change materials (EPCMs) have been identified as effective materials to promote heat transfer and to improve the heat storage performance of the thermal energy systems. The microencapsulation increases the chemical stability and thermal characteristics of solid–liquid PCMs, allowing these materials to be used in a variety of applications. When these components are dispersed in a base fluid (i.e., carrier fluid), they form an encapsulated phase transition material slurry.

DOI: 10.1201/9781003213260-9

The slurry will transfer more heat or cold at the same volume flow rate due to the phase change energy. The heat transfer coefficients are increased further, allowing the use of smaller heat exchangers. Because of these secondary benefits, smaller systems can be attained with the same performance as larger ones, and the system's performance can be improved when a phase change slurry is implemented. This is an intriguing retrofit option for current systems. This chapter addresses the methods available for encapsulations and slurry preparation as well as their most important characteristics for thermal energy storage systems. Furthermore, the applications of these slurries in the domains of thermal energy storage are discussed.

9.2 ENCAPSULATED PCMs

Encapsulation is the micro-/nanoscale coating of PCMs with supporting materials. The confined material is referred to as the core, the coating material is referred to as the shell, and the final product is referred to as micro/nano capsules. In encapsulated PCMs, the core is always PCMs, while the shell can be any supporting material. Depending on the synthetic conditions, capsules can have a mono/poly nuclear, a matrix, or a multi-wall architecture as shown in Figure 9.1 [1].

Structural integrity and stability should be achieved using shell materials. In other words, the coating materials, that are mechanically robust and thermally conductive, not only enclose the PCMs but also increase the rate of heat transmission. The ratio of PCM core to the shell material is a crucial element in determining the composite system's energy storage capacity and structural stability. A relatively high weight percentage of the PCM core results in high storage capacity but poor structural stability and vice versa. As a result, PCM mass loading must be optimized in relation to the mechanical strength of the coating materials. Organic polymers, metal oxides, and hydroxides have all been used as coating materials [2–5]. The coating materials are selected based on their compatibility with the PCM core and the method of synthesis. Polymerization methods are commonly used to encapsulate PCMs with organic coatings, whereas the sol–gel method is used for inorganic coatings. The

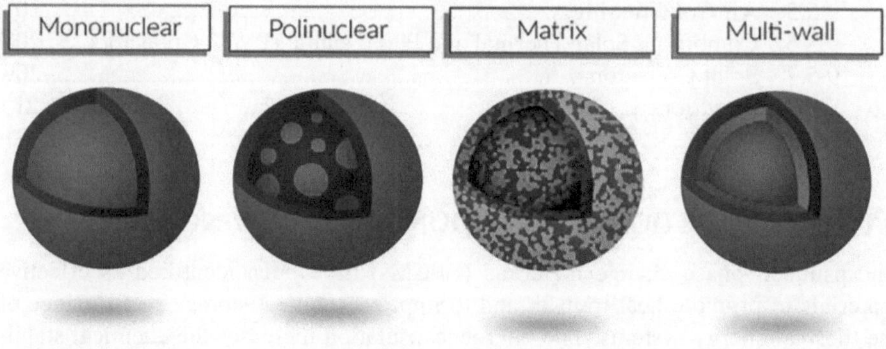

FIGURE 9.1 Synthetic illustration of various architectures for encapsulated PCMs. (Adapted from Jurkowska and Szczygieł [1], reproduced with permission from Elsevier (license number 5177541385586).)

major focus of this chapter is on micro-/nano-encapsulated PCMs (MEPCMs and NEPCMs) since the smaller the microcapsule size, the higher the heat transfer rate, which broadens the field of applications. Capsules come in three sizes: macro, micro, and nano. Smaller capsules enhance the material's surface-area-to-volume ratio, which increases heat transfer. Encapsulating PCMs in 1 mm capsules, for example, is expected to increase the surface area by $300\,m^2/m^3$ when it's compared to the bulk PCM. This impact would be greatly enhanced if their diameter was reduced to the nanometer scale. Other benefits of encapsulating PCMs include reduced leakage and interaction with the external environment, corrosion protection for container materials, control over volume change upon melting, and enhanced thermal cyclic stability. Table 9.1 shows the features of several encapsulation techniques with their particle size ranges and advantages [6].

The materials used as the core of MEPCMs or NEPCMs must be selected based on their applications. Some factors to consider when selecting a capsule's core material are the capsule's working temperature, low supercooling, high latent heat, low reactivity, high thermal conductivity, no toxicity, high density, high availability, and a minimal volume changes during phase transition. Three types of phase change materials exist including organic, inorganic, and eutectic PCMs. Two main kinds of organic materials are characterized as paraffin and non-paraffin. The high stability, non-corrosivity, good availability, low reactivity, and recyclability of these materials are the attractive qualities. The most prevalent and well-known phase change materials in many sectors are paraffin and oleic acid. Mainly salts (e.g., hydrates) and metals are included in nonorganic compounds. Although their energy storage and thermal conductivity are superior to organic materials, they have certain negative effects such as corrosion and excessive supercooling. Eutectics are generated by the combination of one or two components (organic–organic, organic–nonorganic, or nonorganic–nonorganic) that melt and freeze homogeneously and form a mixture of individual crystals as they crystallize. Eutectics melt and freeze nearly without dissociation, and they do not separate. When compared to organic materials, this kind has a higher storage density.

9.2.1 Encapsulation's Benefits and Drawbacks

The most significant characteristic that is affected by encapsulation is the degree of EPCM adhesion to the container wall. Due to the high stickiness in its solid form, PCM will have considerable trouble moving in flow applications (i.e., fluid mechanics). In practice, by encapsulating a PCM and distributing it in the base fluid, the latent heat of PCM may be exploited to alleviate adhesion difficulties. Intimately linked, the encapsulation extends the lifespan of the core PCM by protecting it from corrosion and reactivity. Encapsulation also provides the benefit of a faster phase change rate due to the high area-to-volume ratio. Despite the benefits listed above, there are some disadvantages to this approach, including supercooling increment. Even when the fluid reaches the freezing point, supercooling prevents it from freezing. This situation results in a loss of efficiency, which must be avoided. The temperatures at which solidification and melting happen are different in this situation. Many studies have shown that the supercooling degree of EPCMs increased by reducing the size

TABLE 9.1
The Features of Various Encapsulation Techniques

Encapsulation Techniques	Progress	Advantages	Particle Size Range
Spray drying	• Spraying the core and shell material mixture into the heating chamber. • The solvent is evaporated by contacting the mixture with a hot dry gas. • Separating solid particles.	It's inexpensive and simple.	6–600 μm
Interfacial polymerization	• Making a dispersed oil phase and an aqueous solution. • Adding hydrophilic monomers for polymerization at the water–oil interface.	Encapsulation efficiency is high.	2–2000 μm
In situ polymerization	• Making an aqueous solution containing reactive monomers. • Polymerization is accomplished by adding polymer to an aqueous solution. • The polymer develops on the core material's surface and eventually forms a full shell.	The shell has high quality and is not breakable.	1–2000 μm
Suspension cross-link	• The shell components are dissolved in an aqueous solution. • Adding a dispersed phase containing core components and an initiator to an aqueous solution. • Surfactants are used to encapsulate the core components by the shell materials.	High thermal stability of microcapsules	2–4000 μm
Coacervation	• An aqueous solution is used to dissolve the shell materials and polymers. • To enclose the core, the aqueous solution containing the shell materials and polymers is mixed with the preceding aqueous solution. • Heat treatment or crosslinking to stabilize microcapsule particles.	Microcapsules with high encapsulation efficiency and strong thermal stability.	1–1000 μm
Sol–gel encapsulation	• Preparation of an aqueous solution with core components, shell materials, and complexing substances. • Adding the metal sol to the aqueous soil solution and then polymerizing it into a gel with a 3D network structure. • The gel is dried and microcapsule particles are sintered.	Nanoscale microcapsule particles can be created.	0.1–10 μm

Source: Data from Ran et al. [6].

of the capsules [7]. The shell of the capsule reduces the thermal conductivity and effective volume of the MEPCMs/NEPCMs. To prevent these issues and maintain the high thermal efficiency of the encapsulated PCMs, a careful tuning is required.

9.3 ENCAPSULATED PCM FLUIDS

An encapsulated phase change material slurry (EPCMS) is a liquid that consists of a component that stores heat through phase change and a component which is always in a liquid phase. One example is a mixture of water and microencapsulated PCM. The PCM in the microcapsules stores the heat, and the water ensures fluid behavior even when the PCM in the microcapsules is in the solid form. Because slurry is always liquid, it can be stored in a standard tank, discharged directly, pumped through pipes, and heated or cooled by a heat exchanger. There are even more benefits:

- Due to the phase change, the slurry may carry more heat or cold at the same volume flow rate, therefore, the pipe system already functions as a storage system.
- Heat transfer coefficients are increased even further, allowing for the use of smaller heat exchangers.
- Smaller systems can attain the same performance as larger systems, or a system's performance can be improved by using a phase change slurry instead of a standard heat transfer fluid without a PCM component. This is a fascinating option for retrofitting existing systems.

9.3.1 Concept of Phase Change Slurries

Heat transport systems rely heavily on fluid heat capacity. The larger a fluid's heat capacity, the more energy it can absorb and release. Because these materials have a high energy storage density (i.e., latent heat) in comparison to the specific heat of the base fluid, adding PCMs to the base fluid can enhance the capacity of energy absorption in the base fluid. In fact, by changing their phase, these materials may absorb a significant amount of input energy while maintaining the temperature of the base fluid constant. Figure 9.2 shows a storage tank containing the slurry. It should be noted that the slurry within the pipe system also functions as a storage section, which is an often overlooked but critical aspect.

The phase transition properties of slurries, thermal conductivity, viscosity, and density are all crucial elements to consider while designing the thermal energy storage (TES) systems. Phase change properties of the generated slurry are important factors in determining its TES capability. The PCM and the microcapsule production process both have a significant impact on this parameter. When it comes to heat transport, thermal conductivity is crucial.

Dispersing the microcapsules with inorganic conductive material resulted in a higher heat transfer rate and enhanced heat transfer performance when the thermal conductivity values were higher. Besides, the viscosity of the slurry is an essential consideration when determining pumping power. The flow resistance produced

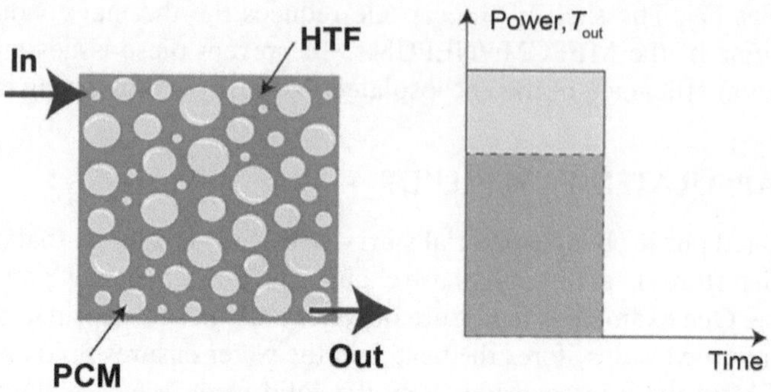

FIGURE 9.2 The general overview of slurry system. The dashed line simply represents the output power for a sensible heat storage. The slurry at the same temperature as sensible heat storage delivers higher power due to the presence of PCM.

by the microcapsules dispersed in the carrier fluid increases the slurries' viscosity. Commercialized slurries are currently available on the market and BASF is the leading manufacturer of PCM microcapsules as well as MEPCM slurries. BASF produces microcapsules by polymerizing a PCM emulsion with diameters of approximately 5 μm and disperses them in water. However, most of the studies have demonstrated the higher structural stability of nanocapsules compared to macro- and microcapsules, therefore it has higher potential to be applied in the heat storage applications. On the other hand, research on producing nano capsules is still at an early stage. Further research is needed in practical applications before they are actually employed. TES applications have effectively marketed macro and microcapsules. Table 9.2 shows product information for the BASF commercialized PCM slurries [8].

9.4 ENCAPSULATED PCM SLURRY (EPCMS) PRIMARY CHARACTERISTICS

9.4.1 Subcooling, Solidification, and Hysteresis

When measurements of the cooling and heating processes provide different findings, this is referred to as hysteresis; nonetheless, hysteresis is usually regarded as a material characteristic. The effects caused by the measuring circumstances are referred to as apparent hysteresis. The hysteresis effect is significantly more evident in inorganic PCMs such as salt hydrates than in organic PCMs like paraffins. Subcooling is the most typical action that causes hysteresis because of the material characteristics. Subcooling occurs when a liquid must be chilled to a temperature lower than its melting point for the crystallization process to begin. Subcooling is always a major issue in PCM research and application sectors, as it expands the operating temperature range of storage devices, reducing energy efficiency. Although many PCMs do not exhibit the subcooling phenomena in macroscopic geometry, the problem exists in microscopic geometries. The effect of hysteresis and subcooling is demonstrated in Figure 9.3 [9].

TABLE 9.2
The BASF Commercial Micronal® PCM Slurries Characteristics [8]

Product Designation	Melting Point Approx.	Application	Integration Range	Overall Storage Capacity Approx.	Latent Heat Capacity Approx.	Solid Content	Density	Viscosity
DS5000	26°C	Summertime excessive heating protection	10°C–30°C	59 kJ/kg	45 kJ/kg	Approx 42%	Approx 0.98	Approx 200–600 mPas
DS5007	23°C	Stabilization of the indoor temperature in the comfort zone Passive and active application	10°C–30°C	55 kJ/kg	41 kJ/kg	Approx 42%	Approx 0.98	Approx 200–600 mPas
DS5030	21°C	Surface cooling systems	10°C–30°C	51 kJ/kg	37 kJ/kg	Approx 42%	Approx 0.98	Approx 200–600 mPas

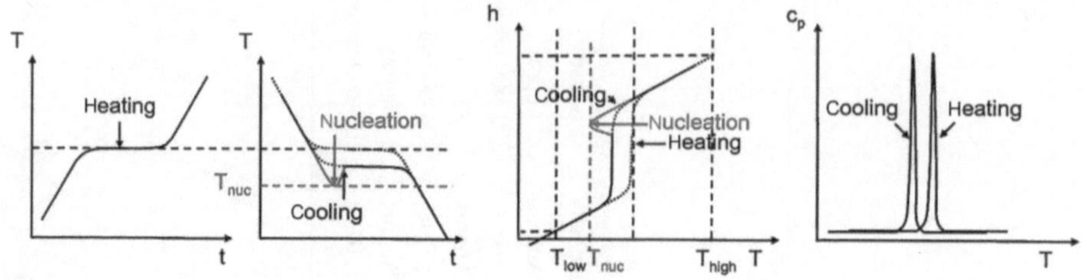

FIGURE 9.3 Hysteresis incl. subcooling, here with a phase change temperature range; heating and cooling $T(t)$, $h(T)$, and $C_p(T)$. In $C_p(T)$, the effect of subcooling cannot be depicted. (Based on Mehling et al. [9].)

Many scholars have investigated the topic of subcooling in PCM emulsions and PCM slurries, as well as how to eliminate it. The value of subcooling is defined as the difference between the peak temperatures recorded with the differential scanning calorimetry (DSC) in the heating and cooling curves. Encapsulated organic-based PCM slurries got more attention due to their self-nucleating characteristics, the absence of subcooling and congruent melting, and their availability in a wide range of melting temperatures appropriate for various applications. The degree of subcooling of inorganic-based EPCM slurries is significantly influenced by encapsulation (e.g., encapsulated salt hydrates). The higher specific surface area of micro-/nano-encapsulated PCMs promotes the nucleation of hydrated salts, which results in less subcooling. Such subcooling effects should be controlled or forecasted in EPCMS applications. Several studies have demonstrated that the size of the encapsulation may have a significant impact on the subcooling effect, causing enthalpy hysteresis within the EPCMS. Subcooling broadens the temperature range by decreasing the freezing point, diminishing cooling energy efficiency. Due to subcooling, which lowers heat capacity, the encapsulated PCM may not be entirely in the solid phase at the end of the tube during the heat transfer process. Several studies have been conducted on strategies for reducing subcooling. The most frequent way of suppressing subcooling effects is to change homogenous nucleation to heterogeneous nucleation using nuclear agents. However, the mass fraction of the nucleating agents should be carefully managed to prevent the loss in EPCMS thermal capacity if this technique is utilized.

9.4.2 Stability and Durability

As a heat storage and a heat transfer medium, EPCMS should be stable during long-term storage and under mechanical–thermal stresses. Suspension stability is more critical for long-term storage, while microcapsule durability is the major element determining stability under mechanical–thermal stresses. Suspension instability of EPCMS can be related to creaming, sedimentation, or agglomeration (Figure 9.4) [10]. Due to gravity force, the density difference between the working fluid and the dispersed EPCMs will induce creaming or sedimentation while the surface charges and adhesivity of the EPCMs cause agglomeration. Surfactants, by generating electrostatic repulsive contact between EPCM and the base fluid, may be utilized to avoid these effects during long-term storage.

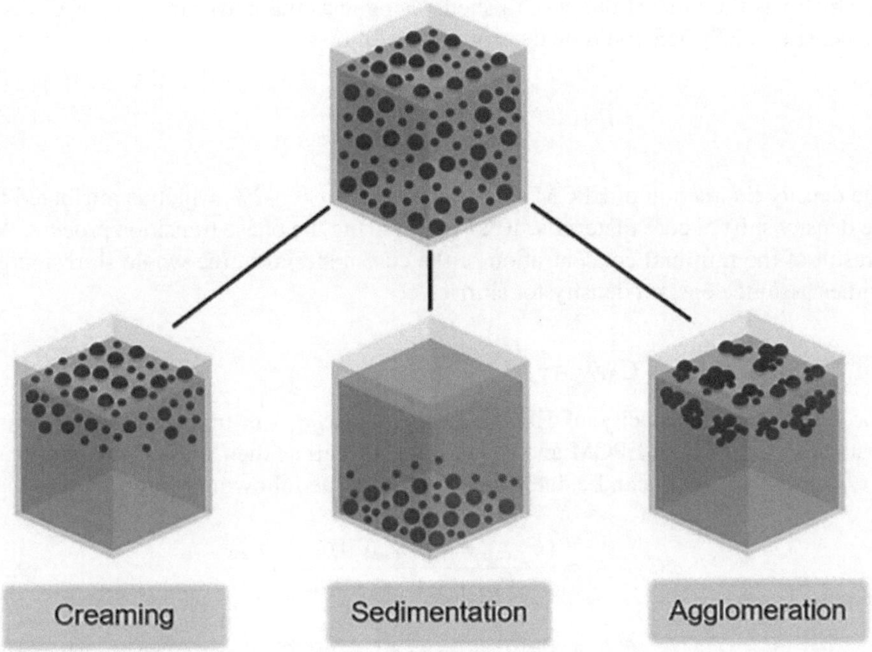

FIGURE 9.4 Processes of Instability in EPCMS. (Based on Shao et al. [10], reproduced with permission from Elsevier (license number 5177551010490).)

Changes in EPCMS characteristics such as the shapes and sizes of the microcapsules, the thermophysical properties, and viscosity can be used to assess their stability. Continuous pumping tests are commonly used to assess the durability of EPCMS. The size distribution is one of the most important factors determining the lifespan of EPCMS. Pumping was revealed to be the most common cause of microcapsule rupture, and when the EPCM diameter was lowered to 5 μm, the capsules rupture risk was minimized, giving stability even after 5000 pumping cycles. Another element influencing EPCMS durability is the core to shell ratio where a greater diameter to thickness ratio increases the possibility of breaking. These findings suggest that the microencapsulation process and parameters used have an impact on the thermal stability of EPCMS.

9.4.3 Density

The encapsulated PCM consists of two primary components: the core and the shell. Therefore, the overall density can be calculated according to conversion of mass theory ([11]):

$$\rho_{\text{EPCM}} = \frac{(1+y)\cdot\rho_{\text{core}}\cdot\rho_{\text{shell}}}{\rho_{\text{core}} + y\cdot\rho_{\text{shell}}} \tag{9.1}$$

In which y is the ratio of the core to shell. Using w as the masse fraction of EPCM, the density of EPCMS could be determined as follows:

$$\rho_{\text{EPCMS}} = \frac{\rho_{\text{EPCM}} \cdot \rho_f}{w \cdot \rho_f + (1-w) \cdot \rho_{\text{EPCM}}} \tag{9.2}$$

The density fluctuation of EPCM slurries is less than 1%–2%, which is attributed to the density shift of core materials (10%–15%) during the phase transition process. As a result of the minimal concentration of the core material in the whole slurry, most studies assume constant density for slurries.

9.4.4 Specific Heat Capacity

The specific heat capacity of EPCM slurry ($C_{p\,(\text{EPCMS})}$) is influenced by specific heat capacities of both EPCM and based fluid, as well as their mass concentrations. $C_{p\,(\text{EPCM})}$ and $C_{p\,(\text{EPCMS})}$ can be determined by using the following formulas:

$$C_{p,\text{EPCM}} = \frac{\left(C_{p,\text{core}} + y \cdot C_{p,\text{shell}}\right) \rho_{\text{core}} \cdot \rho_{\text{shell}}}{\left(y \cdot \rho_{\text{core}} + \rho_{\text{shell}}\right) \cdot \rho_{\text{EPCM}}} \tag{9.3}$$

$$C_{p,\text{EPCMS}} = w \cdot C_{p,\text{EPCM}} + (1-w) \cdot C_{p,f} \tag{9.4}$$

Typically, the EPCM has a lower specific heat than pure phase change materials due to significantly lower specific heat capacities of the shell materials. The specific heat capacity of the EPCM in different phases is shown in Figure 9.5.

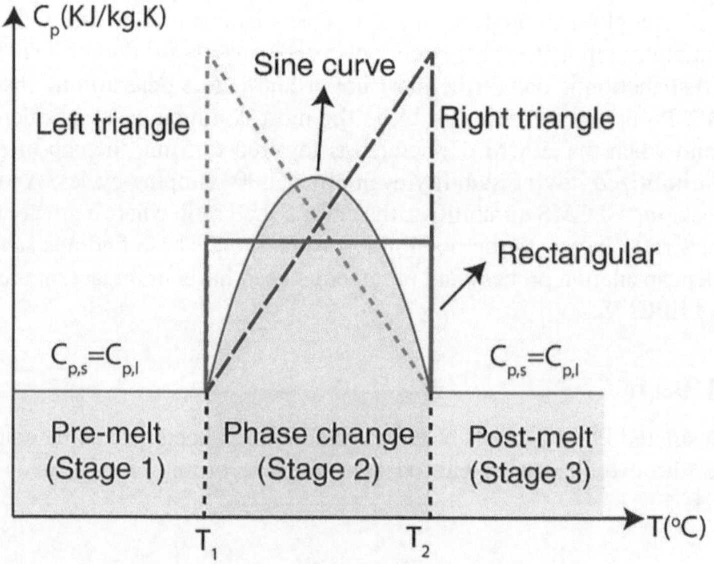

FIGURE 9.5 The temperature-dependent fluctuation of EPCMS specific heat capacity.

For the pre-melt and post-melt stages, the specific heat capacity value is constant and equal to $C_{p,\mathrm{EPCMS}}$. The specific heat capacity during the phase change stage between T_1 and T_2 comprises not only the EPCM's latent heat capacity, but also the sensible heat for both the EPCM and the base fluid which is clearly higher than the other two stages. The enthalpy model, which consists of four types of functions between temperature and specific heat capacity including left triangle, sine curve, rectangle, and right triangle, is widely used to simplify calculations, especially in numerical simulations. The specific heat capacity of the bulk EPCM slurry may also be estimated using the model established by Chen et al. [11].

9.4.5 Thermal Conductivity

Thermal conductivity is an essential characteristic for evaluating EPCMS's performance as a heat transfer medium. EPCMS must have a high heat transport rate when used as a heat transfer medium in order to ensure fast charging and discharging rates. To enhance the thermal conductivity of EPCMS, highly conductive materials should be utilized when encapsulating PCMs. Inorganic materials, such as silicon dioxide and titanium dioxide, have a greater heat conductivity than organic materials. When using low-thermal-conductivity organic materials as microcapsule shells, high-thermal-conductivity additions like graphene or other carbon compounds can be employed. The composite sphere method can be used to determine the thermal conductivity of EPCM particles [12]:

$$\frac{1}{k_{\mathrm{EPCM}} d_{\mathrm{EPCM}}} = \frac{1}{k_{\mathrm{core}} d_{\mathrm{core}}} + \frac{d_{\mathrm{EPCM}} d_{\mathrm{core}}}{k_{\mathrm{shell}} d_{\mathrm{EPCM}} d_{\mathrm{core}}} \tag{9.5}$$

$$\left(\frac{d_{\mathrm{EPCM}}}{d_{\mathrm{core}}}\right)^3 - 1 + \frac{\rho_{\mathrm{shell}}}{\rho_{\mathrm{shell}} + y\rho_{\mathrm{core}}} \tag{9.6}$$

Intimately linked, the Maxwell relations can be implemented to calculate the thermal conductivity of the EPCMS [13]:

$$k_{\mathrm{EPCMS}} = k_f \frac{2k_f + k_{\mathrm{EPCM}} + 2\varphi\left(k_{\mathrm{EPCM}} - k_f\right)}{2k_f + k_{\mathrm{EPCM}} - \varphi\left(k_{\mathrm{EPCM}} - k_f\right)} \tag{9.7}$$

Due to the interaction between the EPCM and working fluid, the effective thermal conductivity of EPCMS while it flows is higher than what is anticipated by Maxwell's equation. As a result, Charunyakorn et al. [14] suggested more precise correlations to estimate the effective thermal conductivity of EPCMS ($k_{e,\mathrm{EPCMS}}$):

$$\frac{k_{e,\mathrm{EPCMS}}}{k_f} = 1 + B \cdot \omega \cdot Pe_{\mathrm{EPCM}}^m \tag{9.8}$$

$$Pe_{\mathrm{EPCM}} = \frac{\vec{e} \cdot d_{\mathrm{EPCM}}^2}{\alpha_f} \tag{9.9}$$

$$\begin{cases} B=3, m=1.5 & Pe_{EPCM}<0.67 \\ B=1.8, m=0.18 & 0.67<Pe_{EPCM}<250 \\ B=3, m=1/11 & Pe_{EPCM}<250 \end{cases} \quad (9.10)$$

Notably, the EPCMS thermal conductivity is influenced by many factors including the shell material, concentration, particle diameters, etc. However, all the studies reviewed so far are suffering from the fact that the methodologies to calculate thermal conductivity of phase change slurries are inadequate and must be improved.

9.4.6 Hydrodynamic Characteristics

It is important to describe several typical characteristics including viscosity, pressure drop, and pumping power to understand fully the hydrodynamic characteristics of EMPCS.

9.4.6.1 Viscosity

The pressure drops of EPCMS and the pumping power of the hydraulic devices are both influenced by viscosity, which is one of the essential hydrodynamic factors for EPCMS. As the volume fraction of EPCMS grows, so does its viscosity. Despite the fact that different phase change materials were examined, the correlations between viscosity and volume fraction found were similar, indicating that the viscosity of EPCMS increases with increasing volume fraction. The influence of key variables on EPCMS viscosity is summarized in Table 9.3 [7,15–24].

All these investigations show that the higher the viscosity of the slurry, the weaker the degree of the turbulence. As a result, in order to maintain a low flow resistance, the volume fraction of the slurry must be well controlled. The EPCMS is often considered as a homogenous fluid for low volume fractions (i.e., up to 37%), and the apparent viscosity (μ_b) of the slurry may be estimated using the correlation by Vand [25]:

TABLE 9.3
Analyses of the Influence Factors on Viscosity

Variable	Alteration	Effect on Viscosity	Effect on Shear Rate	References
Volume fraction (ω)	Increasing	Increasing	Increasing or decreasing depending on the shear rate ranges	[7, 15–22]
Temperature	Increasing	Decreasing	Decreasing	[7, 15–23]
Particle size	Increasing	Increasing	-	[18]
Phase change stage	Solid to liquid	No changes	-	[21, 23, 24]
Shell material	Polymers or others	Not significant	-	[16]
Surfactant	Addition of anionic	Decreasing	-	[7]

$$\frac{\mu_b}{\mu_f} = \left(1 - \varphi - A\varphi^2\right)^{-2.5} \quad (9.11)$$

where A is a variation parameter that depends on the form, size, and type of microcapsules that may be obtained by experimental methods[14,15,26].

9.4.6.2 Pressure Drop and Pumping Power

Despite having a higher heat transfer rate and heat capacity than based fluid, EPCMS is more difficult to pump due to its high viscosity, which increases flow resistance. When compared to a base fluid, the pressure drops and pumping power of EPCMS might be utilized to assess the degree of this difficulty. Numerous studies have been conducted to explore the effects of various variables on EPCMS pressure drop, such as velocity, working temperature, concentration, particle sizes, and flow regime [19,20,22,23]. EPCMS has a considerably greater pressure drop than base fluid (single-phase fluid) in laminar flows [11]. Mechanisms for turbulent flows will be more sophisticated as the pressure drop of EPCMS becomes significantly smaller than that of the carrier fluid because of the drag reduction effect. Considering the complex circumstances in turbulent flow, none of the studies could reach a definite conclusion on the factors generating the drag reduction effect. The most important factor is that, due to the higher heat transfer capacity due to the existence of microcapsules, a less flow rate is required for an equivalent amount of the heat transfer, resulting in a lower pumping power. Consequently, on one hand, higher pumping power is expected due to friction, while on the other hand, it lowers due to the lower necessary flow rate.

9.5 APPLICATIONS OF EPCMS

TES and heat transfer medium in thermal management systems are two areas where EPCMS are frequently used. Figure 9.6 illustrates major examples of EPCMS in various systems, which will be explored in the following sections.

9.5.1 Pipe Flow

Pipe flow is very important due to the necessity of the fluid movement and heat transfer inside pipes for transmission lines in various thermal systems such as air-conditioning and solar systems. As a result, researchers have long sought a method to enhance heat transmission through tubes. The addition of microcapsules to one-phase flows in which heat capacity is a key factor can significantly increase heat transmission.

Many studies investigated flow and heat transmission in pipes under laminar, transient, and turbulent regimes [11,22,27–34]. In general, introducing microcapsules with appropriate operating temperatures should increase specific heat. Nonetheless, a variety of factors influence the decrease and the increase of the heat transfer coefficient [35]. There are a few more features of EPCMS flows within pipes that must be considered. The phase change behavior of EPCMS can be influenced by a variety of factors, including concentration, heat flow, Reynolds number, and pipe length. Therefore, many parameters must be adjusted in such a way that PCM

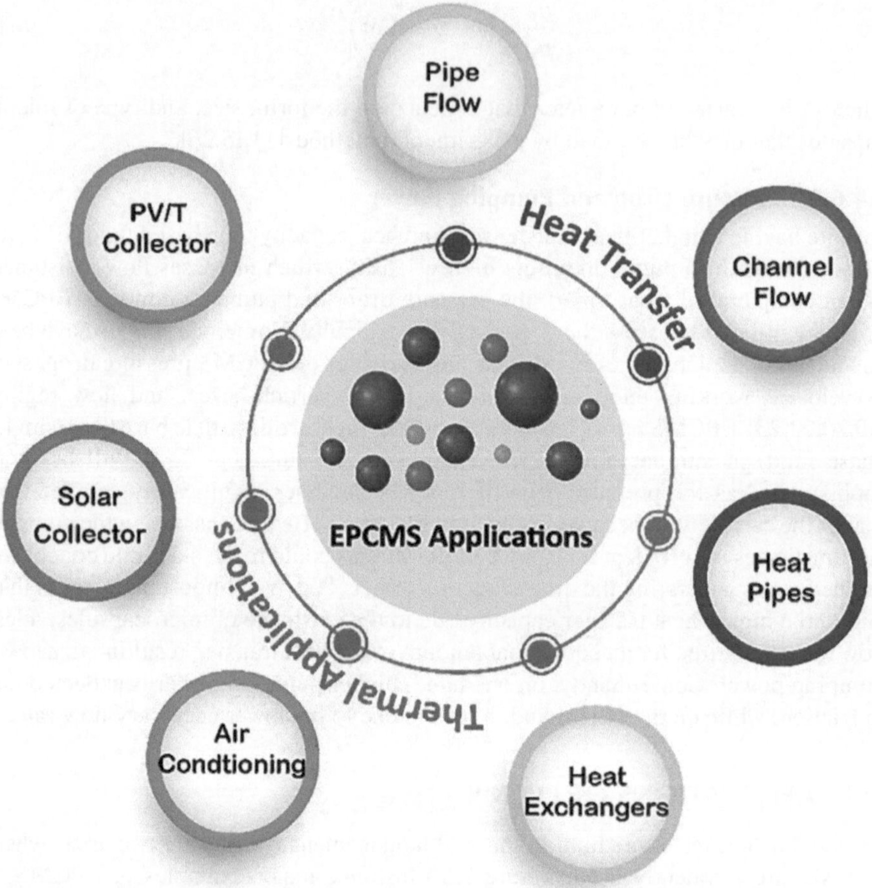

FIGURE 9.6 Heat transfer and thermal applications of EPCMS.

may change phase and function effectively for the appropriate time. For instance, Hashimoto et al. [28] investigated the flow characteristic of EPCMS containing a microcapsule—ethylene glycol/water slurry (see Figure 9.7) [28]. It was observed that when the concentration of EPCM increases, the maximum Reynolds number at which phase transition occurs decreases. In addition, the minimum heat input required to melt or solidify the EPCM increases, as does the minimum time required to accomplish the phase change.

According to the literature, the influence of various factors on the phase change behavior of EPCMS running within the pipe is summarized in Table 9.4.

9.5.2 Channel Flow

One of the most widely used and practical methods of cooling heat sinks, electronic equipment, and other components is the use of micro/mini channels. Due to the improved heat transmission rate they offer, researcher opted to use EPCMS in the

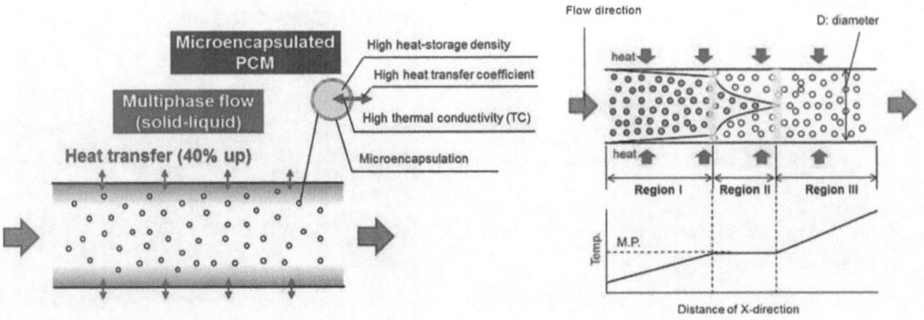

FIGURE 9.7 Heat transfer enhancement in a horizontal circular tube using EPCMS based on three-region melting model. (Adapted from Hashimoto et al. [28], reproduced with permission from Elsevier (license number 5177551484612).)

TABLE 9.4
A Summary of the Parameters' Efficacy in Achieving Complete Phase Change for EPCMS in the Pipe Flow

Variables at Inlet	Effect on Value of Parameters at Output to Maintain the Phase Change Process		
Reynolds number (↑)	Minimum length (↑)	Minimum heat flow (↑)	Minimum concentration (↓)
EPCM concentration (↑)	Minimum length (↑)	Minimum heat flow (↑)	Maximum Reynolds (↓)
Heat flow (↑)	Minimum length (↓)	Maximum Reynolds (↑)	Minimum concentration (↑)

channels. The EPCMS can play a key role in battery system thermal management by combining the better convective heat transfer impact of the liquid cooling technology with the benefit of the latent heat storage. To ensure the superiority of the EPCMS in channel flow systems, the cooling performance of the EPCMS in a channel flow was examined and the findings were compared to the water-cooling performance by many researchers [36–43]. The simulation findings can also be utilized to guide the development of the experimental procedure and technical applications. In a recent study, Bai et al. [36] investigated the effect of different variables including the mass flow rate, EPCMS phase transition temperature, and concentration on the energy consumption and cooling capacity of a mini channel cooling plate as shown in Figure 9.8 [36]. A dimensionless empirical model was developed to evaluate the effect of the EPCMS flow characteristics and physical properties on the cooling performance and the heat transfer. With a rise in the mass flow rate and the EPCMS concentration, the convective heat transfer coefficient was enhanced. Moreover, the cooling performance of the PCS was lower than that of pure water when the mass flow rate exceeded the critical value. It is important for the EPCMS to have enough residence time in the channels for phase change and heat conduction activities while the slurry input temperature should be somewhat lower than the EPCM's theoretical melting point. As current studies do not cover the entire range of operating conditions

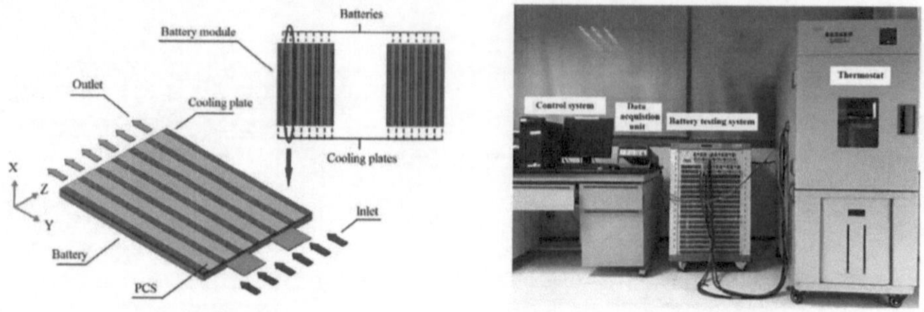

FIGURE 9.8 The microchannel cooling plate construction and experimental setup for evaluating the performance of EPCMS for battery thermal management. (Based on Bai et al. [36], reproduced with permission from Elsevier (license number 5177560122868).)

required for the development of a universal pressure drop and a heat transfer correlation, more effort should be put into conducting experiments and simulations with a broader range of test parameters.

9.5.3 Heat Exchangers

Heat exchangers are frequently utilized in industries to transmit heat. Heat transfer improvement in heat exchangers has been always a key goal, with a plethora of studies dedicated to evaluating these concepts. Improvements in fluid properties can considerably boost the heat transmission in heat exchangers since fluids play such an essential role in heat exchangers. The effect of utilizing EPCMS on a variety of heat exchangers has been studied, including coil [44], double pipe [45], microchannel [39,46,47], shell and tube [48], and square coiled tube [49] heat exchangers. The experimental results demonstrate that EPCMS may enhance both the overall heat transfer coefficient and heat transfer performance of the heat exchangers when compared to a base fluid. However, because of the higher effective viscosity of the fluid, this improvement comes at the cost of higher pressure loss throughout the channel. Furthermore, it has been shown that the thermal conductivity of the EPCMS is critical for improving heat transfer coefficients, which can give better efficiency with smaller pressure drops [39]. EPCMS would enhance the thermal characteristics of working fluids in general, increasing system efficiency while decreasing system size/weight requirements.

9.5.4 Heat Pipes

A heat pipe is a type of heat transmission device that has a compact size but can transport a large amount of heat. Heat pulse pipes (PHPs) are one of the latest ways for high-tech heat transmission that are frequently employed in solar collectors, solar water heaters, and other applications. Working fluid, filling ratio, and inclination angle are all factors that influence pulsing heat pipe performance. Among the variables listed, the working fluid and its thermophysical characteristics have a significant impact on the thermal performance. Due to the fact that the fluid circulates in a closed loop within the PHP, it is cooled and heated at each cycle. Therefore, EPCMS may be utilized as a working fluid to transmit large amounts of heat when melted and solidified by passing between warm and cold sections. There haven't been many studies on the use of EPCMS in heat pipe systems. Heydarian et al. [50] investigated nano-encapsulated paraffin-based EPCMS with varying mass ratios as a working fluid in two-turn PHP. The use of this mixture significantly decreases the PHP's heat resistance (44% at 50 W) when there is an optimal EPCMS concentration to provide the highest thermal performance. It was discovered by analyzing the thermal resistance of the hot and the cold sections of the PHP that while the specific heat is directly responsive to the rise in concentration, there is an optimal concentration at which the lowest resistance in heat transmission occurs owing to the rise in viscosity.

9.5.5 Air Conditioning

Cities' development, population increase, building modernization, etc. have made air-conditioning systems more vital than ever. Storage of the heating and the cooling energy, and therefore the usage of EPCMS, can be highly beneficial in improving these systems. It has been proved that the thermal capacity of the chilled water systems is increased by using EPCMS. Heat transfer properties, stability, and fluidity have all been thoroughly studied to determine its feasibility for incorporation into a cold storage system [51]. EPCMS is suitable for variety of cooling systems which can be classified into two categories: dynamic and static. In a dynamic form, the EPCMS is delivered directly to air-conditioning terminals or heat exchangers. The system, also known as a latent cooling storage and transport system, generates, stores, and transports slurry [52]. The EPCMS is stored in a thermal tank in the event of a static kind. For instance, a new cooling ceiling design that works directly with an EPCMS thermal storage is proposed by Wang et al. [53] as illustrated in Figure 9.9 [53]. The circulation of EPCMS in the ceiling panels eliminates sensible

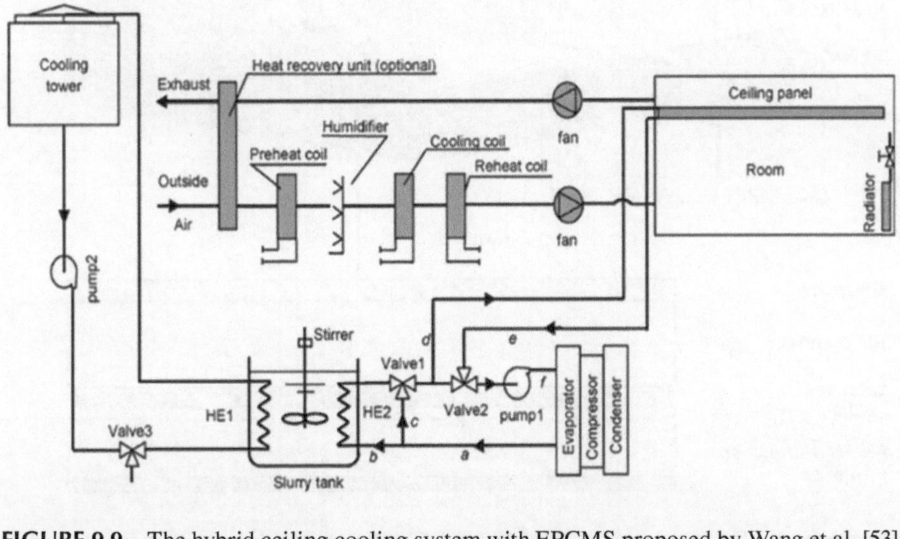

FIGURE 9.9 The hybrid ceiling cooling system with EPCMS proposed by Wang et al. [53], reproduced with permission from Elsevier (license number 5177560246352).)

load. The dehumidified air from an air handling unit eliminates latent load as well as a portion of sensible load. For an office space, the yearly energy consumption of three systems, namely cooling ceiling without thermal storage, with ice storage or EPCMS storage was compared. The combined cooling ceiling with EPCMS storage system was found to be the most energy efficient. PCMs for a cold storage in air-conditioning systems have recently received a lot of attention for two reasons: materials and applications. However, in this segment, most studies focused on composite PCMs rather than EPCMS applications. In the future, the prospect of the cold storage application of EPCMS for air-conditioning systems including solar-driven air-conditioning systems should be studied.

9.5.6 Combining Solar Thermal and Photovoltaic (PV/T) Collectors

Photovoltaic systems are commonly used to convert the solar energy to the electricity. The thermal management of photovoltaic collectors is critical because their efficiency is dependent on temperature of solar panels. Combining solar thermal and photovoltaic collectors might be a way toward increasing the overall efficiency by using the power generation in both electrical and thermal sections of the collectors. Because of the high heat capacity of EPCMS, utilizing these slurries to cool solar panels is becoming an emerging topic. In recent years, there are many modeling and experimental studies on the application of EPCMS for PV/T [54–60]. Liu et al. [61] suggested and investigated a simple arrangement as depicted in Figure 9.10. Because the output temperature was lower when slurry was used, they discovered

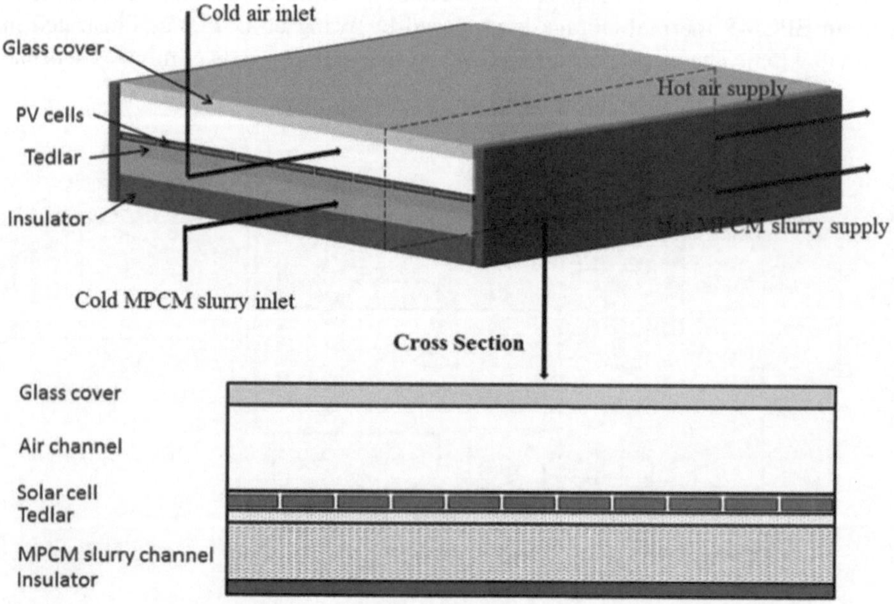

FIGURE 9.10 The integration of EPCMS in PV/T collector. (Based on Liu et al. [61], reproduced with permission from Elsevier (license number 5177560365244).)

that over the course of the day, using slurry gives up to 1.8% higher thermal efficiency than using water. Furthermore, the slurry container had a little lower cell temperature, resulting in a marginally higher electrical efficiency. The slurry case exhibited a greater net efficiency even though it required more pump power owing to the increased viscosity.

Several studies have examined the effects of EPCMS key variables such as EPCM concentration [56,62,63], latent heat [61], melting point [64], Reynolds number, and slurry flow rate [54,56,64] using computational and experimental techniques. They discovered that increasing EPCMS concentration increases the primary energy-saving efficiency at a lower fluid velocity while raising EPCMS concentration increases the total exergy efficiency at a certain fluid velocity. When the intake fluid velocity or the mass flow rate was increased, the thermal and electrical efficiencies increased monotonously, while the exergy efficiency decreased. The thermal and electrical efficiencies were increased when the melting point was reduced, whereas the exergy efficiency was decreased. The high cost of EPCMS, in addition to the high cost of PV/T systems, is one of the most significant hurdles to the implementation of EPCMS in PV/T systems.

9.5.7 Solar Collectors

The most popular devices for conversion of the solar power to the heat are the flat-plate solar thermal collectors. Water-based fluids are frequently used as heat transporters in this technology despite the fact that their efficacy is hindered by thermodynamic and heat storage restrictions. The use of EPCMS as a working fluid instead of typical aqueous solutions allows solar thermal systems to utilize latent heat which results in more flexibility and higher overall performance of flat-plate collectors. Researchers presented several systems designs for implementing EPCMS in solar collectors [65–67]. The isothermal phase change process of EPCMS results in a decrease in thermal energy losses from the working fluid to the environment. To use the EPCMS characteristics, the following conditions must be met: the phase change temperature range for the application must be appropriate, the pumping system must have minimal pressure drops, the EPCMS must be stable under thermal and mechanical stresses to avoid clogging issues. For solar field researchers, improving solar collectors has been always a major concern. The heat loss from solar collectors is reduced by introducing the direct absorption solar collector (DASC) as a replacement for indirect ones. The EPCMS absorbs incoming light directly with a high extension coefficient as illustrated in Figure 9.11. Because of the decreased surface temperature of the solar collector, the heat loss of the DASC is effectively minimized, and the DASC's energy conversion efficiency will be improved. Despite the numerous studies done on fabrication of EPCMS for DASC, only a few studies examine the performance of EPCMS-based DASC [2,68–74]. Ma and Zhang [75] conducted preliminary research on using the EPCM slurry in a DASC. They found that the EPCMS-based DASC increases the collector efficiency. In the early investigations, only the effects of the solar radiation intensity and the flow velocity on the collector efficiency were explored. Despite various studies on nanofluid-based DASC, EPCMS-based DASC research is still in its early stages. Because of variations in solid volume fraction and particle size, the

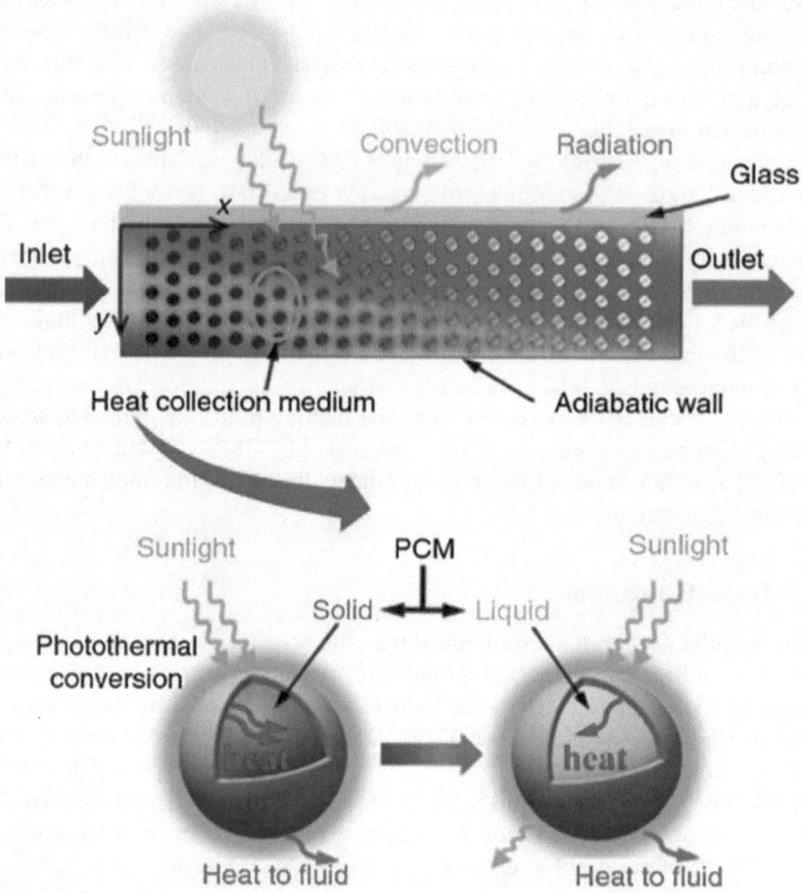

FIGURE 9.11 The concept of direct absorption solar collectors using EPCMS. (Based on Ma et al. [75], reproduced with permission from Elsevier (license number 5177560476408).)

optical property of EPCMS differs from that of nanofluid, prompting more investigation. More research is needed to guide application and research to use slurry in an efficient way by improving the optical and thermal characteristics of heat transfer fluid.

9.6 FUTURE DIRECTIONS

Even though the fundamental applications of micro- and nano-encapsulated PCM fluids have been described in this chapter, there are some intriguing contents that were out of the scope of this chapter [6,76–78]. Despite a high potential for using EPCMS in the thermal applications, there are several major unresolved concerns that must be addressed before commercialization:

- The cost of microencapsulation and EPCMS is predicted to be quite high. A more cost-effective method for slurry production is essential.
- The stability of EPCMS must be improved; this stability refers to as:
  1. EPCM compatibility and stability in the working fluid.
  2. the stability of EPCMS against particle adhesion and deposition on heat transfer components including heat exchangers, channels, pipelines, and so on.
  3. mechanical stability of EPCMs when subjected to hydrodynamic forces in a pump.
- Heat transfer characteristics of EPCM slurries could be enhanced further by improving material properties or adopting designs that offer good contact and minimal heat transfer intervals between the heat source and the impinging micro/nano EPCMs.

As discussed in this chapter, nano-/micro-PCM fluids or EPCMS offer a wide range of potential applications in energy storage and temperature management. Further research on material and system is required to overcome existing issues before their widespread acceptance.

REFERENCES

1. Jurkowska, M.,I. Szczygieł,Review on properties of microencapsulated phase change materials slurries (mPCMS). Applied Thermal Engineering 98 (2016) 365-373.
2. Tahan Latibari, S., J. Eversdijk, R. Cuypers, V. Drosou,M. Shahi,Preparation of Phase Change Microcapsules with the Enhanced Photothermal Performance. Polymers 11 (9) (2019) 1507.
3. Tahan Latibari, S., M. Mehrali, M. Mehrali, A.B.M. Afifi, T.M.I. Mahlia, A.R. Akhiani,H.S.C. Metselaar,Facile synthesis and thermal performances of stearic acid/titania core/shell nanocapsules by sol–gel method. Energy 85 (2015) 635-644.
4. Tahan Latibari, S., M. Mehrali, M. Mehrali, T.M. Indra Mahlia,H.S. Cornelis Metselaar,Synthesis, characterization and thermal properties of nanoencapsulated phase change materials via sol–gel method. Energy 61 (2013) 664-672.
5. Tahan Latibari, S., M. Mehrali, M. Mehrali, T.M.I. Mahlia,H.S.C. Metselaar,Fabrication and Performances of Microencapsulated Palmitic Acid with Enhanced Thermal Properties. Energy & Fuels 29 (2) (2015) 1010-1018.
6. Ran, F., Y. Chen, R. Cong,G. Fang,Flow and heat transfer characteristics of microencapsulated phase change slurry in thermal energy systems: A review. Renewable and Sustainable Energy Reviews 134 (2020) 110101.
7. Yamagishi, Y., T. Sugeno, T. Ishige, H. Takeuchi,A.T. Pyatenko. *An evaluation of microencapsulated PCM for use in cold energy transportation medium. in IECEC 96. Proceedings of the 31st Intersociety Energy Conversion Engineering Conference.* 1996. Washington, DC, USA.
8. BASF,https://www.maisonpassive.be/IMG/pdf/Micronal_EN.pdf
9. Mehling, H., J. Leys, C. Glorieux,J. Thoen,Potential new reference materials for caloric measurements on PCM. SN Applied Sciences 3 (2) (2021) 202.
10. Shao, J., J. Darkwa,G. Kokogiannakis,Review of phase change emulsions (PCMEs) and their applications in HVAC systems. Energy and Buildings 94 (2015) 200-217.

11. Chen, B., X. Wang, R. Zeng, Y. Zhang, X. Wang, J. Niu, Y. Li,H. Di,An experimental study of convective heat transfer with microencapsulated phase change material suspension: Laminar flow in a circular tube under constant heat flux. Experimental Thermal and Fluid Science 32 (8) (2008) 1638-1646.
12. Guyer, E.C., *Handbook of applied thermal design.* 1999, Boca Raton, Florida,USA: CRC press.
13. Maxwell, J.C., *A treatise on electricity and magnetism.* Vol. 1. 1873, Oxford, United Kingdom: Clarendon press.
14. Charunyakorn, P., S. Sengupta,S.K. Roy,Forced convection heat transfer in microencapsulated phase change material slurries: flow in circular ducts. International Journal of Heat and Mass Transfer 34 (3) (1991) 819-833.
15. Yamagishi, Y., H. Takeuchi, A.T. Pyatenko,N. Kayukawa,Characteristics of microencapsulated PCM slurry as a heat-transfer fluid. AIChE Journal 45 (4) (1999) 696-707.
16. Yang, R., H. Xu,Y. Zhang,Preparation, physical property and thermal physical property of phase change microcapsule slurry and phase change emulsion. Solar Energy Materials and Solar Cells 80 (4) (2003) 405-416.
17. Alvarado, J.L., C. Marsh, C. Sohn, G. Phetteplace,T. Newell,Thermal performance of microencapsulated phase change material slurry in turbulent flow under constant heat flux. International Journal of Heat and Mass Transfer 50 (9) (2007) 1938-1952.
18. Zhang, G.H.,C.Y. Zhao,Thermal and rheological properties of microencapsulated phase change materials. Renewable Energy 36 (11) (2011) 2959-2966.
19. Liu, C., Z. Ma, J. Wang, Y. Li,Z. Rao,Experimental research on flow and heat transfer characteristics of latent functional thermal fluid with microencapsulated phase change materials. International Journal of Heat and Mass Transfer 115 (2017) 737-742.
20. Cao, V.D., C. Salas-Bringas, R.B. Schüller, A.M. Szczotok, M. Hiorth, M. Carmona, J.F. Rodriguez,A.-L. Kjøniksen,Rheological and thermal properties of suspensions of microcapsules containing phase change materials. Colloid and Polymer Science 296 (5) (2018) 981-988.
21. Dutkowski, K.,J.J. Fiuk,Experimental investigation of the effects of mass fraction and temperature on the viscosity of microencapsulated PCM slurry. International Journal of Heat and Mass Transfer 126 (2018) 390-399.
22. Zhang, G., G. Cui, B. Dou, Z. Wang,M.A. Goula,An experimental investigation of forced convection heat transfer with novel microencapsulated phase change material slurries in a circular tube under constant heat flux. Energy Conversion and Management 171 (2018) 699-709.
23. Wang, X., J. Niu, Y. Li, X. Wang, B. Chen, R. Zeng, Q. Song,Y. Zhang,Flow and heat transfer behaviors of phase change material slurries in a horizontal circular tube. International Journal of Heat and Mass Transfer 50 (13) (2007) 2480-2491.
24. Dutkowski, K.,J.J. Fiuk,Experimental research of viscosity of microencapsulated PCM slurry at the phase change temperature. International Journal of Heat and Mass Transfer 134 (2019) 1209-1217.
25. Vand, V.,Theory of Viscosity of Concentrated Suspensions. Nature 155 (3934) (1945) 364-365.
26. Mulligan, J.C., D.P. Colvin,Y.G. Bryant,Microencapsulated phase-change material suspensions for heat transfer in spacecraft thermal systems. Journal of Spacecraft and Rockets 33 (2) (1996) 278-284.
27. Liu, L., C. Zhu,G. Fang,Numerical evaluation on the flow and heat transfer characteristics of microencapsulated phase change slurry flowing in a circular tube. Applied Thermal Engineering 144 (2018) 845-853.
28. Hashimoto, S., K. Kurazono,T. Yamauchi,Thermal–hydraulic characteristics of ethylene glycol aqueous solutions containing microencapsulated paraffin. Experimental Thermal and Fluid Science 99 (2018) 297-303.

29. Li, L., D. Zou, X. Ma, X. Liu, Z. Hu, J. Guo,Y. Zhu,Preparation and flow resistance characteristics of novel microcapsule slurries for engine cooling system. Energy Conversion and Management 135 (2017) 170-177.
30. Yang, J., D. Hutchins,C.Y. Zhao,Melting behaviour of differently-sized micro-particles in a pipe flow under constant heat flux. International Communications in Heat and Mass Transfer 53 (2014) 64-70.
31. Taherian, H., J.L. Alvarado, K. Tumuluri, C. Thies,C.-H. Park,Fluid Flow and Heat Transfer Characteristics of Microencapsulated Phase Change Material Slurry in Turbulent Flow. Journal of Heat Transfer 136 (6) (2014)
32. Scott, D.A., A. Lamoureux,B.R. Baliga,Modeling and Simulations of Laminar Mixed Convection in a Vertical Pipe Conveying Slurries of a Microencapsulated Phase-Change Material in Distilled Water. Journal of Heat Transfer 135 (1) (2012)
33. Sabbah, R., J. Seyed-Yagoobi,S. Al-Hallaj,Heat Transfer Characteristics of Liquid Flow With Micro-Encapsulated Phase Change Material: Experimental Study. Journal of Heat Transfer 134 (4) (2012)
34. Zeng, R., X. Wang, B. Chen, Y. Zhang, J. Niu, X. Wang,H. Di,Heat transfer characteristics of microencapsulated phase change material slurry in laminar flow under constant heat flux. Applied Energy 86 (12) (2009) 2661-2670.
35. Zhang, Y., X. Hu,X. Wang,Theoretical analysis of convective heat transfer enhancement of microencapsulated phase change material slurries. Heat and Mass Transfer 40 (1) (2003) 59-66.
36. Bai, F., M. Chen, W. Song, Q. Yu, Y. Li, Z. Feng,Y. Ding,Investigation of thermal management for lithium-ion pouch battery module based on phase change slurry and mini channel cooling plate. Energy 167 (2019) 561-574.
37. Hao, Y.L.,Y.X. Tao,A NUMERICAL MODEL FOR PHASE-CHANGE SUSPENSION FLOW IN MICROCHANNELS. Numerical Heat Transfer, Part A: Applications 46 (1) (2004) 55-77.
38. Zhang, Y., S. Wang, Z. Rao,J. Xie,Experiment on heat storage characteristic of microencapsulated phase change material slurry. Solar Energy Materials and Solar Cells 95 (10) (2011) 2726-2733.
39. Roberts, N.S., R. Al-Shannaq, J. Kurdi, S.A. Al-Muhtaseb,M.M. Farid,Efficacy of using slurry of metal-coated microencapsulated PCM for cooling in a micro-channel heat exchanger. Applied Thermal Engineering 122 (2017) 11-18.
40. Zhang, X., X. Kong, G. Li,J. Li,Thermodynamic assessment of active cooling/heating methods for lithium-ion batteries of electric vehicles in extreme conditions. Energy 64 (2014) 1092-1101.
41. Rao, Y., F. Dammel, P. Stephan,G. Lin,Flow frictional characteristics of microencapsulated phase change material suspensions flowing through rectangular minichannels. Science in China Series E: Technological Sciences 49 (4) (2006) 445-456.
42. Rao, Y., F. Dammel, P. Stephan,G. Lin,Convective heat transfer characteristics of microencapsulated phase change material suspensions in minichannels. Heat and Mass Transfer 44 (2) (2007) 175-186.
43. Dammel, F.,P. Stephan,Heat Transfer to Suspensions of Microencapsulated Phase Change Material Flowing Through Minichannels. Journal of Heat Transfer 134 (2) (2011)
44. Kong, M.-S., K. Yu, J.L. Alvarado,W. Terrell, Jr.,Thermal Performance of Microencapsulated Phase Change Material Slurry in a Coil Heat Exchanger. Journal of Heat Transfer 137 (7) (2015)
45. Doruk, S., O.N. Şara, A. Karaipekli,S. Yapıcı,Heat transfer performance of water and Nanoencapsulated n-nonadecane based Nanofluids in a double pipe heat exchanger. Heat and Mass Transfer 53 (12) (2017) 3399-3408.
46. Hasan, M.I.,Numerical investigation of counter flow microchannel heat exchanger with MEPCM suspension. Applied Thermal Engineering 31 (6) (2011) 1068-1075.

47. Wu, W., H. Bostanci, L.C. Chow, Y. Hong, C.M. Wang, M. Su,J.P. Kizito,Heat transfer enhancement of PAO in microchannel heat exchanger using nano-encapsulated phase change indium particles. International Journal of Heat and Mass Transfer 58 (1) (2013) 348-355.
48. Deshmukh, R.S.,Data Center Cooling Augmentation Using Micro-Encapsulated Phase Change Material. (2014)
49. Kurnia, J.C., A.P. Sasmito, S.V. Jangam,A.S. Mujumdar,Heat Transfer in Coiled Square Tubes for Laminar Flow of Slurry of Microencapsulated Phase Change Material. Heat Transfer Engineering 34 (11-12) (2013) 994-1007.
50. Heydarian, R., M.B. Shafii, A. Rezaee Shirin-Abadi, R. Ghasempour,M. Alhuyi Nazari,Experimental investigation of paraffin nano-encapsulated phase change material on heat transfer enhancement of pulsating heat pipe. Journal of Thermal Analysis and Calorimetry 137 (5) (2019) 1603-1613.
51. Zhai, X.Q., X.L. Wang, T. Wang,R.Z. Wang,A review on phase change cold storage in air-conditioning system: Materials and applications. Renewable and Sustainable Energy Reviews 22 (2013) 108-120.
52. Diaconu, B.M., S. Varga,A.C. Oliveira,Experimental assessment of heat storage properties and heat transfer characteristics of a phase change material slurry for air conditioning applications. Applied Energy 87 (2) (2010) 620-628.
53. Wang, X., J. Niu,A.H.C. van Paassen,Raising evaporative cooling potentials using combined cooled ceiling and MPCM slurry storage. Energy and Buildings 40 (9) (2008) 1691-1698.
54. Ali, S.,M. Mustafa,Barriers facing Micro-encapsulated Phase Change Materials Slurry (MPCMS) in Photovoltaic Thermal (PV/T) application. Energy Reports 6 (2020) 565-570.
55. Eisapour, M., A.H. Eisapour, M.J. Hosseini,P. Talebizadehsardari,Exergy and energy analysis of wavy tubes photovoltaic-thermal systems using microencapsulated PCM nano-slurry coolant fluid. Applied Energy 266 (2020) 114849.
56. Jia, Y., C. Zhu,G. Fang,Performance optimization of a photovoltaic/thermal collector using microencapsulated phase change slurry. International Journal of Energy Research 44 (3) (2020) 1812-1827.
57. Zhou, J., W. Zhong, D. Wu, Y. Yuan, W. Ji,W. He,A Review on the Heat Pipe Photovoltaic/Thermal (PV/T) System. Journal of Thermal Science 30 (5) (2021) 1469-1490.
58. Eisapour, A.H., M. Eisapour, M.J. Hosseini, A.H. Shafaghat, P. Talebizadeh Sardari,A.A. Ranjbar,Toward a highly efficient photovoltaic thermal module: Energy and exergy analysis. Renewable Energy 169 (2021) 1351-1372.
59. Yazdanifard, F., M. Ameri,R.A. Taylor,Numerical modeling of a concentrated photovoltaic/thermal system which utilizes a PCM and nanofluid spectral splitting. Energy Conversion and Management 215 (2020) 112927.
60. Ali, H.M.,Recent advancements in PV cooling and efficiency enhancement integrating phase change materials based systems – A comprehensive review. Solar Energy 197 (2020) 163-198.
61. Liu, L., Y. Jia, Y. Lin, G. Alva,G. Fang,Performance evaluation of a novel solar photovoltaic–thermal collector with dual channel using microencapsulated phase change slurry as cooling fluid. Energy Conversion and Management 145 (2017) 30-40.
62. Qiu, Z., X. Zhao, P. Li, X. Zhang, S. Ali,J. Tan,Theoretical investigation of the energy performance of a novel MPCM (Microencapsulated Phase Change Material) slurry based PV/T module. Energy 87 (2015) 686-698.
63. Qiu, Z., X. Ma, X. Zhao, P. Li,S. Ali,Experimental investigation of the energy performance of a novel Micro-encapsulated Phase Change Material (MPCM) slurry based PV/T system. Applied Energy 165 (2016) 260-271.

64. Yu, Q., A. Romagnoli, R. Yang, D. Xie, C. Liu, Y. Ding,Y. Li,Numerical study on energy and exergy performances of a microencapsulated phase change material slurry based photovoltaic/thermal module. Energy Conversion and Management 183 (2019) 708-720.
65. Huang, M.J., P.C. Eames, S. McCormack, P. Griffiths,N.J. Hewitt,Microencapsulated phase change slurries for thermal energy storage in a residential solar energy system. Renewable Energy 36 (11) (2011) 2932-2939.
66. Serale, G., Y. Cascone, A. Capozzoli, E. Fabrizio,M. Perino,Potentialities of a Low Temperature Solar Heating System Based on Slurry Phase Change Materials (PCS). Energy Procedia 62 (2014) 355-363.
67. Serale, G., E. Fabrizio,M. Perino,Design of a low-temperature solar heating system based on a slurry Phase Change Material (PCS). Energy and Buildings 106 (2015) 44-58.
68. Wang, Z., J. Qu, R. Zhang, X. Han,J. Wu,Photo-thermal performance evaluation on MWCNTs-dispersed microencapsulated PCM slurries for direct absorption solar collectors. Journal of Energy Storage 26 (2019) 100793.
69. Ma, X., Y. Liu, H. Liu, L. Zhang, B. Xu,F. Xiao,Fabrication of novel slurry containing graphene oxide-modified microencapsulated phase change material for direct absorption solar collector. Solar Energy Materials and Solar Cells 188 (2018) 73-80.
70. Liu, J., L. Chen, X. Fang,Z. Zhang,Preparation of graphite nanoparticles-modified phase change microcapsules and their dispersed slurry for direct absorption solar collectors. Solar Energy Materials and Solar Cells 159 (2017) 159-166.
71. Xu, B., J. Zhou, Z. Ni, C. Zhang,C. Lu,Synthesis of novel microencapsulated phase change materials with copper and copper oxide for solar energy storage and photothermal conversion. Solar Energy Materials and Solar Cells 179 (2018) 87-94.
72. Ma, F.,P. Zhang,Heat Transfer Characteristics of a Volumetric Absorption Solar Collector using Nano-Encapsulated Phase Change Slurry. Heat Transfer Engineering 39 (17-18) (2018) 1487-1497.
73. Gao, G., T. Zhang, S. Jiao,C. Guo,Preparation of reduced graphene oxide modified magnetic phase change microcapsules and their application in direct absorption solar collector. Solar Energy Materials and Solar Cells 216 (2020) 110695.
74. Yuan, K., J. Liu, X. Fang,Z. Zhang,Crafting visible-light-absorbing dye-doped phase change microspheres for enhancing solar-thermal utilization performance. Solar Energy Materials and Solar Cells 218 (2020) 110759.
75. Ma, F.,P. Zhang,Performance investigation of the direct absorption solar collector based on phase change slurry. Applied Thermal Engineering 162 (2019) 114244.
76. Yu, Q., X. Chen,H. Yang,Research progress on utilization of phase change materials in photovoltaic/thermal systems: A critical review. Renewable and Sustainable Energy Reviews 149 (2021) 111313.
77. Qiu, Z., P. Li, Z. Wang, H. Zhao,X. Zhao, *PCM and PCM Slurries and Their Application in Solar Systems*, in *Advanced Energy Efficiency Technologies for Solar Heating, Cooling and Power Generation*, X. Zhao and X. Ma, Editors. 2019, Springer International Publishing: Cham. p. 101-141.
78. Omara, A.A.M.,A.A.A. Abuelnour,Improving the performance of air conditioning systems by using phase change materials: A review. International Journal of Energy Research 43 (10) (2019) 5175-5198.

10 Structural Classification of PCM Heat Exchangers

Moghtada Mobedi
Shizuoka University

Chunyang Wang
Institute of Engineering Thermophysics
Chinese Academy of Sciences

CONTENTS

10.1 INTRODUCTION

The rapid development of modern industries requiring high demand for energy, and the continuous increase of greenhouse gas emissions causes the issues of energy to become an important issue and a serious topic for researchers. The energy topic is one of the major concerns of the present century gathering many researchers all around the world together. Many scientists attempt to find effective methods to use renewable energy sources as well as utilize waste heat for solving energy problems. The unstable and/or transient behavior of renewable energy and waste heat sources forces

DOI: 10.1201/9781003213260-10

researchers to find solutions for energy storage. Thermal energy storage becomes an important part of research performing on the use of renewable energy technology systems. Thermal energy storage reduces the mismatch between supply and demand. It can improve the performance and reliability of renewable energy systems and can play an important role in energy conservation. In many industrial applications, the heat which is thrown into the environment can be recovered by thermal energy storage systems and not only the heating of environment can be reduced but also the thermal performance of the systems can be improved. Thermal storage systems can also be used to balance the time gap between the production and consumption of energy such as storage of low-temperature energy during the night and use it during the day when electricity consumption is generally high. There are many methods for energy storage (such as electric or mechanical energy storage); however, due to the wide application of thermal energy, it has attracted the attention of researchers, considerably. Thermal energy storage can be sensible (due to change of temperature), latent type (due to phase change), or chemical (by adsorption or chemical reaction). Among different methods of thermal energy storage, the solid–liquid phase change thermal energy storage (a kind of phase change thermal storage) becomes popular in recent years since the storage of energy in low, moderate, and high temperatures is possible. The amount of stored thermal energy is considerable due to the latent heat and it has a simple principle of operation and long lifetime. These advantages of solid–liquid energy storage have caused many investigations done on the various applications such as:

- Air conditioning industry (to store energy when the electricity price is low and use it when it is expensive),
- Car industry especially for cooling of batteries,
- Cooling of electronic equipment (for sudden power outage),
- Space industry (stabilization of temperature of electronic equipment in satellites).

In spite of many studies on the solid–liquid thermal storage, the design and thermal improvement of solid–liquid thermal energy storage still need further research. A suitable solid–liquid thermal storage system should have high specific thermal storage power, quick response for sensitive applications such as cooling of electronic equipment, and stable outlet temperature for working fluid. Therefore, studies on phase change material (PCM) heat exchangers continue to satisfy the required conditions.

A PCM heat exchanger is a new concept that has become popular in recent years for storing heat and utilization whenever it is required. It is not only used for heating but also cooling applications. Many studies on PCM heat exchangers have been designed and reported in the literature. The suggested PCM heat exchangers have different configurations from single double-pipe heat exchangers to multi-pipe multi-circuit shell and tube heat exchangers. The aim of this chapter is to classify PCM heat exchangers systematically and give hints about the application of each group as well as their advantages and disadvantages. This classification provides the following advantages:

a. It provides an easy way to give a name to a PCM heat exchanger and call a PCM heat exchanger by its name helping to image it.

b. It provides opportunities to compare PCM heat exchangers in a group among themselves and find the best design as well as better methods for improvement of thermal storage.
c. It provides to specify the advantages and disadvantages of each group of PCM heat exchangers and discuss them.

These advantages motivated authors to review the reported PCM heat exchangers and attempt to classify the designed PCM heat exchangers properly.

10.2 DEFINITION OF A PCM HEAT EXCHANGER

There are important differences between the PCM and classical heat exchangers as can be seen in Figure 10.1:

a. Classical heat exchanges are used to transfer heat from a fluid (or fluids) to another fluid (or other fluids). However, the aim of a PCM heat exchanger is to receive heat from a fluid (or fluids), store it efficiently, and release it to another fluid (or other fluids) when it is required.
b. Most of the classical heat exchangers operate simultaneously meaning that the releasing and receiving of heat by two fluids occur at the same time. However, PCM heat exchangers operate intermittently. Heat is received from a fluid and then it is stored and later it is released to another fluid.
c. In the classical heat exchangers, when the constant inlet temperatures for both fluids exist the outlet temperatures of both fluids are also steady and do not change over time. However, in PCM heat exchangers, even the inlet temperatures are constant, the outlet temperatures change over time.

All the above differences show that the classification and design methods employed for a classical heat exchanger cannot be used for a PCM heat exchanger and special methods must be developed for these kinds of heat exchangers.

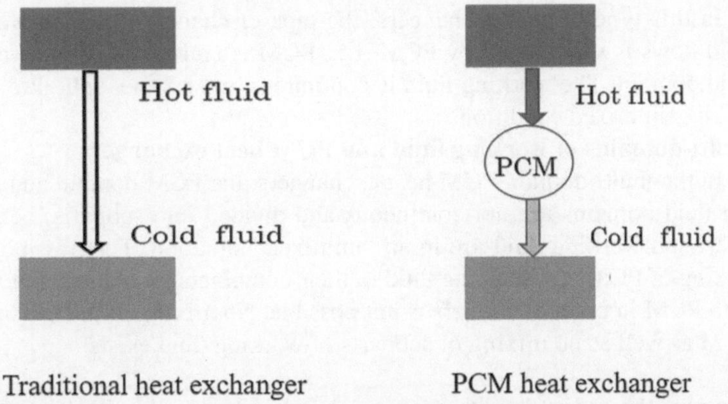

FIGURE 10.1 Schematic principle working of a traditional and a PCM heat exchanger.

10.3 REPORTED REVIEWS ON PCM HEAT EXCHANGERS

Due to the importance of thermal energy storage and advantages of solid–liquid phase change, many reviews were done and reported in the literature. The reported reviewers were illustrated in Table 10.1 and summarized in this section.

Among many reviews that have been done on the solid–liquid thermal storage, those who discuss PCM heat exchangers somehow are selected and presented in Table 10.1. These are the review studies found by authors and there might be more reported studies in the literature. As it can be seen, most of the studies discuss PCM heat exchangers as well as enhancement of heat transfer due to slow heat transfer in most of the phase change materials. The number of references in the studies of Table 10.1 shows the number of studies in this area, and it is very high referring to the importance of the solid–liquid phase change. The interesting point is that many studies have been done on PCM heat exchangers for cooling or heating of the building since a large amount of energy in many countries is consumed in heating or air condition systems.

10.4 STRUCTURAL CLASSIFICATION OF PCM HEAT EXCHANGER

In this study, around 150 papers were reviewed on solid–liquid phase change thermal storage and an attempt was done to classify the studied and reported PCM heat exchangers. The result of the performed review is shown in Figure 10.2. PCM heat exchangers can be classified into three groups as described below:

a. **Working fluid-embedded PCM heat exchangers**

   In this type of heat exchangers, the channel or pipe of the working fluid (i.e., a container of the working fluid) is embedded inside the PCM domain, i.e., the working fluid is surrounded by PCM and PCM is in "mixed" situation. Generally, the PCM domain is continuous and is not divided into parts or subsections. The possibility of mixing of PCM in the entire container of PCM with each other exists; however, the working fluid can be in mixed or unmixed conditions.

b. **PCM-embedded PCM heat exchangers**

   In this type of heat exchangers, the pipe or channel in which working fluid flows is surrounded by PCM, i.e., PCM is embedded in the working fluid domain. The working fluid is continuous and it is not split into parts, i.e., it is in mixed condition.

c. **Multi-domains of working fluid and PCM heat exchangers**

   In the multi-domain PCM heat exchangers, the PCM domain and working fluid domains are not continuous and divided into subparts, i.e., both PCM and working fluid are in an "unmixed" situation. The possibility of mixing of PCM and working fluid in their containers (or channels of pipes) with PCM in other containers is not possible. No mixing of the subparts of PCM as well as no mixing of subparts of working fluid exists.

Each group of the above classification can be classified within itself and they are explained in the following sections.

TABLE 10.1
The List of the Review Studies on PCM Solid–Liquid Thermal Storage in Which PCM Heat Exchanger is Included

Researcher (Publication Year)	PCM Materials	PCM Heat Exchangers	PCM Heat Exchangers for Buildings	Different Applications	Enhancement Heat Transfer	Number of Referenced Studies
Sharma et al. [1] (2009)	X	X		X		155
Jegadheeswaran and Pohekar [2] (2009)		X			X	104
Raj and Verlaj [3] (2010)		X	X	X		68
Agyenim et al. [4] (2010)	X	X			X	135
Fan and Khodadadi [5] (2011)		X			X	40
Tiat and del Barrio [6] (2011)		X		X	X	45
Zhou et al. [7] (2012)		X	X			111
Oró et al. [8] (2012)	X	X		X	X	150
Al-Abidi et al. [9] (2012)	X	X		X	X	79
Osterman et al. [10] (2012)	X	X	X		X	69
Liu et al. [11] (2012)	X	X		X	X	76
Waqas and Din [12] (2013)		X	X			93
Kuravi et al. [13] (2013)		X		X	X	266
Jankowski and McCluskey [14] (2014)		X		X	X	175
Anuar Sharif et al. [15] (2015)		X	X		X	81
Akeiber et al. [16] (2016)		X	X			121
Souayfane et al. [17] (2016)		X	X			151
Kapsalis and Karamanis [18] (2016)	X	X	X		X	154
Gracia and Cabeza [19] (2017)		X				82
Tay et al. [20] (2017)			X		X	87
Dinker et al. [21] (2017)	X	X				115

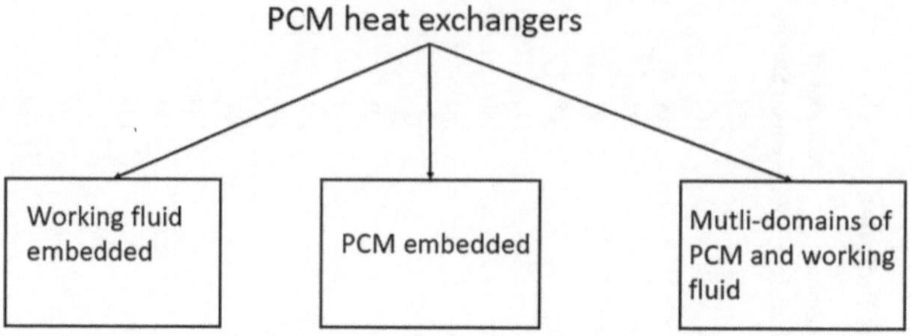

FIGURE 10.2 General classification of PCM Heat exchangers.

10.4.1 Working Fluid-Embedded PCM Heat Exchangers

In this type of PCM heat exchangers, PCM is in a container, and the working fluid flows in the container of PCM. The PCM is in "mixed" condition while working fluid can be in "mixed" or "unmixed" condition. Our review of the designed and reported PCM embedded heat exchangers showed that it is possible to classify it into two main groups as shell and tube and double-plate PCM heat exchangers as shown in Figure 10.3. The shell and tube type heat exchanger can be classified into four groups as shown in Figure 10.3. All of these classifications are described in the following sections.

10.4.1.1 Shell and Tube PCM Heat Exchanger

As it is well known, this type of PCM heat exchangers consists of a shell and a tube (or tubes). PCM is placed in the shell side, i.e., the tube in which working fluid flows is embedded in the shell side. Shell and tube PCM heat exchangers are very popular among PCM heat exchangers and many researchers designed and investigated them. Our review on the PCM shell and tube heat exchangers showed that it is possible to classify them into four groups as (a) single circuit single pass (or double-pipe PCM heat exchangers), (b) single circuit multi-pass, (c) multi-circuit single pass and (d) multi-circuit multi-pass. These kinds of PCM heat exchangers are described below:

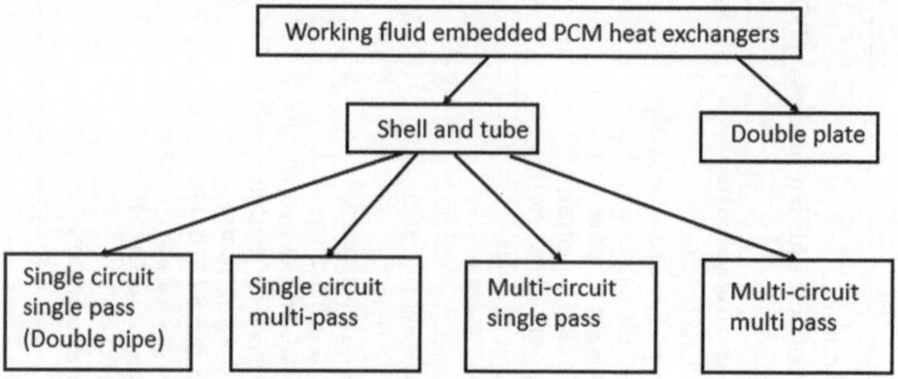

FIGURE 10.3 Classification of working fluid-embedded PCM heat exchangers

a. Single circuit—single pass (double pipe)

In double-pipe PCM heat exchangers, the PCM materials are located in the shell side and the working fluid flows through a pipe placed within the shell side. There is one path for working fluid which means that the charge and discharge fluid paths are the same. The outer surface of the pipe side is generally supported by fins to enhance heat transfer between the pipe and PCM. The outer surface of the shell side is thermally insulated. Figures 10.4 and 10.5 show some of the designed and studied double-pipe PCM heat exchangers. In a simple double-pipe PCM heat exchanger, both the PCM in the container and working fluid are in "mixed" condition as seen in Figure 10.4a. However, using special fins for enhancement of heat transfer may cause "unmixed" conditions for the PCM side. This "unmixed" condition is not due to different domains, and the PCM domain is unique.

There are studies in which designers changed the shape of shells or tubes to improve the storage power of PCM heat exchangers. For instance, the shape of the shell can be prism or conical as it is shown in Figure 10.5a and b. Furthermore, the special shapes of tubes were also designed as can be seen in Figure 10.5c in order to enhance heat transfer both in working fluid and PCM sides.

b. Single circuit–multi-pass

In the single-path (double-pipe) working fluid-embedded heat exchangers, generally the length of the pipe and shell are the same. In the single-circuit multi-pass working fluid-embedded PCM heat exchanger, the length of pipe is longer than shell side since the working fluid travels for more than one tour in the shell side. In other words, the heat transfer area between the working fluid and PCM is greater than single-circuit single-pass PCM heat exchangers. It should be mentioned that in this kind of PCM heat exchanger, one inlet and one outlet only exist and the path of charge and discharge

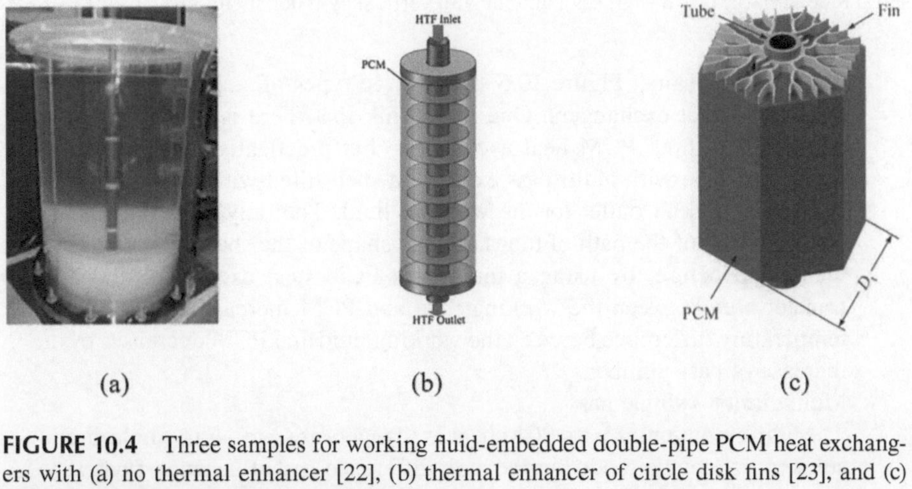

FIGURE 10.4 Three samples for working fluid-embedded double-pipe PCM heat exchangers with (a) no thermal enhancer [22], (b) thermal enhancer of circle disk fins [23], and (c) thermal enhancer of branch axial fin [24].

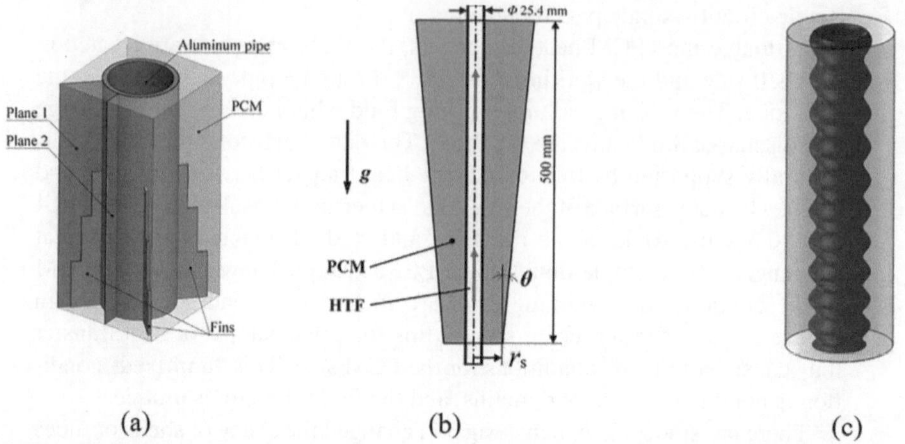

FIGURE 10.5 Three samples for working fluid-embedded double-pipe PCM heat exchangers with different shell and/or pipe configurations: (a) prism sell [25], (b) conical shell [26], and (c) wavy pipe [27].

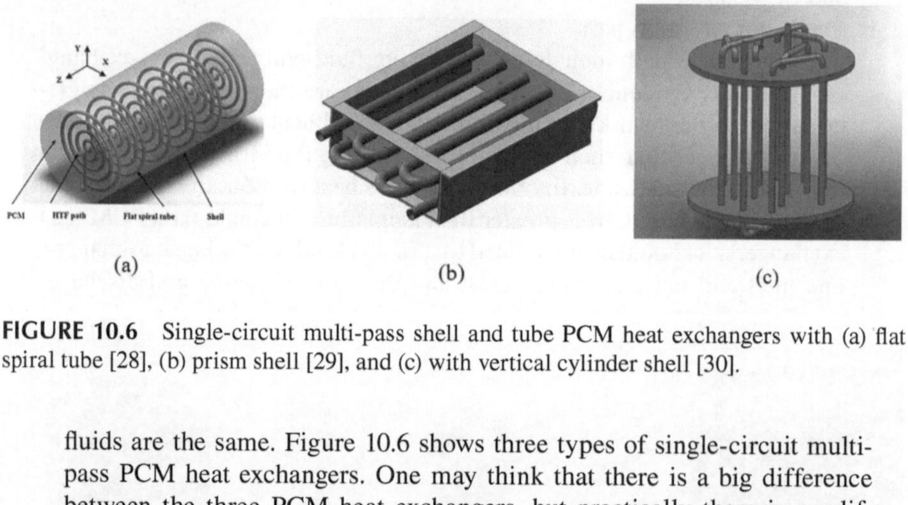

FIGURE 10.6 Single-circuit multi-pass shell and tube PCM heat exchangers with (a) flat spiral tube [28], (b) prism shell [29], and (c) with vertical cylinder shell [30].

fluids are the same. Figure 10.6 shows three types of single-circuit multi-pass PCM heat exchangers. One may think that there is a big difference between the three PCM heat exchangers, but practically there is no difference. A pipe with multi-pass exists in a shell filled with PCM and there is an inlet and an outlet for the working fluid. The only difference is the configuration of the path of tubes and the shape of the shell side. As it was mentioned before, by using a multi-pass PCM heat exchanger, the heat transfer area between the working fluid and PCM increases; however, the temperature difference between the working fluid and PCM decreases by an increase of pass number.

c. Multi circuit—single pass

Multi-circuit multi-pass PCM heat exchangers are similar to single circuit single pass (double pipe), but the only difference is that instead of one tube many tubes exist in the shell side. There are two or more inlets and outlets; so,

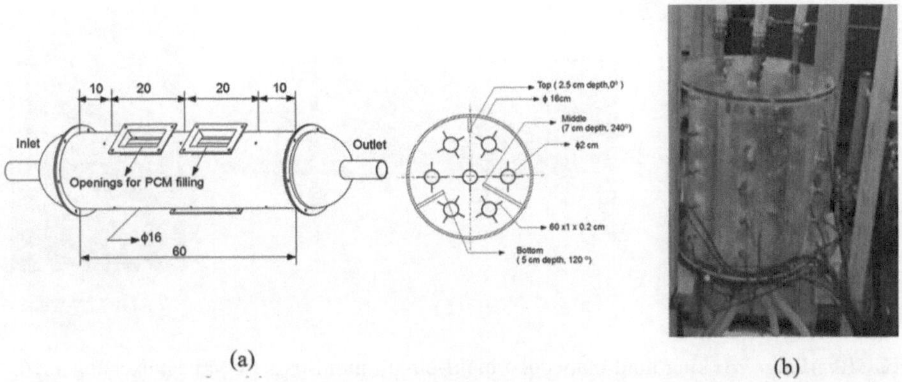

FIGURE 10.7 Two samples of multi-circuit single pass shell and tube PCM heat exchangers: (a) assisted with fin [31] and (b) no thermal enhancer [32].

it is called multi circuits. The total mass flow rate of the working fluid flowing in a pipe is divided into the number of pipes. The outer side of the tubes is generally supported by fins to increase the heat transfer area between the working fluid and PCM. The expectation for a multi-circuit single-pass PCM heat exchanger is more uniform melting and freezing processes in the shell side due to increase of heat transfer area and distributing of pipes in entire shell side. The most important point for a multi-circuit PCM heat exchanger is that the possibility of two working flows (one for charge and the other for discharge) exists. Hence, if the remaining (molecules) of charging working on the internal surface of a pipe effects the discharge fluid (or vice versa), it is possible to split pipes for charge and discharge. Figure 10.7 shows two types of multi-circuit single-pass PCM heat exchangers.

d. Multi circuit—multi pass

For the multi-circuit multi-pass shell and tube PCM heat exchangers, the tubes pass through the shell for more than one. The fluid flowing distance is longer than the length of the PCM heat exchanger. More heat transfer area is achieved; however, the temperature difference between the working fluid and PCM is reduced. Since it is multi-circuit, the total mass flow rate of working fluid in a pipe is divided into the number of circuits. Figure 10.8 shows three multi-circuits multi-pass PCM heat exchangers. Although there are differences between the shell sides, their structure is thermally almost the same. The shell of Figure 10.8a and b is a prism, while it is a cylinder for Figure 10.8b. Figure 10.8a and c have two circuits, while Figure 10.8b has four circuits.

10.4.1.2 Double-Plate Heat Exchanger

Double-plate heat exchanger is similar to a double-pipe heat exchanger; however, the working fluid flows in a channel sandwiched by PCM plate containers as discussed in Ref. [35]. There are not many studies on the double-plate heat exchanger and most designers prefer double-pipe PCM heat exchangers rather than double-plate PCM heat exchangers.

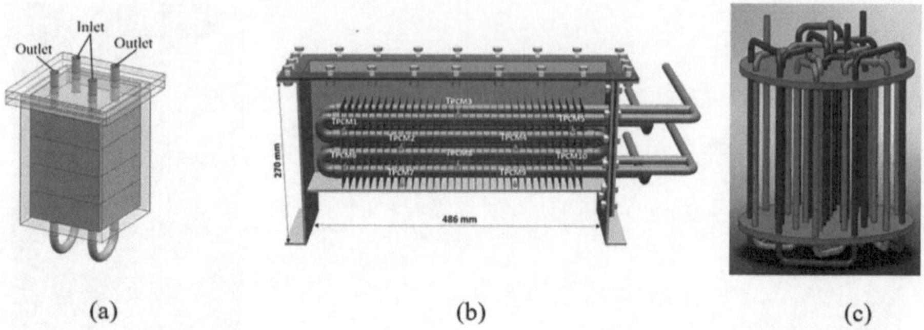

FIGURE 10.8 Working fluid-embedded multi-circuit multi-pass PCM heat exchangers: (a) two circuits [33] (b) four circuit [34], and (c) two circuits [30].

10.4.2 PCM-Embedded-Type PCM Heat Exchanger

In PCM-embedded-type PCM heat exchangers, PCM is in a container which is surrounded by a working fluid (or fluids). The working fluid flows around the PCM container and heat is transferred from the working fluid to the outer surface of the container and then to PCM or vice versa. The classification of PCM embedded type in working fluid is shown in Figure 10.9. This kind of PCM heat exchangers can be classified into shell and tube, cross flow, triplex, and packed bed. Similar to working fluid-embedded-type PCM heat exchangers, the shell and tube type PCM heat exchanger can be divided into double-pipe or multi-pipe heat exchangers. These kinds of PCM heat exchangers are described in the following sections.

10.4.2.1 Shell and Tube Type PCM Heat Exchangers

In this kind of shell and tube PCM heat exchangers, the PCM is in a closed pipe while working fluid flows in the shell side. The beginning and end sides of the pipe in which PCM exists are closed. As it was mentioned above, it can be classified into two types as double pipe and multi-pipe which are explained in the following sections.

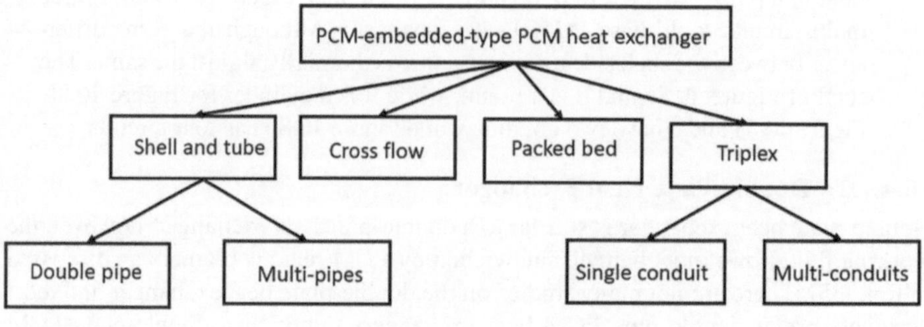

FIGURE 10.9 Classification for PCM-embedded-type PCM heat exchangers.

10.4.2.1.1 Double Pipe

PCM embedded double-pipe PCM heat exchanger is a simple structure in which PCM is in a closed pipe and fluid flows longitudinally around the pipe. The number of studies on this kind of PCM heat exchangers is too limited. Figure 10.10 shows two samples of PCM embedded double-pipe PCM heat exchangers. Figure 10.10a was studied for comparison of PCM embedded double-pipe PCM heat exchanger with working fluid-embedded double-pipe heat exchanger. Figure 10.10b shows this type of PCM heat exchangers when the flow path is zig-zag. In this kind of PCM heat exchanger (similar to working fluid-embedded double-pipe heat exchangers) both PCM and working fluid are in "mixed" condition.

10.4.2.1.2 Multi-Pipe

In the PCM embedded multiple PCM heat exchangers, there are many closed cylindrical containers in which PCM exists and fluid flows around the pipes longitudinally. The number of studies on this kind of PCM heat exchangers is too limited, and Figure 10.11 shows a sample of PCM embedded multiple pipe PCM heat exchangers. As it can be seen from Figure 10.11, the PCM is in "unmixed" condition while the working fluid is in a mixed situation.

10.4.2.2 Cross Flow

In PCM-embedded cross-flow PCM heat exchangers, PCM is in the closed pipes, and working fluid flows perpendicularly to the pipes. The arrangement of pipes may be inline or staggered. Since there is one channel for working fluid, the path for charging and discharging fluids are identical. Figure 10.12 shows samples of cross-flow PCM heat exchangers. Similar to of PCM embedded multi-pipe shell and tube PCM heat exchanger, the PCM is "unmixed" while the working fluid is mixed in this kind of PCM heat exchanger.

10.4.2.3 Capsule PCM Packed Bed

In this type of PCM heat exchangers, a large number of capsules containing PCM are located in a cylindrical or rectangular container as shown in Figure 10.13. The working fluid flows in the voids between the PCM capsules from an inlet to an outlet. The walls of the container are insulated thermally. Heat transfer in the capsules is by conduction and natural convection during melting and solidification periods.

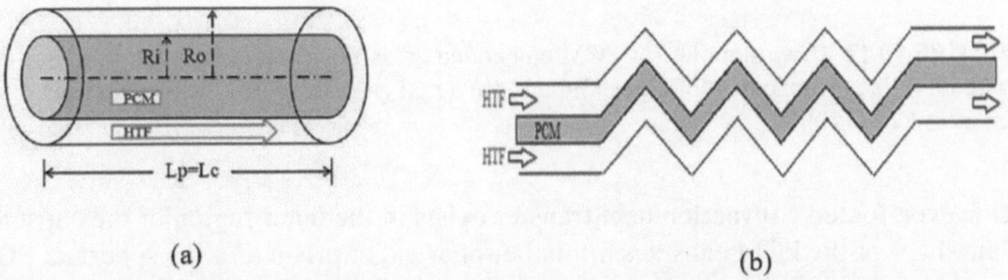

FIGURE 10.10 PCM-embedded-type double-pipe PCM heat exchanger: (a) straight type [36] and (b) zig-zag type [37].

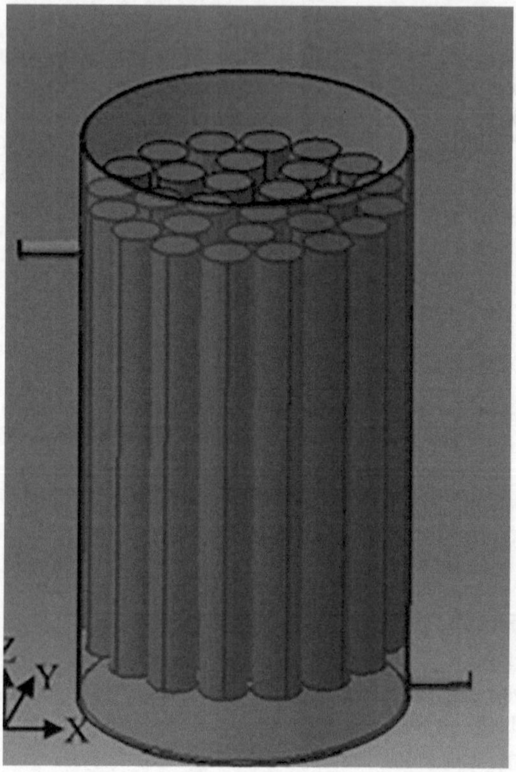

FIGURE 10.11 A view of PCM embedded multi-pipe shell and tube PCM heat exchanger [38].

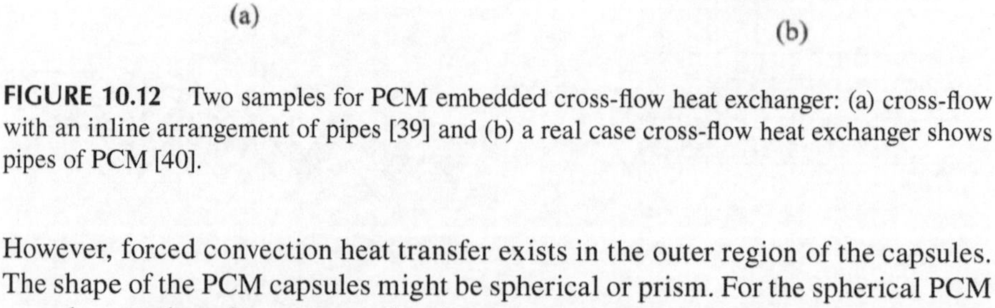

FIGURE 10.12 Two samples for PCM embedded cross-flow heat exchanger: (a) cross-flow with an inline arrangement of pipes [39] and (b) a real case cross-flow heat exchanger shows pipes of PCM [40].

However, forced convection heat transfer exists in the outer region of the capsules. The shape of the PCM capsules might be spherical or prism. For the spherical PCM capsules, a point-wise contact exists between the capsules. However, for the prism capsules, there is no contact between the PCM capsules. The passage of working fluid is identical for the charge and discharge process; however, the direction

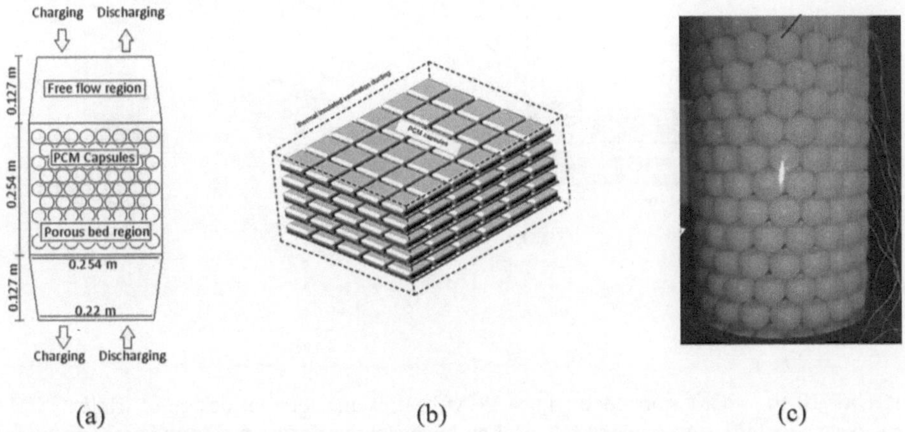

FIGURE 10.13 Packed bed heat exchangers with PCM capsules: (a) a schematic view of a bed with spherical capsules [41], (b) a bed with prism capsules [42], and (c) a view of a real bed with spherical capsules [43].

of flow might be opposite. As a typical behavior of the packed bed, the phase change process in the capsules is started from the inlet and continuous to the outlet region over time. PCM is in "unmixed" condition while working fluid is in "mixed condition".

10.4.2.4 Triplex

The name of Triplex for a PCM heat exchanger may not be very technical, but it is used by many researchers. In this kind of PCM heat exchanger, PCM is between in a conduit which is in contact with working fluid from inner and outer surfaces. Therefore, the PCM in the conduit is melted from both sides which are faster than double-pipe PCM heat exchangers. Triplex PCM heat exchangers can be divided into the following two types:

a. Single-conduit triplex
b. Shell and tube triplex

These kinds of triplex PCM heat exchangers are described in the following sections.

10.4.2.4.1 Single-Conduit Triplex

Samples of the single-conduit triplex PCM are shown in Figure 10.14. As it can be seen, the PCM layer can melt both from inside and outside surfaces. Although it is possible to use a single working fluid for this kind of PCM heat exchangers, the possibility of the use of two working fluids also exists. For instance, the freezing process of PCM can be done by a refrigerant at low temperature and melting can be achieved by water flow in the outer channel. Since heat transfer occurs from both inner and outer surfaces of the PCM conduit, researchers support both surfaces by fins (Figure 10.14b). The PCM in the conduit is in "mixed" condition.

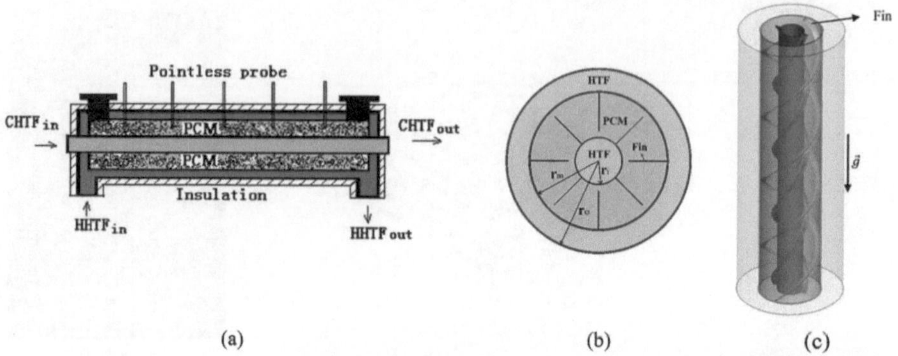

FIGURE 10.14 PCM embedded triplex PCM heat exchangers: (a) design of triplex PCM heat exchanger [44], (b) a triplex PCM heat exchanger supported by fins from inner and outer surfaces [45], and (c) a triplex vertical PCM heat exchange with twisted fins [46].

10.4.2.4.2 Multi-Conduits Triplex

Two samples of shell and tube triplex PCM heat exchanger are shown in Figure 10.15. As it can be seen, there is more than one tube in the PCM conduit and working fluid can flow in those tubes as well as in the shell side and fast melting of PCM with high heat storage (due to many triplex tubes) can be provided. In addition to the design of Figure 10.15a, the design of Figure 10.15b was also reported in the literature. There are many PCM conduits in which the working fluid flows in the center of the pipe. The banks of those conduits are in a shell with a double pass. Hence, a fast melting or freezing is expected in these kinds of triplex PCM heat exchangers.

10.4.2.5 Multi-Domains

In the multi-domains PCM heat exchanger, both the PCM boxes and fluids are unmixed. The working fluid flows between the plates in which PCM exists. The PCMs in the boxes are also unmixed since they are in separate containers. One

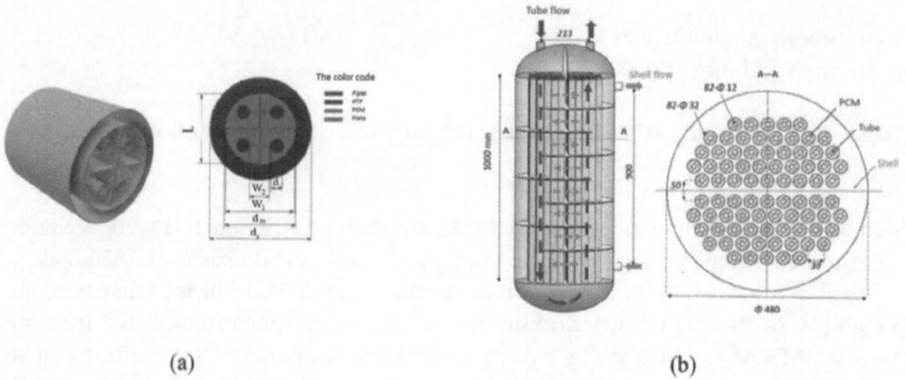

FIGURE 10.15 Two types of triplex shell and tube heat exchangers: (a) one conduit triplex shell and tube [47] and (b) multi-conduit triplex shell and tube [48].

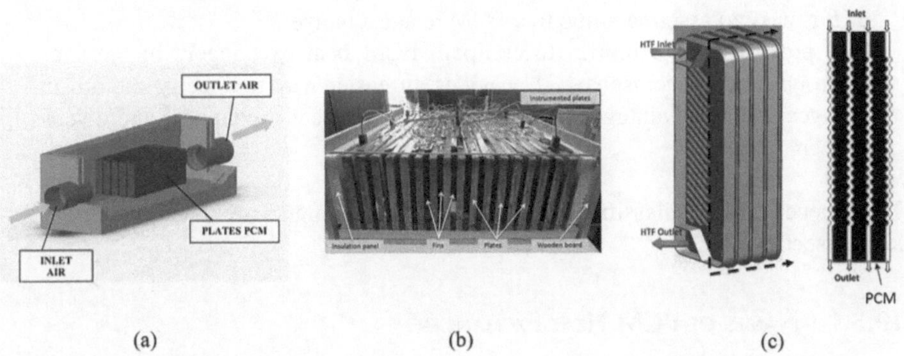

FIGURE 10.16 Three samples for multi-domain PCM heat exchangers: (a) and (b) multi-plate heat exchanger [49,50] and (c) application of traditional plate type heat exchanger [51].

interesting application of a multi-domain heat exchanger is the use of a traditional multi-plate heat exchanger. It can be especially appropriate when the working fluid is liquid. It flows between the plated filled with PCM and is located in parallel. Figure 10.16 shows three types of multi-domain PCM heat exchangers. Both PCM and working fluids are not in mixed condition, and PCM in the containers is melted/frozen separately.

10.5 RESULT AND DISCUSSION

The full classification of the PCM heat exchanger is shown in Figure 10.17. This classification of PCM heat exchangers provides two advantages for the researchers in this field as follows:

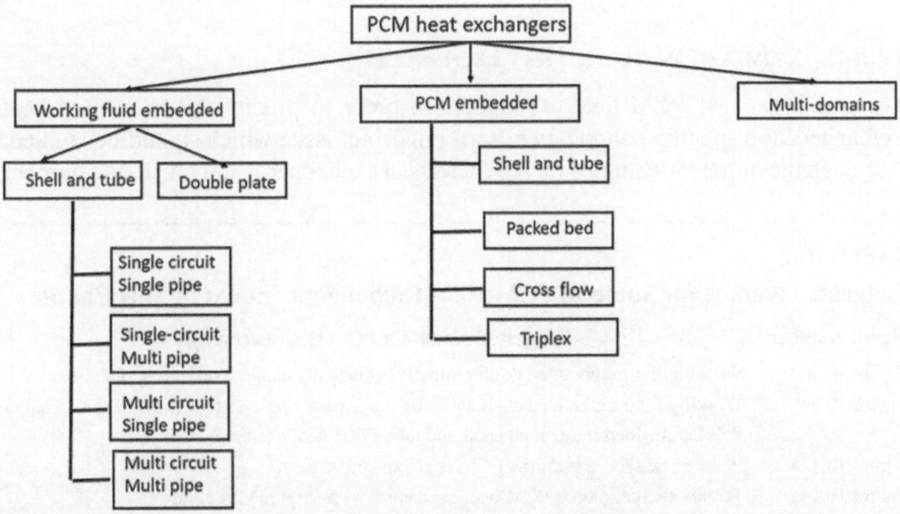

FIGURE 10.17 Classification of PCM heat exchangers.

a. It is easy to assign a name to a PCM heat exchanger.
b. It provides opportunities to compare PCM heat exchangers in the same group among themselves. It provides an easier way to specify the advantages and disadvantages of each type of PCM heat exchangers and discuss them.

These benefits of the classification of PCM heat exchangers are discussed in the following sections.

10.5.1 Names of PCM Heat Exchangers

Based on the classification shown in Figure 10.17, it is possible to assign names to most of the designed PCM heat exchangers. Table 10.2 shows the suggested names for some of the PCM heat exchangers shown in different figures of this study. As it can be seen, the given names can successfully represent the structure of the PCM heat exchanger. For instance, if the name of a heat exchanger as "Fluid-embedded shell and tube single-pipe single-circuit PCM heat exchanger with rectangular shell" is considered, the statement of "fluid embedded" creates an image of a PCM heat exchanger in which the working fluid flowing a pipe and it is surrounded by PCM. The statement of "shell and tube" makes an image of a pipe or pipes which are in a shell filled with PCM. Similar to the traditional single-phase heat exchangers, the statement of "single circuit" provides an image that the PCM heat exchanger has an inlet and an outlet for the working fluid. Finally, a rectangular shell gives an idea about the shape of a shell. Hence, the reader can easily imagine a heat exchanger from its name. Our literature survey showed that for the same PCM heat exchangers, many names are given by researchers, and standardization for the names of PCM heat exchangers is needed. This is one of the reasons motivated authors to do the present study.

10.5.2 Comparison of the Heat Exchangers

The comparison of PCM heat exchangers in order to predict the best PCM heat exchanger for a specified application is an important issue which should be studied. Although the number of studies on the PCM heat exchanger is too much, the number

TABLE 10.2
Suggested Names for Some of PCM Heat Exchangers Shown in This Chapter

Figure Number	Suggested Name for PCM Heat Exchanger
Figure 10.4b	Working fluid-embedded double-pipe PCM heat exchanger with circle disk fins
Figure 10.6c	Working fluid-embedded shell and tube multi-pass single-circuit PCM heat exchanger
	PCM embedded multi-pipe shell and tube PCM heat exchanger
Figure 10.12a	PCM embedded cross-flow PCM heat exchanger
Figure 10.14a	Triplex shell and tube PCM heat exchanger with single PCM conduit
Figure 10.15a	Multi-domains single-circuit single-pass PCM heat exchanger

of studies on the comparison of different types of PCM heat exchangers is limited. Furthermore, topics on the advantages and disadvantages of various PCM heat exchangers have not been discussed in the literature. Some of the comparison studies between two or more PCM heat exchangers reported in the literature are mentioned in Table 10.3. Our literature survey shows that more studies for comparison of different PCM heat exchangers under the same amount of PCM, the same volume of the heat exchanger as well as the same operation condition should be done. This will help designers to know the weak and strong points of a PCM heat exchanger and find a solution for improving their thermal efficiency.

The comparing of the PCM heat exchangers can be done also according to the number of studies reported in the literature to find out the most popular type of PCM heat exchangers and investigate the reasons for attracting high attention compared to other types. It should be mentioned that the charts plotted in this section are the results of reviewing around 150 papers. These results may be changed if further studies will be found and added. Figure 10.18 shows the number of studies for fluid-embedded, PCM-embedded, multi-domain PCM heat exchangers. Most of the studies have been done on the working fluid-embedded PCM heat exchangers. Researchers prefer to put PCM on the shell side rather than inside the tubes or pipes. The number of studies on the multi-domain PCM heat exchanger is less than working

TABLE 10.3
Some of Studies Performed Studies for Comparison of Different PCM Heat Exchangers

Researcher	Compared Heat Exchangers	Results
Gorzin et al. [52]	a. PCM embedded double pipe b. Working fluid embedded double pipe c. Triplex	Based on their geometrical parameters, the freezing time of (b) is shorter than (a) and (c) is shorter than (a) and (b).
Han et al. [36]	a. PCM embedded double pipe b. Working fluid embedded double pipe	Based on their geometrical parameters and for horizontal case: the melting time of (a) is shorter than (b).
Medrano et al. [53]	a. Working fluid embedded double pipe b. Working fluid embedded double pipe with graphite matrix c. Working fluid embedded double pipe with a circular disk d. Working fluid-embedded multi-pass single-circuit shell and tube with prism shell.	Based on their geometrical parameters: (d) has the highest average thermal power (above 1 kW), as it has the highest ratio of heat transfer area to an external volume.
Belusko et al. [54]	a. Working fluid-embedded single-pass multi-circuit shell and tube with prism shell b. Working fluid-embedded multi-pass multi-circuit shell and tube with prism shell	Based on their geometrical parameters: (b) has better thermal performance compared with (a).

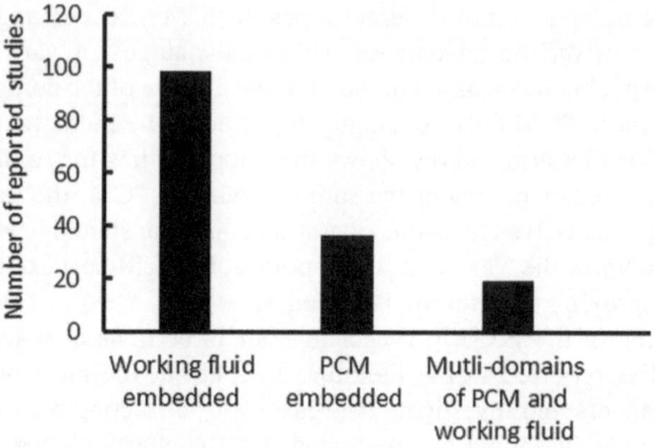

FIGURE 10.18 Number of studies on the main groups of PCM heat exchangers.

fluid-embedded and PCM-embedded-type heat exchangers since for the same volume of a PCM heat exchanger the amount of PCM in the multi-domain PCM heat exchanger may be smaller.

For working fluid-embedded heat exchangers, 97% is the shell and tube heat exchangers, only three studies for the double-plate heat exchangers were found. Figure 10.19a shows the number of studies on working fluid-embedded shell and tube heat exchangers. Single circuit single pass (which is a double-pipe heat exchanger) has attracted the attention of many researchers and around 58 studies (numerically and/or experimentally) were done on it. The number of multi-circuit or multi-pass is considerably smaller than double-pipe heat exchangers which are the simplest heat exchanger. Many studies on different methods for the enhancement of heat transfer in PCM were investigated by application onto the double-pipe heat exchanger. Different fin shapes were designed and applied onto the single-pass single-circuit PCM heat exchangers. This may be another reason for a high number of studies on working fluid-embedded double-pipe heat exchangers. Figure 10.19b shows studies on PCM-embedded-type PCM heat exchangers. Packed bed PCM heat exchanger has attracted

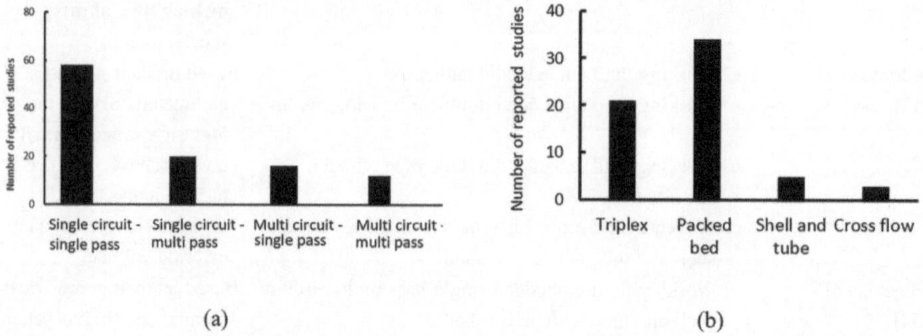

FIGURE 10.19 Number of studies on (a) different kinds of working fluid embedded and (b) different kinds of PCM-embedded-type PCM heat exchangers.

the highest attention and around 34 studies were performed. Simple construction and high convective heat transfer coefficient for the fluid side may be the reasons for this high interest. Triplex PCM heat exchanger is in the second rank from studies numbers point of view among other types of PCM embedded heat exchangers. The advantages of fast melting and freezing and the use of two different working fluids make this kind of PCM heat exchangers attractive. Shell and closed tube as well as cross-flow PCM heat exchangers could not receive too much attention compared to Triplex and Packed bed PCM heat exchangers. The small amount of PCM which can be filled in the tubes may be the reason for the limited performed studies on them.

10.6 CONCLUSION

Due to the importance of thermal storage and advantages of solid–liquid phase change number, many studies have been performed on the PCM heat exchangers. The number of review papers, in which PCM heat exchangers are discussed somehow, is around 23 with around 2400 references in total. In these studies, thermal analysis of PCM heat exchangers as well as innovative methods for the increase of thermal storage power is discussed. The classification of PCM heat exchangers is important since it helps thermal designers to know the advantages and disadvantages of each group as well as to know their best applications. Therefore, a thermal designer can predict the best type of PCM heat exchangers for a required application. To the best of our knowledge, there is no detailed study on the classification of PCM heat exchangers from a structure point of view and so the present study is performed to classify PCM heat exchangers. The following remarks can be concluded from the present study:

- It is possible to classify PCM heat exchangers into three main groups: (a) working fluid-embedded PCM heat exchangers, (b) PCM-embedded-type PCM heat exchangers, (c) multi-domains PCM heat exchangers. Our literature survey on 150 studies showed that most of the studies were done on the working fluid-embedded PCM heat exchanger with 98 studies. The smallest number of studies is multi-domains PCM and working fluid heat exchangers with 19 studies.
- Working fluid-embedded PCM heat exchangers can be also classified into shell and tube and double-plate PCM heat exchangers. The number of studies on double-plate PCM heat exchangers is too limited (three studies were found). Shell and tube heat exchangers can be (a) single circuit single pass (or double pipe), (b) single circuit multi-pass, (c) multi-circuit single pass, and (d) multi-circuit–multi-pass PCM heat exchangers. Single circuit single pass with 58 studies has attracted the highest rate of attention among four types of shell and tube PCM heat exchangers. Many studies on single circuit single pass were done to check the effect of fins performance.
- PCM-embedded-type PCM heat exchangers can be grouped as (a) shell and tube, (b) triplex, (c) cross-flow, and (d) packed bed type PCM heat exchangers. Among them, the highest number of studies was done for the packed bed PCM heat exchangers due to simple design and high heat transfer coefficient in the working fluid side.

- Finally, multi-domains PCM and working fluid PCM heat exchangers attracted attention, and around 22 studies on this kind of PCM heat exchangers could be found.
- Although many studies were reported on the PCM heat exchanger, the number of studies on the comparison of PCM heat exchangers with each other is limited. The present classification helps researchers to compare PCM heat exchangers of each group among themselves as well as do comparison for PCM heat exchangers of different main groups to obtain more meaningful results.
- The classification of PCM heat exchangers provides a methodology to name a PCM heat exchanger correctly. The given name can provide an image of the PCM heat exchanger for the reader (or audience).

Although the classification of PCM heat exchangers was done successfully in this study and briefly the advantages and disadvantages of each group are discussed, more detailed studies on the comparison between groups and their applications are required to be studied, and the present study should be continued by researchers in this field.

REFERENCES

1. A. Sharma, V. V. Tyagi, C. R. Chen, D. Buddhi, Review on thermal energy storage with phase change materials and applications, *Renewable and Sustainable Energy Reviews* 13 (2009) 318–345.
2. S. Jegadheeswaran, S. D. Pohekar, Performance enhancement in latent heat thermal storage system: A review, *Renewable and Sustainable Energy Reviews* 13 (2009) 2225–2244.
3. V. A. A. Raj, R. Velraj, Review on free cooling of buildings using phase change materials, *Renewable and Sustainable Energy Reviews* 14 (2010) 2819–2829.
4. F. Agyenim, N. Hewitt, P. Eames, M. Smyth, A review of materials, heat transfer and phase change problem formulation for latent heat thermal energy storage systems (LHTESS), *Renewable and Sustainable Energy Reviews* 14 (2010) 615–628.
5. L. Fan, J. M. Khodadadi, Thermal conductivity enhancement of phase change materials for thermal energy storage: A review, *Renewable and Sustainable Energy Reviews* 15 (2011) 24–46.
6. V. H. K. Tiat, E. P. del Barrio, Recent patents on phase change materials and systems for latent heat thermal energy storage, *Recent Patents on Mechanical Engineering* 4 (2011) 16–28.
7. D. Zhou, C. Y. Zhao, Y. Tian, Review on thermal energy storage with phase change materials (PCMs) in building applications, *Applied Energy* 92 (2012) 593–605.
8. E. Oró, A. de Gracia, A. Castell, M. M. Farid, L. F. Cabeza, Review on phase change materials (PCMs) for cold thermal energy storage applications, *Applied Energy* 99 (2012) 513–533.
9. A. A. Al-Abidi, S. B. Mat, K. Sopian, M. Y. Sulaiman, C. H. Lim, Abdulrahman, T., Review of thermal energy storage for air conditioning systems, *Renewable and Sustainable Energy Reviews* 16 (2012) 5802–5819.
10. E. Osterman, V. V. Tyagi, V. Butala, N. A. Rahim, U. Stritih, Review of PCM based cooling technologies for buildings, *Energy and Buildings* 49 (2012) 37–49.

11. M. Liu, W. Saman, F. Bruno, Review on storage materials and thermal performance enhancement techniques for high temperature phase change thermal storage systems, *Renewable and Sustainable Energy Reviews* 16 (2012) 2118–2132.
12. A. Waqas, Z. U. Din, Phase change material (PCM) storage for free cooling of buildings—A review, *Renewable and Sustainable Energy Reviews* 18 (2013) 607–625.
13. S. Kuravi, J. Trahan, D. Y. Goswami, M. M. Rahman, E. K. Stefanakos, Thermal energy storage technologies and systems for concentrating solar power plants, *Progress in Energy and Combustion Science* 39 (2013) 285–319.
14. N. R. Jankowski, F. P. McCluskey, A review of phase change materials for vehicle component thermal buffering, *Applied Energy* 113 (2014) 1525–1561.
15. M. K. Anuar Sharif, A. A. Al-Abidi, S. Mat, K. Sopian, M. H. Ruslan, M. Y. Sulaiman, M. A. M. Rosli, Review of the application of phase change material for heating and domestic hot water systems, *Renewable and Sustainable Energy Reviews* 42 (2015) 557–568.
16. H. Akeiber, P. Nejat, M. Z. Abd. Majid, M. A. Wahid, F. Jomehzadeh, I. Z. Famileh, J. K. Calautit, B. R. Hughes, S. A. Zaki, A review on phase change material (PCM) for sustainable passive cooling in building envelopes, *Renewable and Sustainable Energy Reviews* 60 (2016) 1470–1497.
17. F. Souayfane, F. Fardoun, Pascal-Henry Biwole, Phase change materials (PCM) for cooling applications in buildings: A review, *Energy and Buildings* 129 (2016) 396–431.
18. V. Kapsalis, D. Karamanis, Solar thermal energy storage and heat pumps with phase change materials, *Applied Thermal Engineering* 99 (2016) 1212–1224.
19. A. de Gracia, L. F. Cabeza, Numerical simulation of a PCM packed bed system: A review, *Renewable and Sustainable Energy Reviews* 69 (2017) 1055–1063.
20. N. H. S. Tay, M. Liu, M. Belusko, F. Bruno, Review on transportable phase change material in thermal energy storage systems, *Renewable and Sustainable Energy Reviews* 75 (2017) 264–277.
21. A. Dinker, M. Agarwal, G. D. Agarwal, Heat storage materials, geometry and applications: A review, *Journal of the Energy Institute* 90 (2017) 1–11.
22. S. Seddegh, M. Mastani Joybari, X. Wang, F. Haghighat, Experimental and numerical characterization of natural convection in a vertical shell-and-tube latent thermal energy storage system, *Sustainable Cities and Society* 35 (2017) 13–24.
23. X. Yang, Z. Lu, Q. Bai, Q. Zhang, L. Jin, J. Yan, Thermal performance of a shell-and-tube latent heat thermal energy storage unit: Role of annular fins, *Applied Energy* 202 (2017) 558–570.
24. J. Vogel, M. Keller, M. Johnson, Numerical modeling of large-scale finned tube latent thermal energy storage systems, *Journal of Energy Storage* 29 (2020) 101389.
25. M. Gürtürk, B. Kok, A new approach in the design of heat transfer fin for melting and solidification of PCM, *International Journal of Heat and Mass Transfer* 153 (2020) 119671.
26. G. Shen, X. Wang, A. Chan, F. Cao, X. Yin, Study of the effect of tilting lateral surface angle and operating parameters on the performance of a vertical shell-and-tube latent heat energy storage system, *Solar Energy* 194 (2019) 103–113.
27. A. Shahsavar, J. Khosravi, H. I. Mohammed, P. Talebizadehsardari, Performance evaluation of melting/solidification mechanism in a variable wave-length wavy channel double-tube latent heat storage system, *Journal of Energy Storage* 27 (2020) 101063.
28. S. Saedi Ardahaie, M. J. Hosseini, A. A. Ranjbar, M. Rahimi, Energy storage in latent heat storage of a solar thermal system using a novel flat spiral tube heat exchanger, *Applied Thermal Engineering* 159 (2019) 113900.

29. J. Gasia, J. Miguel Maldonado, F. Galati, M. De Simone, L. F. Cabeza, Experimental evaluation of the use of fins and metal wool as heat transfer enhancement techniques in a latent heat thermal energy storage system, *Energy Conversion and Management* 184 (2019) 530–538.
30. N. H. S. Tay, M. Belusko, F. Bruno, An effectiveness-NTU technique for characterising tube-in-tank phase change thermal energy storage systems, *Applied Energy* 91 (2012) 309–319.
31. R. Anish, V. Mariappan, M. M. Joybari, Experimental investigation on the melting and solidification behavior of erythritol in a horizontal shell and multi-finned tube latent heat storage unit, *Applied Thermal Engineering* 161 (2019) 114194.
32. M. M. Joybari, S. Seddegh, X. Wang, F. Haghighat, Experimental investigation of multiple tube heat transfer enhancement in a vertical cylindrical latent heat thermal energy storage system, *Renewable Energy* 140 (2019) 234–244.
33. Z. N. Meng, P. Zhang, Experimental and numerical investigation of a tube-in-tank latent thermal energy storage unit using composite PCM, *Applied Energy* 190 (2017) 524–539.
34. S. Pakalka, K. Valancius, G. Streckien, Experimental and theoretical investigation of the natural convection heat transfer coefficient in Phase Change Material (PCM) based fin-and-tube heat exchanger, *Energies* 14 (3) (2021) 716.
35. P. Talebizadehsardari, J. M. Mahdi, H. I. Mohammed, M. A. Moghimi, A. H. Eisapour, M. Ghalambaz, Consecutive charging and discharging of a PCM-based plate heat exchanger with zigzag configuration, *Applied Thermal Engineering* 193 (2021) 116970.
36. G. Han, H. Ding, Y. Huang, L. Tong, Y. Ding, A comparative study on the performances of different shell-and-tube type latent heat thermal energy storage units including the effects of natural convection, *International Communications in Heat and Mass Transfer* 88 (2017) 228–235.
37. R. Babaei Mahani, H. I. Mohammed, J. M. Mahdi, F. Alamshahi, M. Ghalambaz, P. Talebizadehsardari, W. Yaïci, Phase change process in a zigzag plate latent heat storage system during melting and solidification, *Molecules* 25 (20) (2020) 4643.
38. X. Xiao, P. Zhang, Numerical and experimental study of heat transfer characteristics of a shell-tube latent heat storage system: Part I e Charging process, *Energy* 79 (2015) 337–350.
39. V. Dubovsky, G. Ziskind, R. Letan, Analytical model of a PCM-air heat exchanger, *Applied Thermal Engineering* 31 (2011) 3453–3462.
40. P. Promoppatum, S. Yao, T. Hultz, D. Agee, Experimental and numerical investigation of the cross-flow PCM heat exchanger for the energy saving of building HVAC, *Energy and Buildings* 138 (2017) 468–478.
41. S. Bellan, T. E. Alam, J. Gonzalez-Aguilar, M. Romero, M. M. Rahman, D. Y. Goswami, E. K. Stefanakos, Numerical and experimental studies on heat transfer characteristics of thermal energy storage system packed with molten salt PCM capsules, *Applied Thermal Engineering* 90 (2015) 970–979.
42. M. Zukowski, Mathematical modeling and numerical simulation of a short term thermal energy storage system using phase change material for heating applications, *Energy Conversion and Management* 48 (2007) 155–165.
43. C. Arkar, S. Medved, Influence of accuracy of thermal property data of a phase change material on the result of a numerical model of a packed bed latent heat storage with spheres, *Thermochimica Acta* 438 (2005) 192–201.
44. J. Long, Numerical and experimental investigation for heat transfer in triplex concentric tube with phase change material for thermal energy storage, *Solar Energy* 82 (2008) 977–985.

45. M. M. Hosseini, A. B. Rahimi, Improving heat transfer in a triplex tube heat exchanger containing phase-change materials by modifications of length and position of fins, *Scientia Iranica B* 27 (1) (2020) 239–251.
46. M. Ghalambaz, J. M. Mahdi, A. Shafaghat, A. Hossein Eisapour, O. Younis, P. Talebizadeh Sardari, W. Yaïci, Effect of twisted fin array in a triple-tube latent heat storage system during the charging mode, *Sustainability* 13 (5) (2021) 2685.
47. A. H. N. Al-Mudhafar, A. F. Nowakowski, F. C. G. A. Nicolleau, Performance enhancement of PCM latent heat thermal energy storage system utilizing a modified webbed tube heat exchanger, *Energy Reports* 6 (2020) 76–85.
48. W. Lin, R. Huang, X. Fang, Z. Zhang, Improvement of thermal performance of novel heat exchanger with latent heat storage, *International Journal of Heat and Mass Transfer* 140 (2019) 877–885.
49. B. Zalba, J. M. Marin, L. F. Cabeza, H. Mehling, Free-cooling of buildings with phase change materials, *International Journal of Refrigeration* 27 (2004) 839–849.
50. M. Borcuch, M. Musiał, K. Sztekler, W. Kalawa, St. Gumuła, S. Stefański, The influence of flow modification on air and PCM temperatures in an accumulative heat exchanger, *EPJ Web of Conferences* 180 (2018) 02011.
51. B. Gurel, A numerical investigation of the melting heat transfer characteristics of phase change materials in different plate heat exchanger (latent heat thermal energy storage) systems, *International Journal of Heat and Mass Transfer* 148 (2020) 119117.
52. M. Gorzin, M. J. Hosseini, M. Rahimi, R. Bahrampoury, Nano-enhancement of phase change material in a shell and multi-PCM-tube heat exchanger, *Journal of Energy Storage* 22 (2019) 88–97.
53. M. Medrano, M. O. Yilmaz, M. Nogués, I. Martorell, J. Roca, L. F. Cabeza, Experimental evaluation of commercial heat exchangers for use as PCM thermal storage systems, *Applied Energy* 86 (2009) 2047–2055.
54. M. Belusko, N. H. S. Tay, M. Liu, F. Bruno, Effective tube-in-tank PCM thermal storage for CSP applications, Part 1: Impact of tube configuration on discharging effectiveness, *Solar Energy*, 139 (2016) 733–743.

10.A APPENDIX

Working Fluid-Embedded Double-Pipe PCM Heat Exchanger Studies
Shell and Tube

(a) Single Circuit Single Pass

https://dx.doi.org/10.1016/j.jestch.2015.09.014; https://doi.org/10.1016/j.est.2020.102226; https://doi.org/10.1016/j.egyr.2021.02.034;

https://doi.org/10.1016/j.applthermaleng.2018.07.134; https://doi.org/10.1016/j.ijrefrig.2017.07.014; http://dx.doi.org/10.1016/j.applthermaleng.2016.06.169; https://doi.org/10.1016/j.jobe.2020.101929; http://dx.doi.org/10.1051/epjconf/20146702046; doi:10.1088/1755-1315/354/1/012020; https://doi.org/10.1016/j.energy.2020.119259; https://doi.org/10.1016/j.applthermaleng.2021.116866; https://doi.org/10.1016/j.ijheatmasstransfer.2018.10.126; https://doi.org/10.1016/j.applthermaleng.2020.116046; http://dx.doi.org/10.1016/j.ijthermalsci.2011.06.010; http://dx.doi.org/10.1016/j.apenergy.2016.10.042; https://doi.org/10.1016/j.est.2020.101875; https://doi.org/10.1016/j.est.2018.12.023; https://doi.org/10.1016/j.ijheatmasstransfer.2020.119671; https://doi.org/10.1016/j.est.2021.102513; http://dx.doi.org/10.1016/j.icheatmasstransfer.2013.11.008; https://doi.org/10.1016/j.enconman.2020.112679; http://dx.doi.org/10.1016/j.icheatmasstransfer.2014.02.023; https://doi.org/10.1016/j.est.2020.101396; https://doi.org/10.1016/j.applthermaleng.2020.115898; http://dx.doi.org/10.1016/j.ijthermalsci.2014.03.014; http://dx.doi.org/10.1016/j.apenergy.2013.06.007; https://doi.org/10.1016/j.applthermaleng.2021.117079; https://doi.org/10.1016/j.est.2020.101331; https://doi.org/10.1007/s00231-020-02983-x; doi:10.1016/j.enbuild.2012.02.053; https://doi.org/10.1016/j.est.2020.101319; 10.1016/j.egypro.2019.01.704; https://doi.org/10.1016/j.est.2020.102161; http://dx.doi.org/10.1016/j.ijheatmasstransfer.2015.02.064; https://doi.org/10.1016/j.est.2021.102458; https://doi.org/10.1016/j.renene.2021.04.076; https://doi.org/10.1016/j.applthermaleng.2020.115017; https://doi.org/10.1016/j.renene.2019.10.084; http://dx.doi.org/10.1016/j.scs.2017.07.024; http://dx.doi.org/10.1016/j.scs.2017.07.024; https://doi.org/10.1016/j.est.2019.101063; https://doi.org/10.1016/j.solener.2019.10.077; https://doi.org/10.1016/j.solener.2020.10.003; https://doi.org/10.1016/j.est.2020.102226; https://doi.org/10.1016/j.enconman.2019.03.022; http://dx.doi.org/10.1016/j.apenergy.2015.01.008; http://dx.doi.org/10.1016/j.apenergy.2011.09.039; http://dx.doi.org/10.1016/j.enconman.2013.06.044; https://doi.org/10.1016/j.est.2020.101389; https://doi.org/10.1016/j.apenergy.2019.04.011; http://dx.doi.org/10.1016/j.applthermaleng.2013.04.063; https://doi.org/10.1016/j.est.2019.100802; https://doi.org/10.1016/j.apenergy.2020.115772; https://doi.org/10.1016/j.scs.2019.101764; http://dx.doi.org/10.1016/j.apenergy.2017.05.007; http://dx.doi.org/10.1016/j.apenergy.2017.05.007; http://dx.doi.org/10.1016/j.solener.2014.01.007; doi:10.3390/pr7050266;

(b) Single Circuit Multi Pass

https://doi.org/10.1016/j.jclepro.2020.124238; https://doi.org/10.1016/j.applthermaleng.2019.113900; doi:10.3390/en13236193; https://doi.org/10.1016/j.energy.2019.116083; https://doi.org/10.1016/j.enbuild.2018.03.058; https://doi.org/10.1016/j.enconman.2019.01.085; http://dx.doi.org/10.1016/j.applthermaleng.2016.06.049; https://doi.org/10.1016/j.est.2021.102649; https://doi.org/10.1016/j.applthermaleng.2019.114007; https://doi.org/10.1016/j.applthermaleng.2018.12.024; https://doi.org/10.1016/j.jclepro.2020.120249; https://doi.org/10.1016/j.scs.2018.11.002; http://dx.doi.org/10.1016/j.icheatmasstransfer.2014.02.025; https://doi.org/10.1007/s10973-021-10726-1; https://doi.org/10.1016/j.est.2018.11.006; https://doi.org/10.1016/j.enbuild.2019.109744; https://doi.org/10.1016/j.est.2019.100797; https://doi.org/10.1016/j.applthermaleng.2018.06.087; https://doi.org/10.1016/j.scs.2018.10.012; https://doi.org/10.1016/j.enconman.2017.12.036;

(c) Multi Circuit Single Pass

(*Continued*)

https://doi.org/10.1016/j.applthermaleng.2020.115743; https://doi.org/10.1016/j.renene.2020.08.153; https://doi.org/10.1016/j.applthermaleng.2019.114194; http://dx.doi.org/10.1016/j.solener.2015.09.042; http://dx.doi.org/10.1016/j.solener.2015.09.034; 10.1016/j.proeng.2017.10.308; https://doi.org/10.1016/j.renene.2019.03.037; https://doi.org/10.1016/j.est.2019.03.024; http://doi.org/10.1007/s12206-020-0739-6; DOI: 10.1002/er.5743; https://doi.org/10.1016/j.ijrefrig.2019.03.006; http://dx.doi.org/10.1016/j.ijthermalsci.2011.01.025; https://doi.org/10.1016/j.applthermaleng.2020.115604; 10.1016/j.egypro.2019.01.737; http://dx.doi.org/10.1016/j.apenergy.2011.09.039; https://doi.org/10.1016/j.apenergy.2018.12.057;

(d) Multi Circuit Multi Pass

http://dx.doi.org/10.1016/j.solener.2015.09.042; http://dx.doi.org/10.1016/j.solener.2015.09.034; https://doi.org/10.1016/j.renene.2019.01.110; http://dx.doi.org/10.1016/j.solmat.2017.06.054; https://doi.org/10.1016/j.apenergy.2018.11.041; https://doi.org/10.1016/j.renene.2019.08.053; http://dx.doi.org/10.1016/j.applthermaleng.2016.11.065; http://dx.doi.org/10.1016/j.apenergy.2016.12.163; doi:10.1016/j.cep.2007.02.001; https://doi.org/10.3390/en14030716; https://doi.org/10.1016/j.applthermaleng.2020.115138; http://dx.doi.org/10.1016/j.apenergy.2011.09.039;

Double Plate

https://doi.org/10.1016/j.renene.2019.10.084; https://doi.org/10.1016/j.applthermaleng.2021.116970; http://dx.doi.org/10.1016/j.apenergy.2014.12.050;

PCM-Embedded-Type PCM Heat Exchanger Studies:

Shell and Tube

(a) Single Pass Single Circuit

https://doi.org/10.1016/j.ijheatmasstransfer.2020.119480; http://dx.doi.org/10.1016/j.icheatmasstransfer.2017.09.009; http://dx.doi.org/10.3390/molecules25204643;

(b) Multi-Pass

http://dx.doi.org/10.1016/j.energy.2014.11.020; http://dx.doi.org/10.1016/j.energy.2014.11.061;

Cross Flow

http://dx.doi.org/10.1016/j.applthermaleng.2011.06.031; https://doi.org/10.1007/s00231-018-2462-8; http://dx.doi.org/10.1016/j.enbuild.2016.12.043;

Packed Bed

(a) Plate Type

UDK 536.65:519.61/.64; https://doi.org/10.1016/j.est.2020.102005; https://doi.org/10.1016/j.est.2020.101944; doi:10.1016/j.enconman.2006.04.017;

(b) Spherical Capsules

https://doi.org/10.1016/j.est.2020.101209; https://doi.org/10.1016/j.ijheatmasstransfer.2019.05.093; https://doi.org/10.1016/j.ijheatmasstransfer.2020.120066; http://dx.doi.org/10.1016/j.applthermaleng.2016.02.036; http://dx.doi.org/10.1016/j.apenergy.2014.01.085; http://dx.doi.org/10.1016/j.energy.2017.08.055; http://dx.doi.org/10.1016/j.enbuild.2013.09.045; doi:10.1016/j.renene.2006.04.015; doi:10.1016/j.renene.2008.12.012; https://doi.org/10.1016/j.renene.2020.01.051; http://dx.doi.org/10.1016/j.ijthermalsci.2012.05.010; https://doi.org/10.1016/j.applthermaleng.2020.116473; http://dx.doi.org/10.1016/j.rser.2016.09.092; doi:10.1016/j.tca.2005.08.032; http://dx.doi.org/10.1016/j.cej.2013.06.112; http://dx.doi.org/10.1016/j.applthermaleng.2015.07.056; http://dx.doi.org/10.1016/j.applthermaleng.2012.12.007; https://doi.org/10.1016/j.ijheatmasstransfer.2020.120066; http://dx.doi.org/10.3390/en13236413; https://doi.org/10.1002/er.4254; https://doi.org/10.1016/j.renene.2021.04.022; PII: S 1 3 5 9-4 3 1 1 (0 2) 0 0 0 8 0–7; https://doi.org/10.1016/j.csite.2019.100431; http://dx.doi.org/10.1016/j.enbuild.2007.05.001; http://dx.doi.org/10.1016/j.enconman.2014.02.052; doi:10.1016/j.renene.2008.12.012; http://dx.doi.org/10.1016/j.est.2017.02.003; http://dx.doi.org/10.1016/j.est.2017.02.003;

(*Continued*)

Triplex

(a) Single Conduit

https://doi.org/10.1016/j.est.2020.102227; https://doi.org/10.1016/j.csite.2019.100487; http://dx.doi.org/10.1016/j.enbuild.2013.09.007; http://dx.doi.org/10.1016/j.ijheatmasstransfer.2013.02.030; https://doi.org/10.1016/j.egyr.2020.02.030; https://doi.org/10.1016/j.applthermaleng.2020.114966; https://doi.org/10.3390/su13052685; https://doi.org/10.1016/j.est.2018.12.023; doi: 10.24200/sci.2018.5604.1369; http://dx.doi.org/10.1016/j.enbuild.2017.01.034; doi:10.1016/j.solener.2008.05.006; https://doi.org/10.1016/j.enconman.2018.11.038; http://dx.doi.org/10.3390/en13123254; https://doi.org/10.1016/j.est.2020.102009; https://doi.org/10.1016/j.renene.2020.11.074; https://doi.org/10.1016/j.applthermaleng.2020.115409; https://doi.org/10.1016/j.renene.2019.06.092; https://doi.org/10.1016/j.renene.2021.04.051; http://dx.doi.org/10.1016/j.apenergy.2014.10.051;

(b) Multi-Conduit

https://doi.org/10.1016/j.egyr.2020.02.030; https://doi.org/10.1016/j.enconman.2018.05.086;

Multi-Domains of PCM and Working Fluid PCM Heat Exchangers

https://doi.org/10.1016/j.ijheatmasstransfer.2019.02.047; https://doi.org/10.1016/j.apenergy.2019.05.095; https://doi.org/10.1016/j.est.2020.101705; http://dx.doi.org/10.1016/j.renene.2011.04.008; https://doi.org/10.1051/epjconf/201818002011; https://doi.org/10.1016/j.enbuild.2019.109354; https://doi.org/10.1016/j.ijheatmasstransfer.2019.119117; doi:10.1016/j.enbuild.2005.04.002; DOI 10.1007/s12206-019-1152-x; https://doi.org/10.1016/j.applthermaleng.2020.115630; http://dx.doi.org/10.1016/j.applthermaleng.2016.03.011; http://dx.doi.org/10.1016/j.enbuild.2015.05.023; doi:10.1016/j.apenergy.2009.01.014; https://doi.org/10.1016/j.enbuild.2019.06.022; https://doi.org/10.1016/j.enbuild.2020.109895; http://dx.doi.org/10.1016/j.enbuild.2016.03.078; https://doi.org/10.1016/j.enconman.2018.12.013; https://doi.org/10.1016/j.energy.2021.120527; doi:10.1016/j.ijrefrig.2004.03.015; https://doi.org/10.2298/TSCI170821072Z;

11 Cool Thermal Energy Storage

Water and Ice to Alternative Phase Change Materials

Sandra K. S. Boetcher
Embry-Riddle Aeronautical University

CONTENTS

11.1 INTRODUCTION

Cold thermal energy storage (TES) dates back to ancient times when Hebrews, Greeks, and Romans gathered snow from mountains for various cooling applications. Storing "cold energy" is actually the reverse of adding heat to a material to store energy, since one removes heat from a material in order to "store" the cold. The advent of modern TES began in the mid-nineteenth century when large blocks of ice were cut from frozen lakes in colder climates and transported in insulated railroad cars to other areas. In fact, "a ton of air conditioning" gets its name from one-ton (by weight) blocks of ice measuring 4 ft × 4 ft × 2 ft. These blocks of ice would provide 12,000 Btu/h of cooling for 24 hours. In the early twentieth century, movie theaters were one of the first businesses to implement an early form of air conditioning by blowing a fan over a one-ton block of ice.

By the mid-twentieth century the implementation of cooling commercial buildings was ubiquitous. Electric utilities saw peak daytime demand and increased revenue due to the energy-intensive air conditioning; however, they also sought ways to increase the off-peak demand load by offering lower rates per kWh at night. This

DOI: 10.1201/9781003213260-11

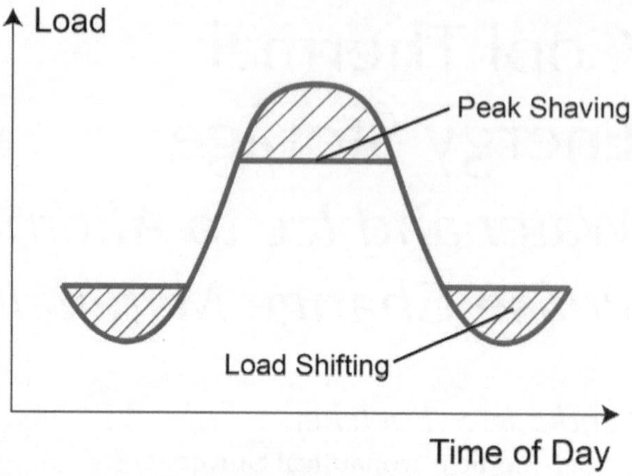

FIGURE 11.1 Diagram illustrating peak shaving and load shifting.

practice is called peak shaving and load shifting (Figure 11.1). Although, due to regulatory policies, the peak-shifting programs had to be revenue neutral, that is, businesses were required to consume twice as much energy during off-peak times if the discounted rate was half of the peak rate. At this time, many pilot cool TES systems were borne; however, due to revenue-neutral requirements, ice-harvesting type storage systems were used [1]. Ice-harvesting systems produce ice on vertical heat transfer surfaces that operate in a cyclical manner with alternating cold and warm fluid to make ice and remove ice, respectively [2]. Many operational and maintenance issues with these dynamic systems caused the static-type ice storage systems to be the prevailing system used moving forward [1].

Like ice, eutectic salts have a long history of being used as a phase change material (PCM). One of the earliest applications of eutectic salts is not for cooling, but for heating. In the late 1800s railroad cars in Britain used this type of PCM for warming seats. Furthermore, so-called eutectic plates were used to store cold for rail and trucking transportation.

Commercial TES systems utilizing ice began to be installed by a small number of manufacturers in the early 1980s. The business grew in the late 1980s and early 1990s with dozens more manufacturers coming on board to take advantge of new demand-side management rebate incentives [3]. Cool TES systems utilizing ice and water are still growing and taking advantage of new market opportunities such as building very large district cooling (DC) systems with a cool TES system servicing multiple buildings, small-scale rooftop TES systems, and emergency backup cooling for mission-critical facilities [1]. However, the push to move ice TES systems into the residential market (or light commercial) has not been as successful with several companies recently going out of buisness [4].

Water/ice has many advantages when used as a PCM. It has an unusually high latent heat of fusion (334 kJ/kg); in fact, it has the highest latent heat among common PCMs. Furthermore, water is free. However, water is limited to a phase-change temperature of is 0°C. Although commercial heating, ventilating, and air-conditioning

(HVAC) chillers can be retrofitted to make ice, residential HVAC systems are not capable of making ice in an efficient manner, and a PCM with a higher phase-change temperature is needed. Even though ice storage works in commercial buildings, there is the potential for energy and cost savings by implementing alternative PCM (such as paraffin wax or salt hydrates) TES systems that melt and freeze at higher temperatures, saving energy and eliminating the need for a sub-zero chiller.

The goal of this chapter is to present different popular types of TES systems based on water and ice and provide an overview of alternative PCMs. Emerging alternative PCM TES systems, their implementation, challenges, and outlook will be presented.

11.2 TYPES OF ICE-BASED THERMAL ENERGY STORAGE SYSTEMS

Ice-based TES systems are classified as either static, where ice is formed directly onto a surface, or dynamic, where ice is produced on the cooling surface and then removed. The majority of ice-based TES systems are of the static type. A typical ice-based TES system for peak shaving and load shifting is shown in Figure 11.2. The following sections outline static and dynamic ice-based systems.

11.2.1 Static Systems

Static types of ice-based TES systems involve producing an ice layer that bonds to a heat exchanger cooling surface (typically coils) in the storage tank itself. Once ice has been formed on the surface, it is available until chilled water is needed for cooling. Warm return water is cooled by the melting ice before it is returned to the building. Other static systems use an alternative fluid, like a refrigerant, that is pumped through tubes. These systems are manufactured as packaged units that are implemented into a building chilled-water system. The ice is either formed inside or

FIGURE 11.2 Ice storage tanks for off-peak load shifting.

outside of the coil. One of the downsides of static systems is that as the layer of ice on the surface increases, the thermal resistance also increases, causing the efficiency of the system to decrease.

11.2.2 Dynamic Systems

Dynamic systems (also referred to as ice shucking or ice harvesting) make crushed ice or slurries and deliver it to a storage tank. These systems typically remove the ice layer via buoyancy [6,7] or fluid flow [8,9]. In theory, dynamic systems offer better efficiency than static systems since the ice layer does not build up, decreasing the thermal performance. Ice slurries are also advantageous because they can cool chilled water to temperatures as low as 1.1°C [5]. However, most TES systems of this type use water solutions, such as ethylene glycol/water and oil/water mixtures, to aid in the ice removal. The efficiency of the system is decreased due to the addition of other fluids into the water which lower both the freezing point and latent heat of fusion.

11.2.3 Static Versus Dynamic Systems

Although dynamic systems offer better efficiency since the ice is cyclically removed, the cost-per-ton of removing the ice is high. Because of this, dynamic ice harvesters were typically configured on a weekly cycle where ice was made on the weekend and then was melted during the rest of the week. The systems were also very complex, which led to many operation and maintenance issues. This caused the static ice storage systems to be the prevailing TES system on the market.

11.3 PHASE CHANGE MATERIALS

Generally, PCMs can be categorized as organic, inorganic, or eutectic (Figure 11.3). Organic PCMs include paraffin waxes (e.g., C_nH_{2n+2} [10]), plant-based materials (e.g., plant fats [11]), or fatty acids (e.g., lauric acid [12]). Organic PCMs are widely used in a number of applications, including emerging TES storage systems for commercial buildings, which will be discussed later. Inorganic PCMs encompass salt hydrates (e.g., $AB{\cdot}nH_2O$ [13]) and metals (e.g., gallium [14]). Finally, eutectics are mixtures of two or more materials that have a lower phase-change temperature than either of the individual materials (e.g., eutectic solution of polyethylene glycol [15]).

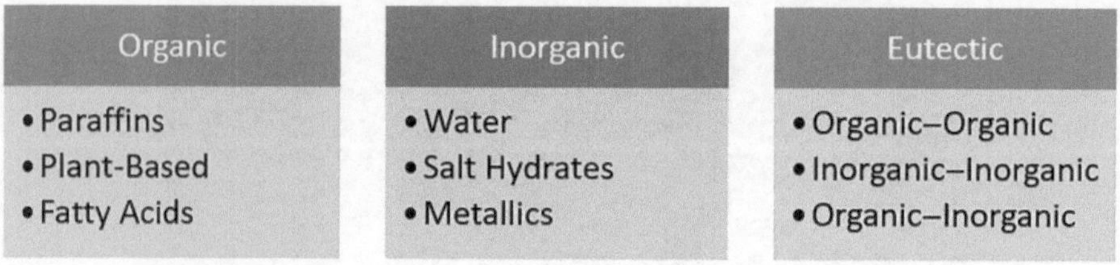

FIGURE 11.3 Classification of different types of PCMs.

Several criteria need to be evaluated before selecting a PCM for a particular application. First, materials with very latent heats of fusion (h_{sl}) are advantageous. The latent heat of fusion is the amount of energy absorbed or released by a material during melting or solidification, respectively. Traditionally, the latent heat of fusion is described on a per-mass basis, and water has the highest latent heat on a per-mass basis of any other common substance used as a PCM at moderate temperatures. Recently, it has become popular to describe PCMs on a per-volume basis for building applications, since space, rather than weight, is more of an issue in HVAC-related applications [16]. By describing PCMs in terms of energy storage density, often described in units kWh/m^3, instead of kJ/kg as for the per-mass basis, other PCMs become more attractive when space-based issues are relevant. Water has an energy storage density of ~92 kWh/m^3; however, some inorganic PCMs such as salt hydrates have energy storage densities well above 100 kWh/m^3. Salt-hydrate PCMs have also been gaining popularity due to new demand for reducing the footprint of TES systems in buildings. Furthermore, although nothing can surpass pure water in terms of cost, salt hydrates are relatively inexpensive.

While salt hydrates have high energy density and low cost, they suffer from a number of disadvantages [17]. These disadvantages include incongruent melting and phase segregation, which is caused by insufficient water needed to dissolve the solid-state material. This causes less-hydrated salts to sink to the bottom, causing the phase transition to become irreversible over long periods of time. Supercooling is also a major issue in which the solidification of the material begins at a much lower temperature than its phase-change temperature. Since PCMs need to be packaged into heat exchangers, another issue that prevents wide-spread implementation is the corrosivity of salt hydrates. Furthermore, salt hydrates have limited phase-change-temperature availability, and environmental control is necessary for maintaining proper hydration of the salts if not in a sealed environment.

Although PCMs have high latent heats of fusion, they are known to have very low thermal conductivity. Materials with low thermal conductivity cannot efficiently distribute the heat. Current research focuses, perhaps too heavily, on trying to increase the thermal conductivity of PCMs through additives such as carbon fibers [18], metal and metal oxide nanoparticles [10,19], and expanded graphite [19]. Implementing additives to increase thermal conductivity can be challenging due to the inability to obtain a homogeneous mixture and the fragility of some fillers such as carbon nanotubes. Furthermore, thermal-conductivity-enhancing additives occupy space, effectively reducing the energy storage density of the composite material. The trade-off of adding particles is that there is less effective latent heat capacity.

There have been a number of attempts to quantify the trade-offs between thermal conductivity and latent heat by developing simplified figures of merit (FOM), such as the one defined by Shamberger [20], shown here.

$$\mathrm{FOM} = \sqrt{kL_{sl}} \tag{11.1}$$

In this equation, k is the thermal conductivity h, and L_{sl} is the volumetric latent heat of fusion (ρh_{sl}). The FOM suggests that the higher the thermal conductivity and volumetric latent heat of fusion, the better. Of course, a material with both high thermal conductivity and capacity is desirable, but materials that are useful TES materials and are high in both properties are rare, or likely to occur in materials where the phase-change temperature range is not desirable for the intended application.

The Ragone framework has also emerged as a popular way to quantify energy versus power for TES systems [21–23]. The potential energy capacitance is material specific, as defined by its latent heat fusion, and the power is dependent on both the thermal conductivity of the material and the geometry of the heat exchanger. Many Ragone analyses focus on simplified thermal resistances and do not consider the overall geometrical configuration when assessing power capabilities. The low thermal conductivity of a particular material can be mitigated by the overall thickness of the material in the heat exchanger.

The total thermal resistance through PCM is not only a function of the conductivity of the material, it is also a function of the thickness. The thermal resistances, in W/K, for a plane wall and annulus are

$$R_{\text{wall}} = \begin{cases} \dfrac{t}{(kA)_w}, & \text{plane wall} \\[2ex] \dfrac{\ln\left(r_o/r_i\right)}{2\pi k_w L}, & \text{annulus} \end{cases} \tag{11.2}$$

In the equation k_w is the thermal conductivity of the material. For the plane wall, t is the thickness, and A is the area perpendicular to the heat transfer. For the annulus, L is the length of the pipe, and r_o and r_i are the outer and inner radii, respectively. As can be seen from Eq. (11.2), there are two ways in which to reduce the total thermal resistance through heat exchanger wall. First, the thermal conductivity can be increased; however, the same effect can be achieved by decreasing the wall thickness as seen in the application of polymer heat exchangers.

Polymer materials have been used for heat exchangers since the 1960s. The first polymer heat exchangers were in the form of flexible tube bundles made of Teflon (polytetrafluoroethylene or PTFE) [24]. In order to overcome the low thermal conductivity of the material, the diameter of the Teflon tubes was made proportional to the wall thickness. These designs took advantage of the high surface area to increase heat transfer. With the advent of additive manufacturing, polymer heat exchanger devices with very thin walls and heat-transfer-enhancing elements can be made [25–27]—making the low thermal conductivity of the material irrelevant. As shown in Figure 11.4, the total thermal resistance of a heat exchanger is dependent on the convective resistances of the fluids, the fouling, and the wall thickness. Another advantage of polymer heat exchangers is that unlike metals, they are resistant to fouling, eliminating the additional thermal resistances due to fouling on both the cold and hot side.

TES systems can also potentially take advantage of geometric design to overcome their inherently low thermal conductivity. If PCM can be 3D printed into compact high surface-area-to-volume heat exchangers, then filling PCMs with thermal

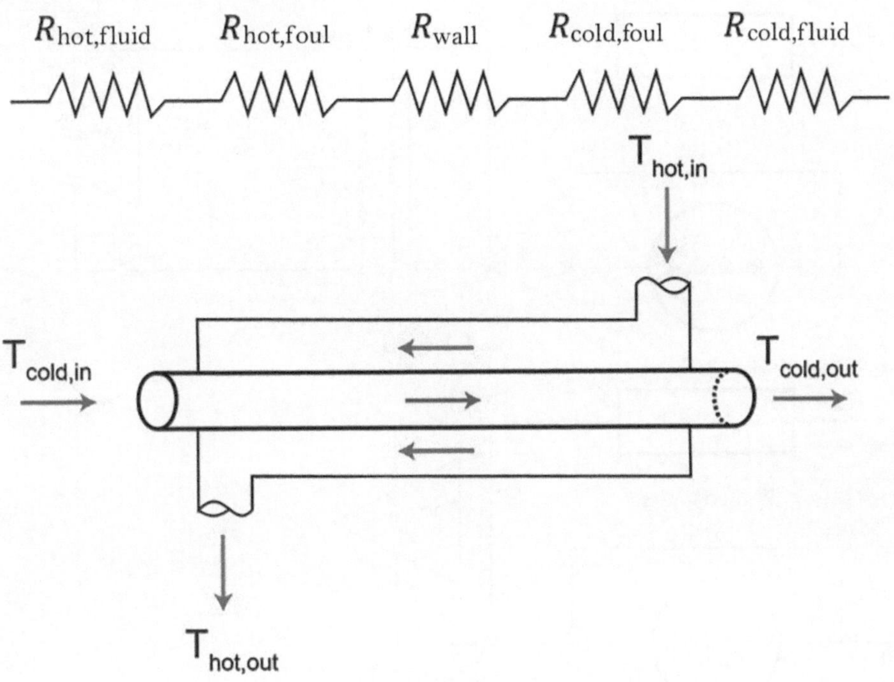

FIGURE 11.4 Schematic diagram of heat exchanger thermal resistances.

conductivity additives may no longer be necessary. Pioneering work by Freeman et al. [28–30] has shown the ability to combine PCM with polymers to additively manufacture compact TES systems.

Whereas TES systems for cooling started out utilizing large blocks of ice cut from lakes, the future of TES relies upon considering both the type of materials used and the arrangement of the materials in a TES system to achieve optimal TES density and power.

11.4 PCM-BASED THERMAL ENERGY STORAGE SYSTEMS

Alternative PCM-based TES systems to replace water- and ice-based systems for cold storage are beginning to be implemented in commercial applications. While ice-based thermal energy systems have been proven to work and are cost effective, there are several disadvantages of such systems when compared to alternative PCM TES systems. First, water melts and freezes at 0°C; this requires a sub-zero ice-making chiller. Often, existing chillers are not configured for making ice. Since a PCM with a higher phase-change temperature can be chosen, a higher set-point can be used and a sub-zero chiller is not necessary, thus saving energy.

Ice TES systems usually require two separate loops. Glycol (typically at a temperature between −7°C and −4°C) is used to freeze the water in the TES system during off-peak hours and transfer heat to the ice during peak hours. A secondary

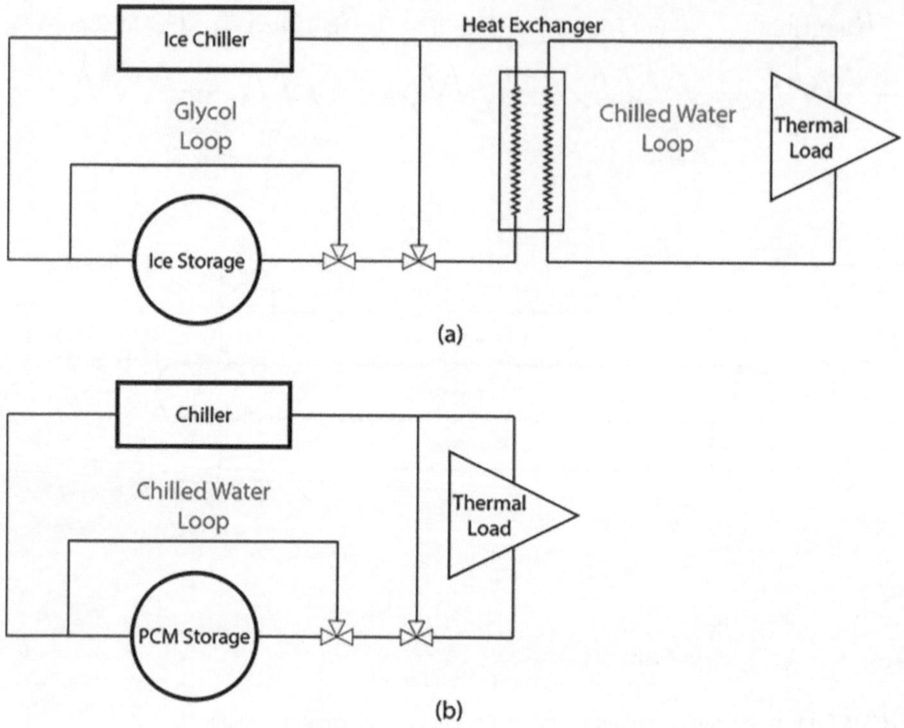

FIGURE 11.5 Comparison between (a) ice storage and (b) PCM thermal energy storage systems.

loop is used to transfer heat from the glycol side to the chilled-water side and vice versa. However, PCM-based TES systems can store energy at higher temperatures and work directly with a building chilled-water system, thus eliminating the need for a separate glycol loop (see Figure 11.5).

Another issue encountered with ice-based systems is that as water expands and contracts during the phase change, it causes thermal stresses and potential problems with thermal contact of the ice and water on the heat transfer pipes. Furthermore, very large tanks are often required to handle the expansion of the water as it freezes. Most other PCMs do not expand when they solidify.

Often, upgrading existing HVAC systems to include ice TES is complex and costly. As previously stated, most existing chillers cannot make ice. Furthermore, the addition of a glycol loop with a heat exchanger also complicates matters. On the other hand, alternative PCM-based systems can be integrated more easily into existing equipment without the need to increase the capacity of the chiller or install a separate glycol flow loop.

Since the technology is nascent, only a few papers available in the open literature detail the components and techno-economic analysis of commercial-scale alternative PCM HVAC TES systems [31–34]. For reference, the technical details regarding the PCM used in each system is detailed in Table 11.1.

TABLE 11.1
Summary of PCMs Used in Commercial-Scale Non-Ice-Based TES Systems Found in the Literature

	Jokiel [31] (2016)	Alam et al. [32] (2019)	Saeed et al. [33] (2018)	Tan et al. [34] (2020)
Type of Study	Case Study: Bergen, Sweden	Case Study: Melbourne, Australia	Commercial-Scale Laboratory Experiment	Case Study: Gothenburg, Sweden
PCM	PCM Products Ltd. Salt Hydrate	PCM Products Ltd. Salt Hydrate	Hexadecane ($C_{16}H_{34}$)	Rubitherm Salt-Hydrate SP11
T_{pc} (°C)	10	15	18.3 (melting) 15.5 (freezing)	11
h_{sl} (kJ/kg)	170	160	238.4 (melting) 234.5 (freezing)	155
ρ (kg/m^3)	1470	1510	828 (solid) 775 (liquid)	1330 (solid) 1320 (liquid)
c_p (J/kg·K)	1900	1900	1925 (solid) 2350 (liquid)	2000
k (W/m·K)	0.43	0.43	0.295 (solid) 0.152 (liquid)	0.6

11.4.1 Commercial-Scale PCM TES Systems

In 2016, the performance of a salt-hydrate-based TES system at the University College of Bergen in Norway was modeled and compared to manufacturer performance data sheets [31]. The TES system at the University College of Bergen, which is the largest system of its kind in Europe, comprises very large tanks filled with containers called FlatICE (PCM Products Ltd., Yaxley, Cambridgeshire, UK) comprising salt-hydrate-based PCM. The projected annual cooling demand for the educational building on campus is 1060 MWh, and the size of the cooling system is reduced through use of the TES system (e.g., the chillers provide 1400 kW for the entire day, although the peak cooling-load demand is 3000 kW). The PCM storage tanks, with a total volume of 228 m^3 and a total TES capacity of 11,200 kWh, have the potential to reduce the chiller load by up to 50%.

After developing a lumped model to analyze the TES, Jokiel [31] determined that while the maximum cooling capacity of 11,200 kWh could be reached when the discharge time was sufficiently long, it was unable to be realized under normal night-time (charging) and daytime (discharging) cycles, and the tanks were oversized for the current mode of operation. Jokiel [31] also pointed out a number of disadvantages of such a system. Since the PCM containers are horizontally aligned, due to natural convection, the heat transfer is higher on the bottom than the top during melting. Furthermore, since a salt hydrate is utilized, subcooling occurs and the long-term efficacy of the PCM is not known.

Alam et al. [32] analyzed the energy savings of an inorganic hydrated-salt PCM TES system designed to minimize the daytime cooling load on the chiller for an 11-story building in Melbourne, Australia. Active and passive chilled beams are used to supply the required cooling and heating load, and the chilled beams are served by a secondary chilled-water system. The insulated PCM tanks are 40 m³ in volume. The TES system also was made of 5120 FlatICE panels, stacked in layers, totaling 15,360 kg of PCM (PCM Products Ltd., Yaxley, Cambridgeshire, UK, same supplier as [31]) encapsulated in HDPE. In this system, water is used (as opposed to glycol) as the heat transfer fluid. It utilizes free cold air during the night to reduce the peak demand during the day. More technically, if there is no cooling call from the building, and ambient temperatures are less than 11°C, then charging of the PCM is activated. Similarly, charging is deactivated if the building calls for cooling, the ambient temperature rises above 11°C, or when the temperature exiting the PCM heat exchanger falls below 13°C for 5 minutes. When discharging the tank during daytime hours, the system takes advantage of PCM cooling until the outlet temperature of the fluid flowing through the PCM is 14°C.

They found that only 15% of the theoretical thermal storage capacity was used. One reason for the poor performance was possibly due to supercooling, which is a known issue with salt hydrates. Interestingly enough, the PCM tank was essentially useless during the summer months (when cooling demand is highest) because the nighttime ambient temperature was not cold enough to charge the PCM. Furthermore, the energy consumed by the pumps utilized in the TES charging and discharging was higher than the actual energy stored in the PCM. The authors of [32] highlight the issue of the difference between the design of the TES tank versus the actual performance of the TES tank.

Saeed et al. [33] designed a plate-type heat exchanger TES tank utilizing hexadecane (Sigma-Aldrich, St. Louis, MO, 99% purity) as the PCM. The investigators in [33] emphasize that the single most important factor in designing PCM TES systems is the heat exchanger geometry. In other words, the device must be compact and comprise thin slabs of PCM. Alternative methods of trying to increase thermal response, such as additives decrease performance, as shown by Marín et al. [35] who added PCM to a porous matrix for the purposes of increasing the thermal conductivity, but then observed a 20% reduction in TES density.

The hexadecane TES system had a net latent heat capacity of 114 MJ and had a volume of 1.5 m³ and a mass of 480 kg. The performance of the commercial-scale TES system was tested in the laboratory. Inlet temperatures between 45°F (7.2°C) and 55°C (12.8°F) were used to "freeze" the PCM, which had a phase-change temperature of 18°C. Inlet temperatures between 75°F (23.9°C) and 95°F (35.0°C), mimicking uncooled-air conditions, were tested in discharging the PCM TES. Heat exchanger plates made of aluminum were used to create channels where the heat exchanger fluid circulated. The authors also noted that due to the aluminum material and very close spacing between aluminum sheets, fouling may result. Overall, they concluded that the latent heat system had the potential to deliver cost savings compared to ice storage systems; however, the TES unit was not tested in a real-life building environment.

Tan et al. [34] performed a techno-economic assessment of a multi-story office building on the Chalmers University of Technology Campus Johanneberg in Gothenburg, Sweden. The air-handling unit (AHU) uses water at $T = 12°C$ from a district cooling system to cool down 16°C return-air flow from the AHU. The system includes a commercial salt-hydrate PCM TES system that is connected by two heat exchangers to the district cooling and the AHU. A direct connection of the PCM TES unit to the district cooling and the AHU was not implemented due to the potential contamination if the PCM leaks. Water at 8°C from the district cooling is used to charge the PCM TES system. Water between 14°C and 16°C is used to discharge the system, and dependent on the cooling load, a fraction of the return air flows through the heat exchanger attached to the PCM TES unit. The storage outlet temperature can vary between 8°C and 16°C during a complete discharge or charge cycle.

The actual PCM TES tank consists of a rectangular steel tank of volume 11.2 m^3. The inside of the tank is filled to a height of 1.6 m with 9380 kg of Rubitherm SP11 salt hydrate. Due to phase separation that was discovered in a laboratory-scale test [36], an additional 3%-by-weight of a cross-linked polymer (sodium polyacrylate) thickening agent was added to the PCM to prevent phase separation. Copolymer polypropylene capillary-tube heat exchangers were submerged into the PCM. Interestingly, this heat exchanger configuration is flexible and can also work with ice TES systems. The reason that a submerged heat exchanger in a vat of PCM was chosen over a design where PCM is encapsulated was to reduce the risk of the PCM contaminating the system.

The initial requirement for the TES capacity of the system was specified to be 190 kWh, which would cover approximately 9.5% of the peak cooling demand of the building over 5 hours. However, as a safety measure the system was oversized, and the capacity ended up as 275 kWh.

Overall, Tan et al. [34] found that the manufacturer of the TES system overestimated the performance. This is likely due to the lack of knowledge in designing these types of systems with salt hydrates. The authors recommend more laboratory-scale tests for validation and improvement of the design. Tan et al. [34] also determined that the charging and discharging rates were the limiting factor for using more storage capacity. The authors also conducted a techno-economic analysis regarding the PCM TES system and found that the current existing TES system is not a viable business because of the high cost of investment and the costs due to cycling losses and auxiliary equipment.

11.5 FUTURE OUTLOOK OF IMPLEMENTATION OF PCM TES SYSTEMS

Alternative PCM systems for cool TES for HVAC applications are in their infancy as witnessed by the nascent technologies describe in this chapter. Many professional societies, such as the American Society of Heating, Refrigerating and Air-Conditioning Engineers (ASHRAE) and the Federation of European Heating, Ventilation and Air Conditioning Associations (REHVA), are working hard to develop design guidelines for PCM TES systems.

As seen from the preceding sections on commercial implementation of cool TES systems, effective use of the total latent heat capacity of the PCM has not been achieved, and many issues have resulted. This is due to both utilization of salt hydrates (since they have many disadvantages) and the configuration of the PCM in the TES

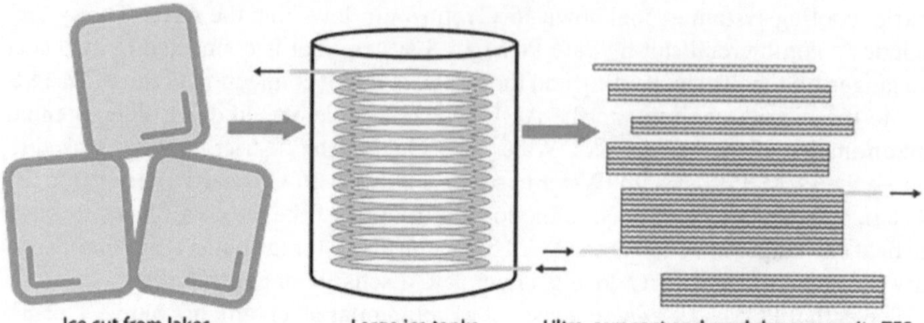

FIGURE 11.6 Evolution of TES systems from the 1800s to the future.

heat exchanger. Furthermore, the use of metals in these types of systems led to fouling due to the close spacing. It is interesting to note that most ice-based TES systems use polyethylene to prevent corrosion; however, polyethylene also helps prevent fouling.

In the future, TES system researchers and designers will need to focus on PCM heat exchanger geometry in order to effectively use the full latent heat capacity of the PCM material. Attempting to mitigate the low thermal conductivity of the PCM by adding thermal-conductivity-enhancing particles may not be a viable solution. Cool TES started by cutting large blocks of ice from lakes and has morphed into the large ice tanks with polyethylene tubes that are seen in commercial applications today (see Figure 11.6). In the future, ultra-compact and modular next-generation TES systems need to be developed. In these compact heat exchangers, PCM is arranged in layers so thin that the impact of the thermal conductivity of the material is negligible (see Eq. (11.2)), and the surface area is very high as to promote effective heat transfer and utilization of the total latent heat capacity of the PCM material. With the advent of advanced manufacturing techniques, such as additive manufacturing, geometries not possible with traditional manufacturing are now able to be realized.

REFERENCES

1. B. B. Lindsay and J. S. Andrepont, "Evolution of thermal energy storage for cooling applications," *ASHRAE J.*, 2019.
2. D. E. Knebel, "Optimal control of ice harvesting thermal energy storage systems," in *Proceedings of the 25th Intersociety Energy Conversion Engineering Conference*, Reno, Nevada, 1990, vol. 6, pp. 209–214. doi: 10.1109/IECEC.1990.748054.
3. M. M. MacCracken, "Thermal Energy Storage Myths," *Energy Eng.*, vol. 101, no. 4, pp. 69–80, 2004.
4. J. Spector, "Ice energy, thermal storage evangelist, files for bankruptcy." Accessed: Dec. 03, 2021. [Online]. Available: Ice Energy, Thermal Storage Evangelist, Files for Bankruptcy.
5. İ. Dinçer and M. Rosen, *Thermal Energy Storage: Systems and Applications*, 2nd ed. Hoboken, NJ: Wiley, 2011.
6. T. Hirata, M. Kato, K. Nagasaka, and M. Ishikawa, "Crystal ice formation of solution and its removal phenomena at cooled horizontal solid surface," *Int. J. Heat Mass Transf.*, vol. 43, no. 5, pp. 757–765, Mar. 2000, doi: 10.1016/S0017-9310(99)00181-7.

7. T. Hirata, Y. Matsuzaki, and M. Ishikawa, "Ice formation of aqueous solution and its removal phenomena on vertical cooled plate," *Heat Mass Transf.*, vol. 40, no. 11, pp. 829–834, Sep. 2004, doi: 10.1007/s00231-003-0422-3.
8. K. Matsumoto, M. Okada, T. Kawagoe, and C. Kang, "Ice storage system with water–oil mixture," *Int. J. Refrig.*, vol. 23, no. 5, pp. 336–344, Aug. 2000, doi: 10.1016/S0140-7007(99)00073-0.
9. K. Matsumoto, Y. Shiokawa, M. Okada, T. Kawagoe, and C. Kang, "Ice storage system using water–oil mixture Discussion about influence of additive on ice formation process," *Int. J. Refrig.*, vol. 25, no. 1, pp. 11–18, Jan. 2002, doi: 10.1016/S0140-7007(01)00024-X.
10. J. Wang, Y. Li, D. Zheng, H. Mikulčić, M. Vujanović, and B. Sundén, "Preparation and thermophysical property analysis of nanocomposite phase change materials for energy storage," *Renew. Sustain. Energy Rev.*, vol. 151, p. 111541, Nov. 2021, doi: 10.1016/j.rser.2021.111541.
11. L. Boussaba, S. Makhlouf, A. Foufa, G. Lefebvre, and L. Royon, "vegetable fat: A low-cost bio-based phase change material for thermal energy storage in buildings," *J. Build. Eng.*, vol. 21, pp. 222–229, Jan. 2019, doi: 10.1016/j.jobe.2018.10.022.
12. C. Cárdenas-Ramírez, M. A. Gómez, F. Jaramillo, A. G. Fernández, and L. F. Cabeza, "Experimental determination of thermal conductivity of fatty acid binary mixtures and their shape-stabilized composites," *Renew. Energy*, vol. 175, pp. 1167–1173, Sep. 2021, doi: 10.1016/j.renene.2021.05.080.
13. M. Casini, *Smart Buildings: Advanced Materials and Nanotechnology to Improve Energy-Efficiency and Environmental Performance*. Amsterdam: Elsevier: Woodhead Pub Ltd, 2016.
14. H. Ge and J. Liu, "Keeping smartphones cool with gallium phase change material," *J. Heat Transf.*, vol. 135, no. 5, p. 054503, May 2013, doi: 10.1115/1.4023392.
15. L. Abdolmaleki, S. M. Sadrameli, and A. Pirvaram, "Application of environmental friendly and eutectic phase change materials for the efficiency enhancement of household freezers," *Renew. Energy*, vol. 145, pp. 233–241, Jan. 2020, doi: 10.1016/j.renene.2019.06.035.
16. I. Gur, K. Sawyer, and R. Prasher, "Searching for a better thermal battery," *Science*, vol. 335, no. 6075, pp. 1454–1455, Mar. 2012, doi: 10.1126/science.1218761.
17. B. K. Purohit and V. S. Sistla, "Inorganic salt hydrate for thermal energy storage application: A review," *Energy Storage*, vol. 3, no. 2, Apr. 2021, doi: 10.1002/est2.212.
18. F. Frusteri, V. Leonardi, S. Vasta, and G. Restuccia, "Thermal conductivity measurement of a PCM based storage system containing carbon fibers," *Appl. Therm. Eng.*, vol. 25, no. 11–12, pp. 1623–1633, Aug. 2005, doi: 10.1016/j.applthermaleng.2004.10.007.
19. C. Ma, Y. Zhang, X. Chen, X. Song, and K. Tang, "Experimental study of an enhanced phase change material of paraffin/expanded graphite/nano-metal particles for a personal cooling system," *Materials*, vol. 13, no. 4, p. 980, Feb. 2020, doi: 10.3390/ma13040980.
20. P. J. Shamberger, "Cooling capacity figure of merit for phase change materials," *J. Heat Transf.*, vol. 138, no. 2, p. 024502, Feb. 2016, doi: 10.1115/1.4031252.
21. M. T. Barako, S. Lingamneni, J. S. Katz, T. Liu, K. E. Goodson, and J. Tice, "Optimizing the design of composite phase change materials for high thermal power density," *J. Appl. Phys.*, vol. 124, no. 14, p. 145103, Oct. 2018, doi: 10.1063/1.5031914.
22. K. Yazawa, P. J. Shamberger, and T. S. Fisher, "Ragone relations for thermal energy storage technologies," *Front. Mech. Eng.*, vol. 5, p. 29, Jun. 2019, doi: 10.3389/fmech.2019.00029.
23. J. Woods, A. Mahvi, A. Goyal, E. Kozubal, A. Odukomaiya, and R. Jackson, "Rate capability and Ragone plots for phase change thermal energy storage," *Nat. Energy*, vol. 6, no. 3, pp. 295–302, Mar. 2021, doi: 10.1038/s41560-021-00778-w.
24. R. E. Githens, W. R. Minor, and V. J. Tosmic, "Flexible tube heat exchangers," *Chem. Eng. Prog.*, vol. 61, no. 7, pp. 55–62, 1965.

25. M. A. Arie, A. H. Shooshtari, R. Tiwari, S. V. Dessiatoun, M. M. Ohadi, and J. M. Pearce, "Experimental characterization of heat transfer in an additively manufactured polymer heat exchanger," *Appl. Therm. Eng.*, vol. 113, pp. 575–584, Feb. 2017, doi: 10.1016/j.applthermaleng.2016.11.030.
26. D. C. Deisenroth, R. Moradi, A. H. Shooshtari, F. Singer, A. Bar-Cohen, and M. Ohadi, "Review of heat exchangers enabled by polymer and polymer composite additive manufacturing," *Heat Transf. Eng.*, vol. 39, no. 19, pp. 1648–1664, Nov. 2018, doi: 10.1080/01457632.2017.1384280.
27. M. A. Arie, D. M. Hymas, F. Singer, A. H. Shooshtari, and M. Ohadi, "An additively manufactured novel polymer composite heat exchanger for dry cooling applications," *Int. J. Heat Mass Transf.*, vol. 147, p. 118889, Feb. 2020, doi: 10.1016/j.ijheatmasstransfer.2019.118889.
28. T. B. Freeman, K. Nabutola, D. Spitzer, P. N. Currier, and S. K. S. Boetcher, "3D-printed PCM/HDPE composites for battery thermal management," in *Volume 8B: Heat Transfer and Thermal Engineering*, Pittsburgh, Pennsylvania, USA, Nov. 2018, p. V08BT10A041. doi: 10.1115/IMECE2018-86081.
29. T. B. Freeman, D. Spitzer, P. N. Currier, V. Rollin, and S. K. S. Boetcher, "Phase-change materials/HDPE composite filament: A first step toward use with 3D printing for thermal management applications," *J. Therm. Sci. Eng. Appl.*, vol. 11, no. 5, p. 054502, Oct. 2019, doi: 10.1115/1.4042592.
30. T. B. Freeman, M. A. Messenger, C. J. Troxler, K. Nawaz, R. M. Rodriguez, and S. K. S. Boetcher, "Fused filament fabrication of novel phase-change material functional composites," *Addit. Manuf.*, vol. 39, p. 101839, Mar. 2021, doi: 10.1016/j.addma.2021.101839.
31. M. Jokiel, "Development and performance analysis of an object-oriented model for phase change material thermal storage." Department of Thermal Energy SINTEF Energy Research, 2016.
32. M. Alam, P. X. W. Zou, J. Sanjayan, and S. Ramakrishnan, "Energy saving performance assessment and lessons learned from the operation of an active phase change materials system in a multi-storey building in Melbourne," *Appl. Energy*, vol. 238, pp. 1582–1595, Mar. 2019, doi: 10.1016/j.apenergy.2019.01.116.
33. R. M. Saeed, J. P. Schlegel, R. Sawafta, and V. Kalra, "Plate type heat exchanger for thermal energy storage and load shifting using phase change material," *Energy Convers. Manag.*, vol. 181, pp. 120–132, Feb. 2019, doi: 10.1016/j.enconman.2018.12.013.
34. P. Tan, P. Lindberg, K. Eichler, P. Löveryd, P. Johansson, and A. S. Kalagasidis, "Thermal energy storage using phase change materials: Techno-economic evaluation of a cold storage installation in an office building," *Appl. Energy*, vol. 276, p. 115433, Oct. 2020, doi: 10.1016/j.apenergy.2020.115433.
35. J. M. Marín, B. Zalba, L. F. Cabeza, and H. Mehling, "Improvement of a thermal energy storage using plates with paraffin–graphite composite," *Int. J. Heat Mass Transf.*, vol. 48, no. 12, pp. 2561–2570, Jun. 2005, doi: 10.1016/j.ijheatmasstransfer.2004.11.027.
36. P. Tan, P. Lindberg, K. Eichler, P. Löveryd, P. Johansson, and A. S. Kalagasidis, "Effect of phase separation and supercooling on the storage capacity in a commercial latent heat thermal energy storage: Experimental cycling of a salt hydrate PCM," *J. Energy Storage*, vol. 29, p. 101266, Jun. 2020, doi: 10.1016/j.est.2020.101266.

12 Evolution of Melt Path in a Horizontal Shell and Tube Latent Heat Storage System for Concentrated Solar Power Plants

Soheila Riahi, Michael Evans, Ming Liu, Rhys Jacob, and Frank Bruno
University of South Australia

CONTENTS

12.1 INTRODUCTION: BACKGROUND

Effective and reliable thermal storage systems are required for the next generation of high-temperature concentrating solar power (CSP) plants. Supercritical CO_2 (sCO_2) as the heat transfer fluid (HTF) in a Brayton power generation cycles was proved to provide higher efficiencies at higher temperatures (greater than 700°C) versus the conventional Rankine cycles with superheated steam up to 550°C (Turchi et al., 2013). Different configurations of shell and tube latent heat thermal storage systems including vertical and horizontal orientations have been a subject of research as potential effective storage systems (Belusko et al., 2016a, 2016b; Riahi et al., 2017).

DOI: 10.1201/9781003213260-12

Apart from the thermal design, the mechanical design involves a highly challenging problem considering the possibility of the failure of the system due to the thermal stresses and corrosion. There are a few studies of thermal stress analysis of tubular solar receivers (Abedini-Sanigy et al., 2015; Logie et al., 2018; Ortega et al., 2016); however, there are a limited number of studies in the thermomechanical scope of phase change material (PCM) systems. The stress/strain analysis are crucial considering the high operating temperature range during the charging and discharging processes (750°C–550°C) and cyclic loading due to the nature of the system. In an experimental and theoretical study, Maruoka and Akiyama (2003) investigated the crack formation in the nickel shell of their proposed encapsulated PCM (lead) system. Experimental work showed that different volume expansion of the PCM and nickel during the melting process results in increased inner pressure and tensile stress.

Nomenclature

C	Specific heat, J/kg K
d	Tube diameter, m
D	Dimension
E	Modulus of elasticity, MPa
h	Heat transfer coefficient, W/m^2 K
H	Latent heat of fusion, J/kg
k	Thermal conductivity, W/m K
P	Pressure, Pa
R	Radius, m
s	second
S	Allowable design stress or strength
t	Time, s
T	Temperature, °C
U	Tensile strength, MPa

Abbreviation

LHTES	Latent heat thermal energy storage
HTF	Heat transfer fluid
PCM	Phase change material
CFD	Computational fluid dynamics
FEA	Finite element analysis
AISI	American Iron and Steel Institute

Greek Letters

α	Thermal expansion coefficient, K^{-1}
Δ	Difference
ε	Strain, %
μ	Dynamic viscosity, Pa s
ν	Poisson ratio, –
ρ	Density, kg/m^3
σ	Stress

Subscript

eq	Equivalent
h	Hot wall, hot HTF or hydraulic
i	Inner
l	Longitudinal
m	Maximum
o	Initial or reference, outer
R	Radial
y	Yield

Using analytical and numerical methods, Dal Magro et al. (2016) assessed thermomechanical viability of a PCM system for temperature smoothing of waste gas where aluminium as the PCM is confined in a stainless-steel shell or in an annulus. The authors suggested that only the annulus can withstand the thermal stresses, which is sensitive to the diameter of the inner tube and the thickness of the outer tube of annulus. Lopez et al. (2010) studied the stress on a composite material of graphite and PCM due to the volume expansion during the melting process. In another work,

Blaney et al. (2013) investigated the stress/strain on spherical nickel and cylindrical stainless-steel shells containing zinc as PCM during a melting process. They found that the welded cylindrical stainless steel is a viable option with only 0.5% strain with an initial volume fraction of 86% for the PCM. In another work by the authors of this study (Riahi et al., 2021), structural integrity of a vertical PCM system under different operating conditions were investigated. Learning from that study has inspired work on a horizontal case, with results presented here.

The thermal and mechanical analyses provide insights into the optimum selection of materials for the PCM, shell, and tubes leading to a cost-effective storage system for CSP plants. In regard to the thermal analysis, a previous study by the authors of this work showed that the horizontal orientation of a shell and tube system provides a more uniform process of melting and solidification, resulting in a higher effectiveness for a longer period of time (Riahi et al., 2017). Therefore, a horizontal shell and tube design is selected for this study. However, a careful stress/strain analysis is required due to the high-temperature gradients, particularly at the start of a charging process. The ASME Boilers and Pressure Vessels Code (BPVC), Section VIII, Division 1 provides guidelines for Design by Rules from BPVC, Section VIII, Division 2 (2013a) of conventional boilers and superheaters. However, for higher temperatures, pressures, and diurnal cyclic operating conditions, a Design by Analysis from BPVC, Section VIII, Division 2 (2013b) is the preferred method.

A critical phenomenon involving the early stages of a melting process is the formation of liquid PCM around the tubes. In a vertical orientation, gravity provides a path for the hot melted PCM to flow, preventing a melt trap in a certain location around a tube. Conversely, in a horizontal orientation gravity is not helpful. A hot-liquid PCM trap is a concern which may result in a hot spot location around a tube (e.g. close to a tube sheet where the HTF flows from a manifold into the tubes) introducing a potential tube failure due to the high thermal stresses.

This study therefore aims to investigate the possibility of a melt trap around a tube in a horizontal orientation of a shell and tube thermal energy storage system, from which a thermal stress analysis has been completed. This is a novel phenomenon to study, which is important in a horizontal shell and tube storage system. A high-temperature PCM with a melting temperature equal to 705.8°C (named PCM705) is considered as the storage medium with sodium as the HTF due to its high thermal conductivity.

12.2 PCM SYSTEM

PCM705 has been developed in the thermal storage laboratory at the University of South Australia where its thermophysical properties, stability and compatibility with austenitic steels have been examined. A lab-scale vertical PCM system was designed in the same laboratory and manufactured by Britannia Jahco (Britannia-Jahco). The system comprised of seven tubes (outer and inner diameters of 17.15 and 12.53 mm, respectively, and 60 mm tube spacing) in a shell with the diameter of 0.275 m (Figure 12.1a). The PCM705 is confined in the shell while sodium as the HTF passes through the tubes. The tube and shell were fabricated from SS347H as an austenitic steel suitable for high temperatures and good corrosion resistance

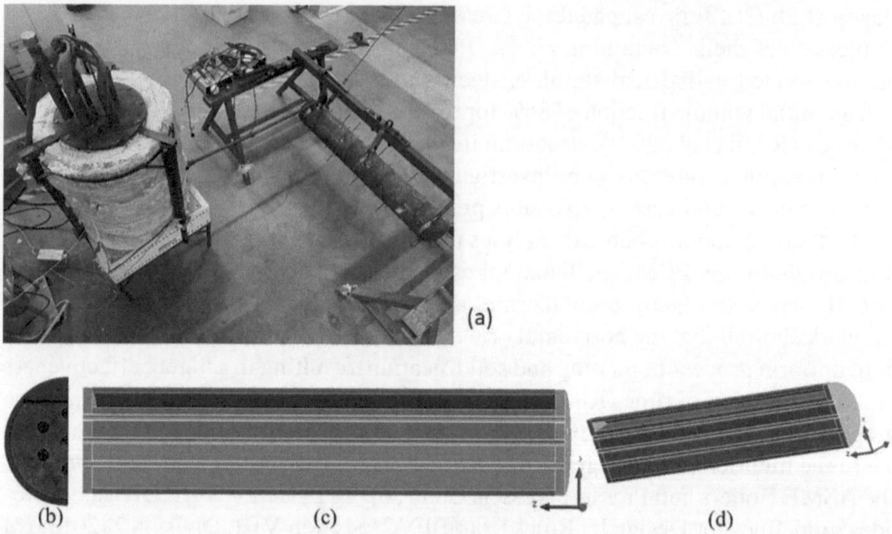

FIGURE 12.1 (a) Picture of the PCM horizontal shell and tube storage tank (right) and the vertical (left) heating and filling system. (b) A cross section of the grid used for modelling. (c) Front view of the symmetrical half geometry of PCM system including the void space at the top of the shell. (d) A three-dimensional view of the model.

TABLE 12.1
Thermophysical Properties of Materials (Boerema et al., 2012; Kenisarin, 2010)

Material	Composition (wt.%)	ρ, kg/m³ Solid–Liquid	C_{pl}, kJ/kg K Solid–Liquid	ΔH, kJ/kg	k, W/m K	μ, Pa s Liquid	T_m, °C
PCM705	Na_2CO_3-K_2CO_3 (47:52)	2400/1919	1.4/1.6	145	0.6/0.5	0.007	705.8
Sodium	Na	783	1.252	–	51.8	1.8e-4	–
Argon	Ar	1.6228	0.5206	–	0.0158	2.1e-5	–
SS347H	Stainless steel	7800	880	NA	16	NA	NA

according to a Designer's Handbook from American Iron and Steel Institute (AISI) (2020). The thermophysical properties of the materials are shown in Table 12.1.

The vertical lab-scale system was considered for transient melting-solidification testing for 1000 hours and 50 cycles. For the replicate horizontal system, the same number of cycles in 1000 hours testing was assumed.

12.3 NUMERICAL MODELLING

12.3.1 Geometry and Grid

A horizontal version of the lab-scale PCM system was considered for the investigation of melt path during the early stages of a melting process. Using ANSYS Design Modeller, a three-dimensional (3D) geometry and grid of a symmetrical half of the shell and tube system was generated with a void space filled with argon at the top of the shell as shown in Figure 12.1b–d. The 3D model comprises of 20 million cells and six domains: three solid domains of tube sheets and tubes, in addition to three fluid domains of sodium as HTF, PCM, and the argon domain at the top. The tube sheets are the interface between the manifolds (inlet and outlet) containing HTF, and the shell containing the PCM and argon. The tube sheet at the inlet side where the HTF flows in the tubes is considered as the most critical section due to the high temperature differences between the HTF and frozen PCM at the initial condition of the melting process.

12.3.2 Thermo-hydraulic Modelling

Three-dimensional computational fluid dynamic (CFD) modelling of the melting process with natural convection was performed using ANSYS FLUENT (2020R1). Time dependent Navier–Stokes equations were solved, assuming a constant and equal density for the solid and liquid phases, except for the buoyancy term in the momentum calculation where a linear density–temperature relation was considered as the Bossinesq approximation. Ensuring zero velocity for the solid phase, a sink term is considered with a mushy zone constant (A_{mush}) which measures how fast the fluid velocity approaches zero as it solidifies. More detail of the governing equations and model validation with experimental data can be found in a previous study by the authors (Riahi et al., 2016).

Two cases with different initial conditions investigated. In Case 1 (T_h = 750°C, T_{ini} = 700°C, ΔT = 50 K), the imposed temperature difference on the system is lower with the frozen PCM705 initially set at T_{ini} = 700°C, and the HTF inlet temperature at T_h = 750°C. In Case 2 (T_h = 750°C, T_{ini} = 650°C, ΔT = 100 K), initial temperature of PCM was assumed at T_{ini} = 650°C with the HTF inlet temperature of T_h = 750°C leading to a higher temperature difference (ΔT = 100 K) through the system.

The boundary condition at the inlet tube sheet in contact with the HTF in the inlet manifold was set at 750°C. The HTF inlet was set as velocity inlet at 750°C and the outlet was set as pressure outlet. The symmetry boundary condition was considered at the symmetrical faces of tubes, HTF, PCM705, and shell. Other boundaries including the outer surface of shell were considered adiabatic. Temperature profiles, melt fraction and heat flux through the tube surfaces were monitored during the melting processes.

12.3.3 Thermomechanical Modelling

Three-dimensional finite element analysis (FEA) was performed using ANSYS workbench to link the results of thermo-hydraulic modelling to thermomechanical modelling, namely Fluent to ANSYS structural. Considering the low operating pressure of the system (less than 3 bar), the thermal driven stresses were investigated. The transient and cyclic nature of the operation of LHTES systems requires to investigate structural behaviour of the metal part of the system at the early stages of a cycle where the temperature differences are high, resulting in high thermal stresses. From a thermomechanical analysis, a structural design and material selection can be performed or assessed to achieve a cost-effective outcome. Such an analysis can be used to calculate the highest rate of heat transfer that a system can operate without failure in short time and also during the expected lifetime of the system.

A FEA model was validated with an analytical model in a previous study by the authors (Riahi et al., 2021). The validated model was used for thermal stress analysis where equivalent thermal (von Mises) stress and strain under two different initial conditions ($\Delta T = 50$ K, $\Delta T = 100$ K) were calculated using Eqs. (12.1) and (12.2).

$$\sigma_{eq} = \sqrt{\frac{(\sigma_1 - \sigma_2)^2 + (\sigma_1 - \sigma_3)^2 + (\sigma_3 - \sigma_2)^2}{2}} \tag{12.1}$$

$$\varepsilon_{eq} = \frac{\sqrt{2}}{3}\sqrt{(\varepsilon_1 - \varepsilon_2)^2 + (\varepsilon_1 - \varepsilon_3)^2 + (\varepsilon_3 - \varepsilon_2)^2} \tag{12.2}$$

In a simple elastic analysis, it is assumed that material undergoes an elastic deformation under the imposed load where stress (σ) is proportional to strain (ε) and the modulus of elasticity (E) of the material (SS347H) as shown in Eq. (12.3). Considering major stresses driven by thermal load in this study, strain is proportional to the imposed temperature difference (ΔT) and the thermal expansion coefficient of material (α) as shown in Eq. (12.4).

$$\varepsilon = \alpha \, \Delta T \tag{12.3}$$

$$\sigma = E \, \varepsilon \tag{12.4}$$

12.4 RESULTS AND DISCUSSION

12.4.1 Thermo-Hydraulic Analysis

Figure 12.2 shows the results of modelling during the first 460 seconds of the melting processes. The profiles present the temperature evolution in time (from bottom to top) and along the length of the tubes and axial coordinate (left to right). After 400 seconds, minor changes were observed in time and along the tubes as the profiles after 460 seconds are very close to those of 400 seconds.

The black profiles for Case 1 ($T_h = 750$°C, $T_{ini} = 700$°C, $\Delta T = 50$ K), present longer plateau in the axial coordinate and closer profiles in time (bottom to top) consistent with lower temperature gradients. For Case 2 with the imposed higher temperature difference ($T_h = 750$°C, $T_{ini} = 650$°C, $\Delta T = 100$ K), the steep decline in grey temperature profiles demonstrate the higher axial temperature gradients. Similarly, the wider distance between the profiles from bottom to top demonstrates the higher temperature gradients in time. While a higher temperature gradient is favourable to deliver a higher heat transfer rate, the consequential thermal stress is a design constraint.

Figure 12.3 shows the evolution of melt front in time, after 60, 100, 200, and 400 seconds and in the axial direction along the outer tube surface where heat transfer occurs. Heat transfer from HTF inside the tubes to PCM in the shell results in the formation of a layer of melt which grows and moves during time. The interface between

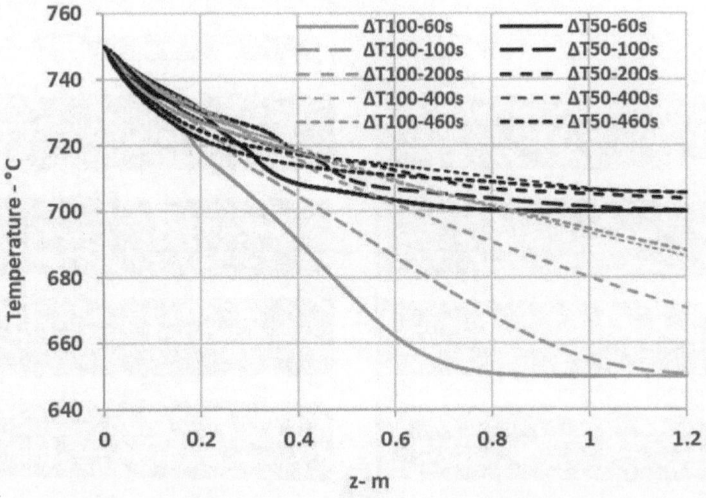

FIGURE 12.2 Comparison of temperature profiles in early stages of melting processes after 60, 100, 200, 400 and 460 seconds. Black profiles represent Case 1 with $\Delta T = 50$ K, and grey profiles represent Case 2 with $\Delta T = 100$ K.

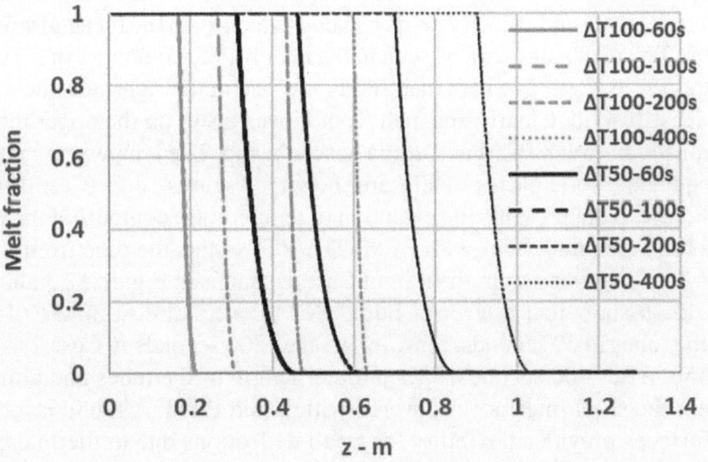

FIGURE 12.3 Evolution of melt front along the tube in the melting processes after 60, 100, 200, and 400 seconds. Black lines represent the case with $\Delta T = 50$ K, and grey lines represent the case with ΔT=100K.

the liquid and solid PCM, namely melt front, becomes the main heat transfer surface area which is called a moving boundary. In Case 1 with $T_{ini} = 700°C$, it takes shorter period of time to form a melt layer of PCM ($T_m = 705.8°C$) around the tube and inlet tube sheet in contact with HTF at 750°C. On the other hand, grey lines in Figure 12.3 show a major delay in the formation of a liquid layer and movement of melt front in Case 2 with $T_{ini} = 650°C$.

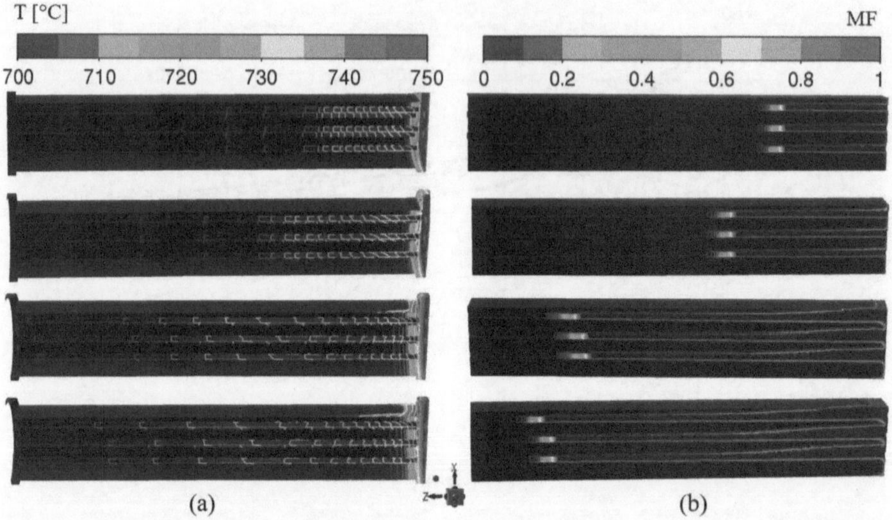

FIGURE 12.4 (a) Temperature and (b) melt fraction (MF) contours in Case 1 with $\Delta T = 50$ K; from top to bottom: at 60, 100, 300, and 400 seconds.

The impact of the delay in melt formation and movement can been seen in the evolution of temperature and velocity fields and consequently in the thermal stress–strain distribution. The melt fronts can be seen in the melt fraction contours in Figures 12.4b (Case 1) and 12.5b (Case 2) where melt fraction is one in the right side and it is zero in the left side, still solid. Clearly, the melt front moves faster on the upper tubes due to the heat transfer between PCM and argon gas at the top. The temperature gradients in both the liquid and solid parts of PCM are shown in Figures 12.4a (Case 1) and 12.5a (Case 2). Figure 12.4a presents the evolution of temperature contours at the tube, tube sheet, and the argon interfaces with PCM. The HTF enters the tube from the right at 750°C and exits the tubes from the left. In Case 1, graphs in Figure 12.3 and contours in Figure 12.4b show that a layer of liquid PCM covers almost 30% (0.4 m) of the length of the tubes in 60 seconds, however, it takes 200 seconds in Case 2 as shown in Figure 12.5b. After 400 seconds, 75% of outer length of the tubes and almost 100% of the inlet tube sheet interface is covered with liquid PCM. A liquid layer covering the tube surfaces provides flexibility for small deflections due to thermal expansion compared with a tube covered by solid PCM, particularly at higher temperature differences between PCM and HTF. This will be shown through the FEA modelling where the cause and effects of thermal stresses are analysed.

The wider bands of temperature contours from top to bottom as presented in Figure 12.4a suggest that the temperature gradient in the axial direction along the tube diminishes in time which is consistent with Figure 12.2. However, closer temperature bands at the shell junction with inlet tube sheet (can be seen in right side of temperature contours in Figure 12.4a) suggest higher temperature gradients compared with the tubes. This is the consequence of higher heat transfer rate between HTF inside the tubes and PCM outside the tube which leads to a layer of melt formation, moving and mixing the fluid and lower temperature gradients. At the shell side,

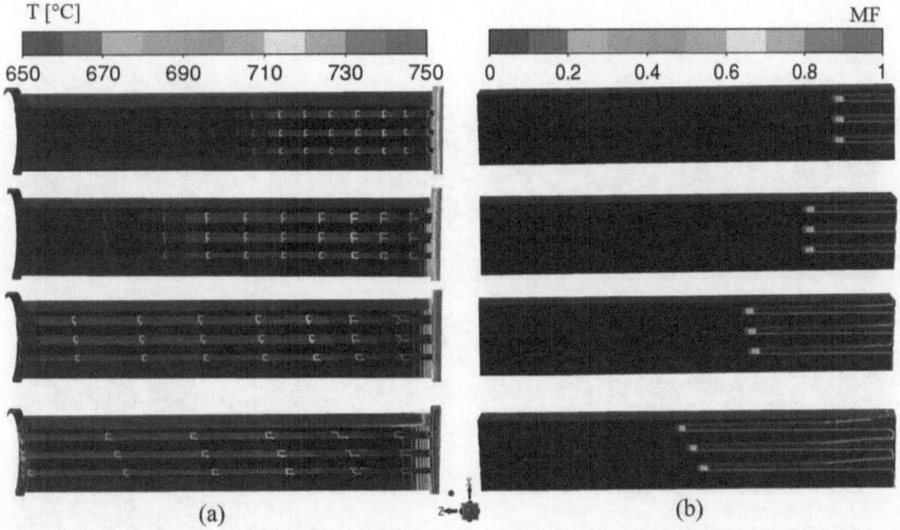

FIGURE 12.5 (a) Temperature and (b) melt fraction (MF) contours in the early stages of the melting process with $\Delta T = 100$ K; from top to bottom: at 60, 100, 200, and 400 seconds.

the PCM remains solid for longer (up to 200 seconds when the wider bands show up) as the main heat transfer mode is conduction between solid PCM, and the inlet tube sheet. At the top of the shell in contact with argon, convection heat transfer plays a role in reducing temperature gradients in 60 seconds as shown in Figure 12.6. The ramifications on the thermal stress and strain are discussed in FEA analysis later in this chapter.

In two previous study from the authors (Riahi et al., 2017, 2018), it has been shown that in the horizontal annulus of tube PCM small eddy cells namely Benard cells (Bejan, 2013) form in the liquid layer which mix and convey the liquid mass resulting in higher rate of convection heat transfer and more uniform temperature distribution. Evolution of the eddy cells are shown in the contours of velocity field in Figure 12.6 (Case 1) and Figure 12.7 (Case 2). While the larger convection cells are apparent in the argon area at the top of the shell in both cases, the small eddy cells form at the PCM-tube sheet interface in 60 seconds in Case 1 and in 100 second in Case 2. In Case 1, after 100 seconds and in Case 2 after 200 seconds, the eddy cells can be seen at the upper side of each tube with increasing numbers towards the 400 seconds. Therefore, despite the slow axial movement of liquid PCM, the lateral movement and permeation of liquid into the solid PCM (the mushy zone) prevents a melt trap and a local hot spot around a tube. In the argon area, maximum velocity increases from 0.37 to 0.57 m/s (red band) as temperature rise in the upper part of the shell, however, velocity of the eddy cells is steady at about 0.02 m/s as the temperature of the liquid PCM remains around the melting point, 705.8°C.

Figures 12.6 and 12.7 show the convection cells which form at the interface of argon with tube sheet (right side) and PCM. The argon gas in contact with the top of the tube sheet absorbs heat from the HTF in the manifold resulting in dense and

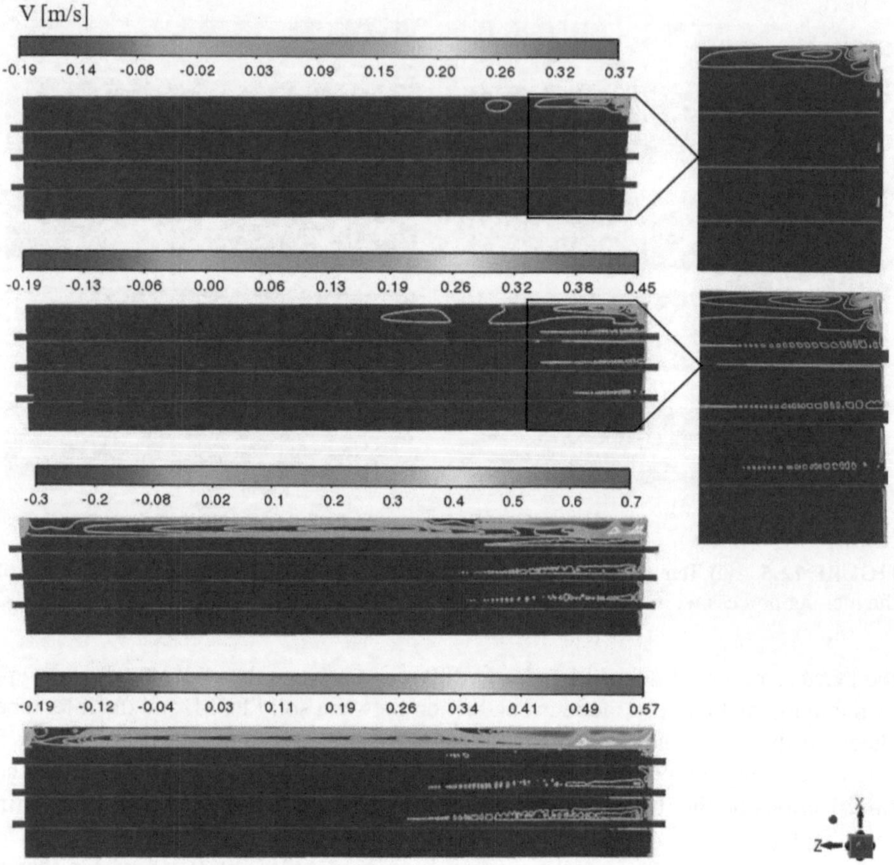

FIGURE 12.6 Velocity contours in the early stages of the melting process with $\Delta T = 50$ K; from top to bottom: at 60, 100, 300, 400 seconds.

high-velocity cells(above 0.3 m/s). The convection cells with lower velocity (below 0.25 m/s) form at the interface with PCM due to the temperature differences between solid PCM and argon. This phenomenon provides a way to mix and transport liquid PCM upward into the void area at the top of the shell which was observed in an experimental work by Shmueli et al. (2010). Moreover, the movement of the melted PCM towards the top of the shell due to the thermosiphon effect was observed in another experimental study by Longeon et al. (2013). This provides more certainty that the hot melted PCM around the tubes connect across the inlet tube sheet and move upwards into the argon area.

Overall, the results of the thermo-hydraulic analysis demonstrate that with the assumed boundary conditions, a melt path forms shortly after the PCM at the tube sheet interface reaches its melting point. It is earlier in Case 1 (60 seconds) and later in Case 2 (100 seconds). The formation of small eddy cells in the liquid layers around the tubes prevents a melt trap and hot spots in the most critical area close to the tube sheet in a horizontal shell and tube thermal storage system. Moreover, the convection

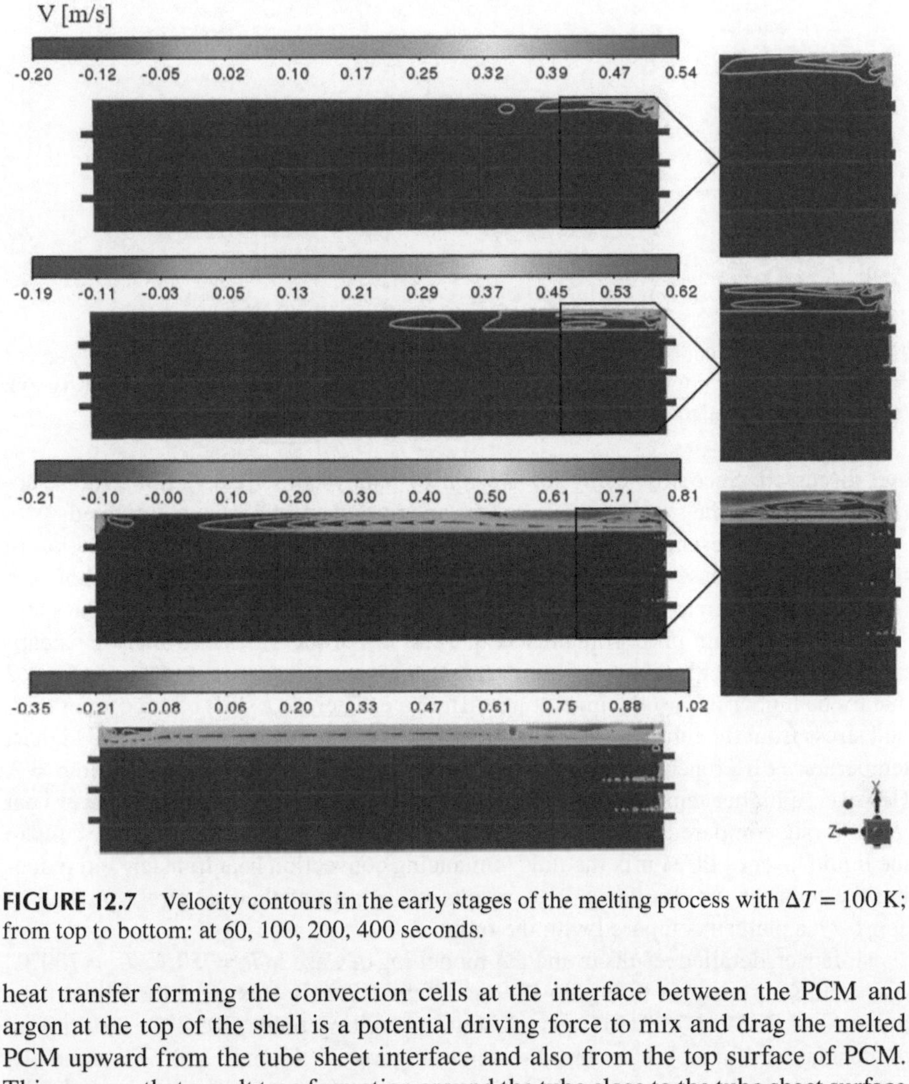

FIGURE 12.7 Velocity contours in the early stages of the melting process with $\Delta T = 100$ K; from top to bottom: at 60, 100, 200, 400 seconds.

heat transfer forming the convection cells at the interface between the PCM and argon at the top of the shell is a potential driving force to mix and drag the melted PCM upward from the tube sheet interface and also from the top surface of PCM. This ensures that a melt trap formation around the tube close to the tube sheet surface is unlikely. The following thermomechanical analyses show the consequential thermal stress–strain distributions through the shell and tube PCM system.

12.4.2 Thermomechanical Analysis

The results of the thermo-hydraulic analysis provided insights into the temperature and velocity distributions and melt path formation in space and time. Using a thermoelastic method and importing the temperature profiles into the FEA model in ANSYS structural, the results and analysis are the subject of this section. Firstly, the equivalent (von Mises) stress and strain distributions in the two cases of ($T_h = 750$°C, $T_{\text{ini}} = 700$°C, $\Delta T = 50$ K) and ($T_h = 750$°C, $T_{\text{ini}} = 650$°C, $\Delta T = 100$ K) are presented

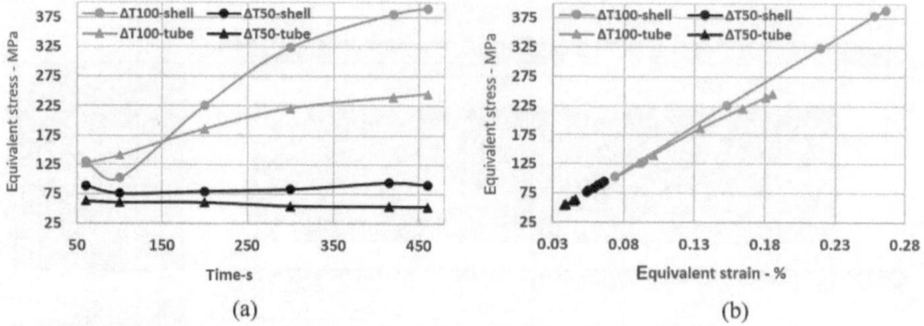

FIGURE 12.8 Evolution of equivalent (von Mises) thermal stress–strain in the early stages of the melting process for Case 1 with $\Delta T = 50$ K (black) and Case 2 with $\Delta T = 100$ K (grey): (a) maximum equivalent stress in 400 seconds and (b) equivalent stress versus strain.

and discussed. Secondly, using the Design-by-Analysis method (2013b), the structural integrity of the system under the two operating conditions is investigated.

Figure 12.8 presents an overall view of the equivalent stress–strain evolution in time for the two cases. In Case 1, the black lines in Figure 12.8a clearly show the trend of thermal stress evolution in tubes and shell with slightly higher stress in shell.

The stress–strain proportionality (Eq. 12.4) in the thermoelastic analysis is captured in Figure 12.8b. The grey lines represent higher stress–strain values in Case 2 due to the higher imposed temperature difference. There is a shift of maximum thermal stress from the tubes to shell after 60 seconds as the tubes experience the highest temperature difference earlier than the shell when HTF flows into tubes at time = 0. However, a higher temperature difference lasts longer in the shell due to lower heat transfer rate compared with the tube sides. In the tube interface, small eddy cells in the liquid layer of PCM mix the fluid, enhancing convection heat transfer, and reducing the temperature gradients. This results in lagging of thermal stress in shell to approach a plateau compared with the tubes.

Moreover, detailed results from FEA modelling of Case 1 ($T_h = 750$°C, $T_{ini} = 700$°C, $\Delta T = 50$ K) are shown in Figure 12.9 and Table 12.2. In this case, the initial temperature of PCM at 700°C is very close to its melting point (705.8°C) resulting in the formation of a melt layer in 60 seconds (Figures 12.4b and 12.6). Consequently, a moderate thermal stress of 63.5 MPa appears at the junction of tubes and inlet tube sheet (in contact with sodium inlet header at 750°C) and declines smoothly to 53.7 MPa (Figure 12.9b, top to bottom) as the temperature gradients decline in 400 seconds (Figure 12.2). The more concentrated stress at the start (top) is recognisable with a contrast between the highest stress (63.5 MPa) and the lowest (12.8 MPa) which becomes more distributed and uniform from 200 seconds (more green colour). Noteworthy, the maximum stress appears inside the tube at the junction to the tube sheet where the highest temperature difference occurs.

On the other hand, Figure 12.9a and Table 12.2 show a higher thermal stress about 90 MPa in the shell joint to inlet tube sheet compared with about 60 MPa in tubes due to the higher temperature gradients (Figure 12.4a). The maximum thermal stress appears at the lower side of the shell in contact with PCM (60 – 70 MPa) and reduces

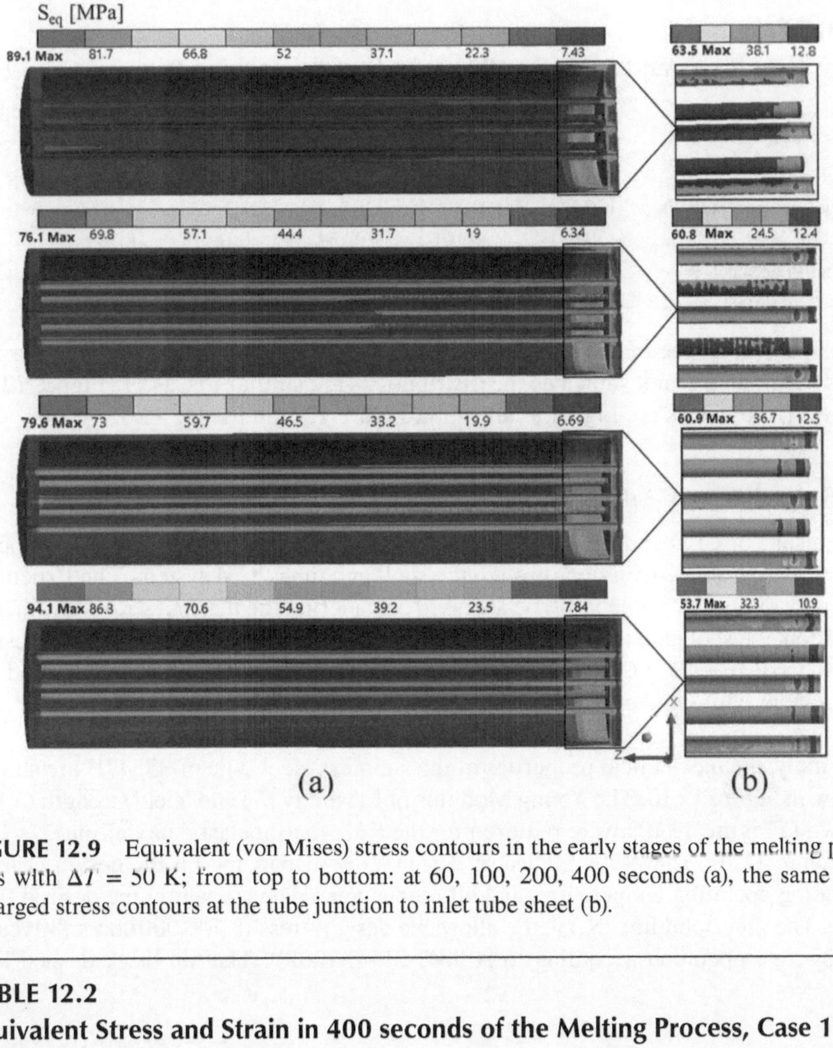

FIGURE 12.9 Equivalent (von Mises) stress contours in the early stages of the melting process with $\Delta T = 50$ K; from top to bottom: at 60, 100, 200, 400 seconds (a), the same but enlarged stress contours at the tube junction to inlet tube sheet (b).

TABLE 12.2
Equivalent Stress and Strain in 400 seconds of the Melting Process, Case 1

$\Delta T = 50$ K	60 seconds	100 seconds	200 seconds	300 seconds	420 seconds	460 seconds
S_{eq}-tubes, MPa	63.5	60.8	60.9	54.7	53.7	53.3
S_{eq}-shell-tube sheet, MPa	89.1	76.1	79.6	83.1	94.1	90.3
ε_{eq}-tubes, %	0.046	0.044	0.045	0.039	0.039	0.038
ε_{eq}-shell-tube sheet, %	0.062	0.054	0.055	0.059	0.066	0.064

at the upper side which is in contact with argon (below 50 MPa). Similarly, the shell at the junction with the inlet tube sheet experiences a higher strain compared with the tubes as thermal strain is proportional to the temperature gradients according to Eq. (12.3). Thermal strain declines during time (Table 12.2) as the melt movement and mixing reduces temperature gradients as shown in Figures 12.2 and 12.4a.

TABLE 12.3
Equivalent Stress and Strain in 400 seconds of the Melting Process, Case 2

$\Delta T = 100$ K	60 seconds	100 seconds	200 seconds	300 seconds	420 seconds	460 seconds
S_{eq}-tubes, MPa	126.9	141.1	186.1	220.7	239.4	245
S_{eq}-shell-tube sheet, MPa	130.1	102.4	226.2	323.2	379.9	390
ε_{eq}-tubes, %	0.091	0.101	0.134	0.164	0.180	0.185
ε_{eq}-shell-tube sheet, %	0.094	0.074	0.153	0.219	0.258	0.266

The results of the FEA modelling for Case 2 are shown in Table 12.3. The thermal stress contours are not shown as the distributions are similar to Case 1 (Figure 12.9), however, the values are higher as shown in Figure 12.8 and Table 12.3.

12.4.3 Thermoelastic Analysis

The results of CFD and FEA analysis provided insight into the temperature, velocity and stress–strain distribution through the shell and tube PCM system. The structural integrity of the system needs to be assessed considering the thermal stress–strain distribution, the shell and tube material properties and standard codes. Applying Design by Analysis from BPVC, Section VIII, Division 2 (2013b) as the preferred method for high-temperature and cyclic operation, the results of the thermoelastic stress–strain calculation are compared with allowable stress values in the BPVC.

Firstly, the mechanical properties of the stainless-steel type of SS347H are shown below in Figure 12.10. The Young Modulus of Elasticity (E) and Yield strength (S_y) as shown in Figure 12.10a were required for the FEA thermoelastic modelling.

Figure 12.10b shows the allowable design stress depending on the design criteria including operating temperature, and allowable creep damage within the desired lifetime. The grey solid line (S_m) is the allowable design stress in 100,000 hours (30 years) steady-state operation according to ASME BPVC (2010). The top black dashed line

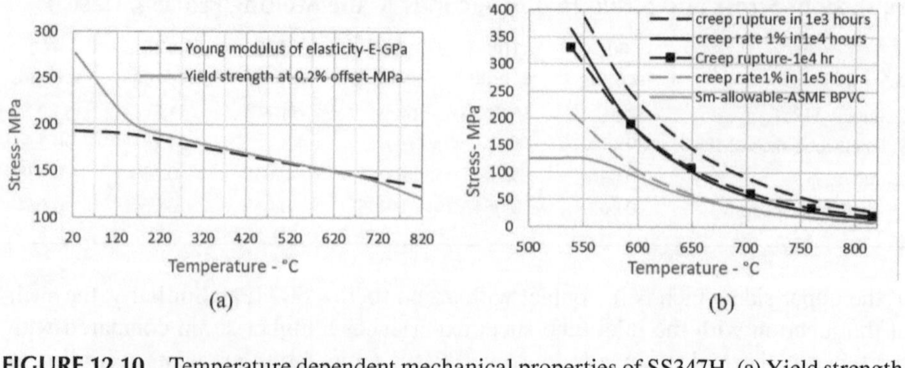

FIGURE 12.10 Temperature dependent mechanical properties of SS347H. (a) Yield strength and Young Modulus of Elasticity. (b) Stress from ASME BPVC (2010) and creep properties from AISI Designer's Handbook (2020).

shows the stress values that can result in creep rupture in 1000 hours and the back dashed line with square markers shows creep rupture in 10,000 hours operation. The grey dashed line shows the stress value which results in 1% strain in 100,000 hours which is very close to the S_m values in temperatures higher than 600°C. The black solid line represents higher values of allowable design stress for strain rate of 1% in 10,000 hours.

For cyclic operation under high temperature and/or pressure, creep-fatigue analysis is required following the guidelines from BPVC, Section VIII, Division 2 (2013b). For the lab-scale system under investigation assuming 50 cycles in 1000 hours, the relatively low number of cycles results in a minor fatigue damage compared with the expected creep damage due to the high operation temperature. Considering safety, the system is not designed (expected) to rupture during the desired 1000 hours, therefore, 1% strain is considered. From the available stress values for rupture in 1000 and 10,000 hours, and 1% creep in 10,000 hours from a Designer's Handbook published by AISI (2020) at 750°C (Figure 12.10b), an estimated allowable design stress (S) for 1% creep in 1000 hours is 45.4 MPa. This estimation is justified assuming the similarity between isochronous curves ((σ, ε) for different lifetimes (t)) (CO, 2012).

According to the Design-by-Analysis guideline from BPVC, Section VIII, Division 2 (2013b), the maximum equivalent stress should be less than the larger quantity of $2S_y$ and/or $3S$. For SS347H at 750°C with S = 45.4 MPa, Eqs. (12.5)–(12.7) can be used for the calculation and comparison of the stresses as follows:

$$3S = 136.3\,\text{MPa}, \quad \text{and} \quad 2S_y = 221.2\,\text{MPa}, \tag{12.5}$$

$$\text{In case one with } \Delta T = 50\,\text{K}, \sigma_{\text{eq}}(\text{tube}) = 63.5\,\text{MPa}, \sigma_{\text{eq}}(\text{shell}) = 94\,\text{MPa} \tag{12.6}$$

$$\text{In case two with } \Delta T = 100\,\text{K}, \ \sigma_{\text{eq}}(\text{tube}) = 239.4\,\text{MPa}, \ \sigma_{\text{eq}}(\text{shell}) = 379.9\,\text{MPa} \tag{12.7}$$

The above examination shows that the equivalent thermal stress in Case 1 is lower than the threshold values of $3S$ and $2S_y$. However, when the imposed temperature difference doubles as in Case 2, the system fails the examination which means the material undergoes plastic deformation and more than 1% creep in 1000 hours testing time and structural failure. Noteworthy, the thermoelastic analyses in cases with plastic deformation are less accurate and result in overestimation of thermal stresses. In Case 2, with ΔT = 100 K, and the imposed initial temperature of 650°C, resulted in a delay of the liquid PCM formation and high thermal stresses beyond the elastic limitation. For a more accurate thermal stress–strain calculation, a bilinear method is an alternative option which is beyond the focus of this study. The focus of this study was to investigate the melting behaviour of the high-temperature PCM under different initial condition and the impact on the structural integrity of the system and its life expectancy.

12.5 CONCLUSION

The liquid path formation during a melting process of a high-temperature PCM with melting point of 705.8°C in a horizontal shell and tube system and the consequential thermal stresses was investigated.

In one case with the initial PCM temperature of 700°C, it takes 60 seconds for a liquid layer to form on the inlet tube sheet and covers 30% of the length of the tubes.

Temperature gradients drive the formation of small eddy cells in the liquid layer which enhances the convection heat transfer and reducing the temperature gradients. This results in a moderate thermal stress of 63.5 MPa in the tubes and 94.1 MPa in the shell which is under an allowable stress of 136.3 MPa for SS347H considering 1% creep in 1000 hours operation time.

In Case 2, however, the initial PCM temperature of 650°C results in a delay in the formation of a liquid layer until 200 seconds. The tubes and shell joint with the inlet tube sheet endure a higher temperature gradient during the early stages leading to higher thermal stresses than the recommended thresholds. From the thermoelastic analysis, 239.4 MPa in inner tube surfaces and 379.9 MPa in the shell were predicted surpassing the 136.3 MPa as the limit for 1% creep in 1000 hours. However, the overestimation of the predicted thermal stresses warrants an alternative method such as bilinear to confirm whether this is the case.

It was also found that heat transfer between the argon area at the top with the inlet tube sheet and the PCM drives the melt formation in the upper side of the shell and tubes. Moreover, the convection cells in hot gas at the top mix and drag the liquid PCM into the void environment. The convection heat transfer reduces the temperature gradients and thermal stresses which resulted in the maximum thermal stress in the lower part of the joint of the shell to inlet tube sheet.

The results of this study shed light on the melting behaviour of a high-temperature PCM in a horizontal shell and tube system and its role in reducing temperature gradients and thermal stresses. The combined CFD and FEA modelling provides a valuable method for the assessment of thermo-hydraulic and mechanical performance of a system operating under harsh conditions with transient (cyclic or diurnal) high temperature and/or pressure.

ACKNOWLEDGEMENTS

The authors gratefully acknowledge that this work was supported by the Australian Solar Thermal Research Institute (ASTRI) and the Australian Government through the Australian Renewable Energy Agency (ARENA).

REFERENCES

2010. Properties of materials, ASME Boiler and Pressure Vessel Code.

2013a. ASME Boiler and Pressure Vessel Code, Section VIII, Division 1, Rules for Construction of Pressure Vessels.

2013b. ASME Boiler and Pressure vessel Code, Section VIII, Division 2, Alternative Rules for Construction of Pressure vessels.

2020. American Iron and Steel Institute, high-temperature characteristics of stainless steel, a designers' handbook series no. 9004.

Abedini-Sanigy, M.H., Ahmadi, F., Goshtasbirad, E., Yaghoubi, M., 2015. Thermal stress analysis of absorber tube for a parabolic collector under quasi-steady state condition. *Energy Procedia* 69, 3–13.

ANSYS. 2020R1. *ANSYS Fluent Users Guide, Release 20. R2.*

Bejan, A., 2013. *Internal Natural Convection, Convection Heat Transfer.* John Wiley & Sons, Inc., Hoboken, NJ, pp. 233–294.

Belusko, M., Tay, N.H.S., Liu, M., Bruno, F., 2016a. Effective tube-in-tank PCM thermal storage for CSP applications, Part 1: Impact of tube configuration on discharging effectiveness. *Solar Energy* 139, 733–743.

Belusko, M., Tay, N.H.S., Liu, M., Bruno, F., 2016b. Effective tube-in-tank PCM thermal storage for CSP applications, Part 2: Parametric assessment and impact of latent fraction. *Solar Energy* 139, 744–756.

Blaney, J.J., Neti, S., Misiolek, W.Z., Oztekin, A., 2013. Containment capsule stresses for encapsulated phase change materials. *Applied Thermal Engineering* 50(1), 555–561.

Boerema, N., Morrison, G., Taylor, R., Rosengarten, G., 2012. Liquid sodium versus Hitec as a heat transfer fluid in solar thermal central receiver systems. *Solar Energy* 86(9), 2293.

Britannia-Jahco, www.britanniajahco.com.au.

Braun C.F.& CO., 2012. Isochronous stress-strain curves for 1 1/4 Cr-1 Mo, Type 304-304H, and Type 316-316-H steels. Technical Report 2012 – Part I.

Dal Magro, F., Benasciutti, D., Nardin, G., 2016. Thermal stress analysis of PCM containers for temperature smoothing of waste gas. *Applied Thermal Engineering* 106, 1010–1022.

Kenisarin, M.M., 2010. High-temperature phase change materials for thermal energy storage. *Renewable and Sustainable Energy Reviews* 14(3), 955–970.

Logie, W.R., Pye, J.D., Coventry, J., 2018. Thermoelastic stress in concentrating solar receiver tubes: A retrospect on stress analysis methodology, and comparison of salt and sodium. *Solar Energy* 160, 368–379.

Longeon, M., Soupart, A., Fourmigué, J.-F., Bruch, A., Marty, P., 2013. Experimental and numerical study of annular PCM storage in the presence of natural convection. *Applied Energy* 112, 175–184.

Lopez, J., Caceres, G., Del Barrio, E.P., Jomaa, W., 2010. Confined melting in deformable porous media: A first attempt to explain the graphite/salt composites behaviour. *International Journal of Heat and Mass Transfer* 53(5), 1195–1207.

Maruoka, N., Akiyama, T., 2003. Thermal stress analysis of PCM encapsulation for heat recovery of high temperature waste heat. *Journal of Chemical Engineering of Japan* 36(7), 794–798.

Ortega, J., Khivsara, S., Christian, J., Ho, C., Dutta, P., 2016. Coupled modeling of a directly heated tubular solar receiver for supercritical carbon dioxide Brayton cycle: Structural and creep-fatigue evaluation. *Applied Thermal Engineering* 109, 979–987.

Riahi, S., Evans, M., Belusko, M., Flewell-Smith, R., Jacob, R., Bruno, F., 2021. Transient thermo-mechanical analysis of a shell and tube latent heat thermal energy storage for CSP plants. *Applied Thermal Engineering* 196, 117327.

Riahi, S., Saman, W.Y., Bruno, F., Belusko, M., Tay, N.H.S., 2017. Comparative study of melting and solidification processes in different configurations of shell and tube high temperature latent heat storage system. *Solar Energy* 150, 363–374.

Riahi, S., Saman, W.Y., Bruno, F., Belusko, M., Tay, N.H.S., 2018. Performance comparison of latent heat storage systems comprising plate fins with different shell and tube configurations. *Applied Energy* 212, 1095–1106.

Riahi, S., Saman, W.Y., Bruno, F., Tay, N.H.S., 2016. Numerical modeling of inward and outward melting of high temperature PCM in a vertical cylinder. *AIP Conference Proceedings* 1734(1), 050039.

Shmueli, H., Ziskind, G., Letan, R., 2010. Melting in a vertical cylindrical tube: Numerical investigation and comparison with experiments. *International Journal of Heat and Mass Transfer* 53(19), 4082–4091.

Turchi, C.S., Ma, Z., Neises, T.W., Wagner, M.J., 2013. Thermodynamic study of advanced supercritical carbon dioxide power cycles for concentrating solar power systems. *Journal of Solar Energy Engineering* 135(4), 041007.

13 Sensible and Latent Thermal Energy Storage in Parallel Channels

B. Buonomo, A. di Pasqua, O. Manca, S. Nardini, and S. Sabet
Università degli Studi della Campania "Luigi Vanvitelli"

CONTENTS

13.1 INTRODUCTION

To meet global energy demands, one should start from the data related to the growth of world population, householders and urbanization and their forecasting. These aspects are related to the built environment, which is one of the greater aspects responsible for global energy consumption and greenhouse emissions (Borri et al., 2021). Furthermore, more than 75% of the energy demand employed in heating and cooling systems is still based on fossil fuels (REN21, 2020). Consequently, reducing energy consumption and a more efficient and sustainable energy conversion is a key to reduce gas emission and preserve environmental health. This leads to a transition into the energy sector and the need for policies that include strategies to develop a sustainable built environment based on high-efficiency buildings and a high share of energy produced from renewable sources as well as the need in energy saving as an energy source. In this scenario, the thermal energy storage (TES) can play a strategic role in energy recovering (Borri et al., 2021; Buonomo et al., 2020).

Energy storage technologies are strategic and necessary components for the efficient utilization of renewable energy sources and energy conservation. The traditional types of energy storages are mechanical, chemical, biological, magnetic and

DOI: 10.1201/9781003213260-13

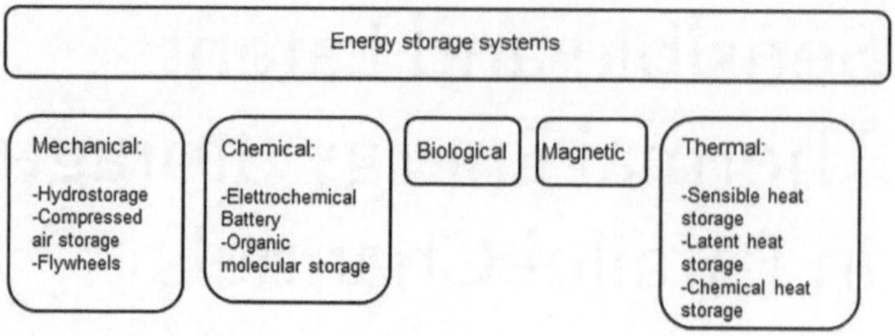

(a)

<table>
<tr><th colspan="8">Thermal Energy Storage</th></tr>
<tr><th colspan="5">Thermal</th><th colspan="3">Chemical</th></tr>
<tr><th colspan="2">Sensible Heat</th><th colspan="3">Latent Heat</th><th rowspan="2">Thermal Chemical Pipeline</th><th rowspan="2">Heat of Reaction</th><th rowspan="2">Heat Pump</th></tr>
<tr><td>Liquids</td><td>Solids</td><td>Solid-Liquid</td><td>Liquid-Gaseous</td><td>Solid-Solid</td></tr>
</table>

(b)

FIGURE 13.1 Classification of (a) energy storage systems and (b) different types of thermal storage of solar energy.

thermal energy storages. Some of them have different forms of energy as shown in Figure 13.1a (Dincer and Rosen, 2002).

Energy conservation and management are needed in several industrial and commercial applications in order to supply thermal energy. Different systems are employed to satisfy the energy demand which can vary on daily, weekly and seasonal bases. TES is very useful for energy conservation and it allows to align energy production with consumer demand. It is expanding mainly within the renewable energy technologies and it is very important to deliver energy to eliminate or reduce the intermittence mainly in solar energy systems.

13.2 THERMAL ENERGY STORAGE SYSTEM CLASSIFICATION

TES systems have the potential of increasing the effective use of thermal energy equipment and of facilitating large-scale switching. They are useful for correcting the mismatch between the supply and demand of energy. TES systems can be classified using different criteria, and a more detailed subdivision of TES systems is depicted in Figure 13.1b. However, the most common is the classification by the TES technology, with sensible heat storage, latent heat storage, and sorption and chemical reactions (Borri et al., 2021). In the following sections, the attention is focused on sensible heat thermal energy storage (SHTES) and latent heat thermal energy storage (LHTES).

In sensible heat storage the energy is stored in the change of temperatures of substances that experience a change in internal energy. Besides the density and the specific heat of the storage material other properties are important for sensible heat storage and they are operational temperatures, thermal conductivity and diffusivity, vapor pressure, compatibility among materials stability, heat loss coefficient as a function of the surface areas to volume ratio, and cost. Sensible TES consists of a storage medium, a container and inlet/outlet devices. Tanks must both retain the storage material and prevent losses of thermal energy. Sensible heat storage can be made by solid media or liquid media. Each medium presents advantages and disadvantages and various solid matrices are used as the storage medium, such as packed beds, foams, and parallel channels (honeycomb) (Cabeza, 2014; Gil et al., 2010; Hanchen et al., 2011; Andreozzi et al., 2014; Luo et al., 2014; Janchen et al., 2015; Li and Chan, 2017).

In terms of porosity, the solid matrix with honeycomb appears more versatile with respect to the other two matrices, as indicated by Andreozzi et al. (2013) and Afrin et al. (2013), and as reported by Luo et al. (2014). Medrano et al. (2010) observed that honeycomb ceramic storages, with respect to concrete storages, have 1.2 times the storage capacity and 1.35 times the thermal conductivity. In honeycomb systems the porosity can be chosen or selected as a function of TES in terms of charge and/or discharge time. The gas flows in each channel of storage material and it is well distributed in the transversal cross section because of the symmetric honeycomb structure (Luo et al., 2014). Moreover, the TES system is easy to design due to the simple geometric units, as indicated by Gil et al. (2010).

In latent heat storage media, the energy is stored nearly isothermally in some substances as the latent heat of phase change, as heat of fusion (solid–liquid transition) or heat of vaporization (liquid–vapor transition). Mainly the solid–liquid transition is used, and substances used under this technology are called phase change materials (PCM). Storage systems utilizing PCM can be reduced in size compared to single-phase sensible heating systems. Heat transfer design and media selection are more difficult, and experience with low-temperature salts has shown that the performance of the materials can degrade after moderate number of freeze melt cycles.

PCM allow large amounts of energy to be stored in relatively small volumes, resulting in some of the lowest storage media costs of any storage concepts. One of the worse drawbacks of using PCM is its very low value of thermal conductivity, therefore some enhancements are required. In literature there are many improvements technique as addition of fins (Zhang and Faghri, 1996), metal foam (Zhao et al., 2010) or honeycomb structure. The honeycomb structure has the possibility to improve the PCM thermal conductivity inside an LHTES system due to the large surface area with PCM, besides softening the mechanical stress decreasing the consequences of the PCM expansion during the phase change process (Farid et al., 2004). There are several applications of the honeycomb as thermal building (Lai and Hokoi, 2014) and thermal management (Kim et al., 2013).

13.3 LITERATURE REVIEW

The thermal storage systems with honeycomb matrix either for sensible or for latent heat seem very promising mainly for the possibility to choose the ratio between the flow area and the storage volume. The parallel channels have different unit channel sections such as parallel, squared, rectangular, hexagonal, triangular, circular, and so on. However, the complexity of the honeycomb geometry could be a problem for numerical simulation due to the increase of the computational cost for complex and regular geometry with numerous channels. Therefore, in literature there are some works where the honeycomb is modeled as a porous media.

A two-dimensional numerical model to determine the dynamic temperature and velocity profiles of gases and solid heat-storing materials in a composite honeycomb regenerator was developed by Rafidi and Blasiak (2005). Low-temperature TES were studied by Liu et al. (2013), Afrin et al. (2013), Liu and Nagano (2014), and Liu et al. (2015).

A numerical study on high-temperature TES systems with honeycomb solid matrix, with different porosity values, was accomplished by Andreozzi et al. (2014). An investigation on a TES with ceramic honeycombs was numerically and experimentally carried out by Luo et al. (2014). A transient analysis of a high-temperature thermal storage honeycomb system with parallel squared channels is numerically studied by Andreozzi et al. (2015). One-dimensional dynamic models of the air receiver and TES were developed using the Modelica language with a graphical user interface by Li et al. (2016). The sensible heat TES was a single honeycomb ceramic block with parallel square channels. An array of parallel flat plates were considered with air as working fluid by Andriotty et al. (2016) to minimize the system heat storage mass. A simplified analysis to handle a parallel plate system with parallel channels as a porous medium was accomplished by Andreozzi et al. (2017a). A ceramic honeycomb system for high-temperature thermal storage application was investigated experimentally and numerically by Srikanth et al. (2017). The study was accomplished on parallel channels with hexagonal transversal sections. A numerical investigation on a honeycomb ceramic TES in a high-temperature solar thermal power plant with air as working fluid was provided by Li et al. (2018). A simplified one-dimensional TES model was proposed. A heat storage system with ceramic or concrete structure of modular symmetric components forming parallel air channels was studied numerically and experimentally by Sacharczuk and Taler (2019). The study was extended (Taler et al., 2019) considering the combined heat transfer problem conductive–convective with the air flow inside parallel squared channels. A numerical study on honeycomb SHTES employed in concentrated solar power micro-turbine was carried out by Iaria et al. (2019). A one-dimensional model was assumed to simulate the thermal storage with parallel circular tubes. A honeycomb system with triangular transversal section channels with external heat losses was numerically investigated by Andreozzi et al. (2019). A methodology to achieve the optimal design of an SHTES system modeled by a simplified lumped element was proposed by Andriotty et al. (2019). The design methodology was set up on correlations between the number of transfer unit and time. A system with parallel channels was studied as an SHTES under periodic thermal conditions by Andriotty et al. (2020) by comparing a proposed lumped element model with a three-dimensional model.

As a honeycomb system in LHTES, the parallel plate configuration is widely employed with the PCM inside flat containers with external heat transfer fluid (HTF). Very few studies or applications have used a parallel channel system with the different transversal section of the channels such as squared, triangular or hexagonal combined with an HTF. It should be underlined that in most cases of LHTES systems, the PCM is enclosed into boxes to contain the liquid phase PCM and also to avoid the

PCM contacting with the environment, as underlined by Mehling and Cabeza (2008). Farid et al. (2004) reviewed the first investigations on the parallel plate systems. An experimental and numerical investigation on a parallel plate stack of ice was carried out by Ismail et al. (1999). A parallel plate phase change energy storage system with air as HTF was numerically studied by Vakilaltojjar and Saman (2001). Zalba et al. (2004) studied a thermal storage with flat plates with encapsulated PCM, and it was found that for melting process the inlet temperature affected more that the thickness of the plates. A similar system to the one studied by Vakilaltojjar and Saman (2001) was investigated by Saman et al. (2005) with a two-dimensional model and including the sensible heat part. A one-dimensional model to analyze the heat transfer inside the configuration given by Vakilaltojjar and Saman (2001) and Saman et al. (2005) was used by Halawa et al. (2005). Hed and Bellander (2006) developed a mathematical model for PCM parallel plates with air as HTF. A stack of parallel plates filled with PCM inside polyethylene film bags were modeled by Zukowski (2007) by a three-dimensional analysis was accomplished. Lazaro et al. (2009) presented an empirical model starting from experimental results. A simplified algorithm to solve one-dimensional parallel plate phase change problems with air as HTF in forced convection was provided by Halawa et al. (2010). A parallel plate phase change thermal storage system was studied employing a one-dimensional numerical model by Liu et al. (2011). An LHTES made up of a stack of parallel slabs was investigated by Dolado et al. (2011) considering a one-dimensional conduction model. A study on one and two-dimensional models related to an LHTES was performed by Belusko et al. (2012). A parametric study on an LHTES with PCM encapsulated in plates was accomplished by Amin et al. (2012). An LHTES with PCM slabs and air as HTF was studied numerically by Lopez et al. (2013). A phase change thermal storage unit for high temperature with PCM in parallel slabs was studied by Liu et al. (2014). A numerical two-dimensional model of an LHTES with parallel slabs was investigated by Zsembinszki et al. (2014) during the discharging process. A parallel plate channel system with PCM as TES was numerically studied by Andreozzi et al. (2017b). A numerical study on a high-temperature flat plate LHTES is proposed by Liao et al. (2018) to evaluate the cyclic performance. A 2D transient numerical model was carried out by Farah et al. (2019) to investigate of an LHTES in a heat pump system. An approximate analytic technique was used by Ding et al. (2020) to examine the thermal behavior of an LHTES with a stack of parallel slab. A numerical investigation on a cascaded LHTES was proposed by Nekoonam and Roshandel (2021). A two-dimensional numerical model to study an LHTES with a stack of slabs was accomplished by Crespo et al. (2021).

The honeycomb LHTES systems with duct transversal sections different from the parallel plate channels were also studied. In fact, a ceramic honeycomb with PCM in squared transversal sections was experimentally studied by Li et al. (2015) and numerically by Andreozzi et al. (2018a), whereas different transversal sections such as triangular (Andreozzi et al., 2018b) or circular (Mahdi et al., 2021) were examined.

It is important to note that the TES honeycomb systems can also be modeled as a system of heat transfer fluid flowing through a porous medium, where the physical relationship between HTF and the porous thermal storage material becomes a general phenomenon. In fact, the so-called solid medium can also be another fluid for energy storage, as long as it is separated from the HTF. However, for adiabatic boundary conditions, the use of the porous model is trivial and to employ the periodicity with the direct model is enough. But for a system with thermal losses toward the ambient, the periodicity of the direct model is lost and, the discretization must be done with a very high number of nodes. The use of an analogous porous model allows realizing a discretization with a lower node number obtaining similar results of the direct model and the computational times are significantly reduced. In the following sections, some different porous media models are presented to simulate the honeycomb system related to both SHTES and LHTES systems.

13.4 GEOMETRY CONFIGURATIONS AND MATHEMATICAL MODELING

This section illustrates the approaches employed to model SHTES and LHTES systems considered as a container with a honeycomb solid matrix inside. In the following sections, different forms of tanks with various honeycomb structures used as TES system will be examined. Two geometric configurations of the container under investigation: (1) cylindrical tank, with diameter D and length L, and (2) rectangular parallelepiped tank, with transversal area $H \times H$ and length L.

The storage medium is made of PCM for the latent heat TES and cordierite for the sensible heat TES and it is analyzed in three different honeycomb matrix configurations: (1) parallel channels with transversal square section, (2) parallel channels with triangular section and (3) parallel plates. The geometrical characteristics of the honeycomb are depicted in Figure 13.2a–c. It is designed with parallel channels, half of them are filled with PCM or solid and in the others the air passes through. The height of a single elementary channel is H_n, the thickness is δ_n and the length is L. The solid walls of the shell of the honeycomb structure are made of the ceramic cordierite. The PCM is encapsulated inside the channels of the honeycomb in checkerboard way as shown in Figure 13.2, and the convective effects related to melting processes are neglected.

The two main geometric parameters involved in the honeycomb structure are porosity, ε, and the number of channels per unit of length (CPLs). Various honeycomb system for different CPLs are studied at the same porosity and volume. Various honeycomb structures with reference to the parallelepiped-shaped latent heat storage system are shown in Figure 13.2. Starting from the base channel, it is repeatedly divided into four equal parts to obtain the different configurations of the honeycomb matrix. Similar procedure is employed to obtain various honeycomb systems with different cross-sectional shape.

The porosity is defined:

$$\varepsilon = \frac{V_{\text{Air}}}{V_{\text{tot}}} = \eta\left(\frac{H}{H+2\delta}\right)^q \tag{13.1}$$

where V_{Air} is f volume of air passing the honeycomb matrix, V_{tot} is the packaging volume, $q = 1$ for parallel plate channel and $q = 2$ for squared and triangular channel.

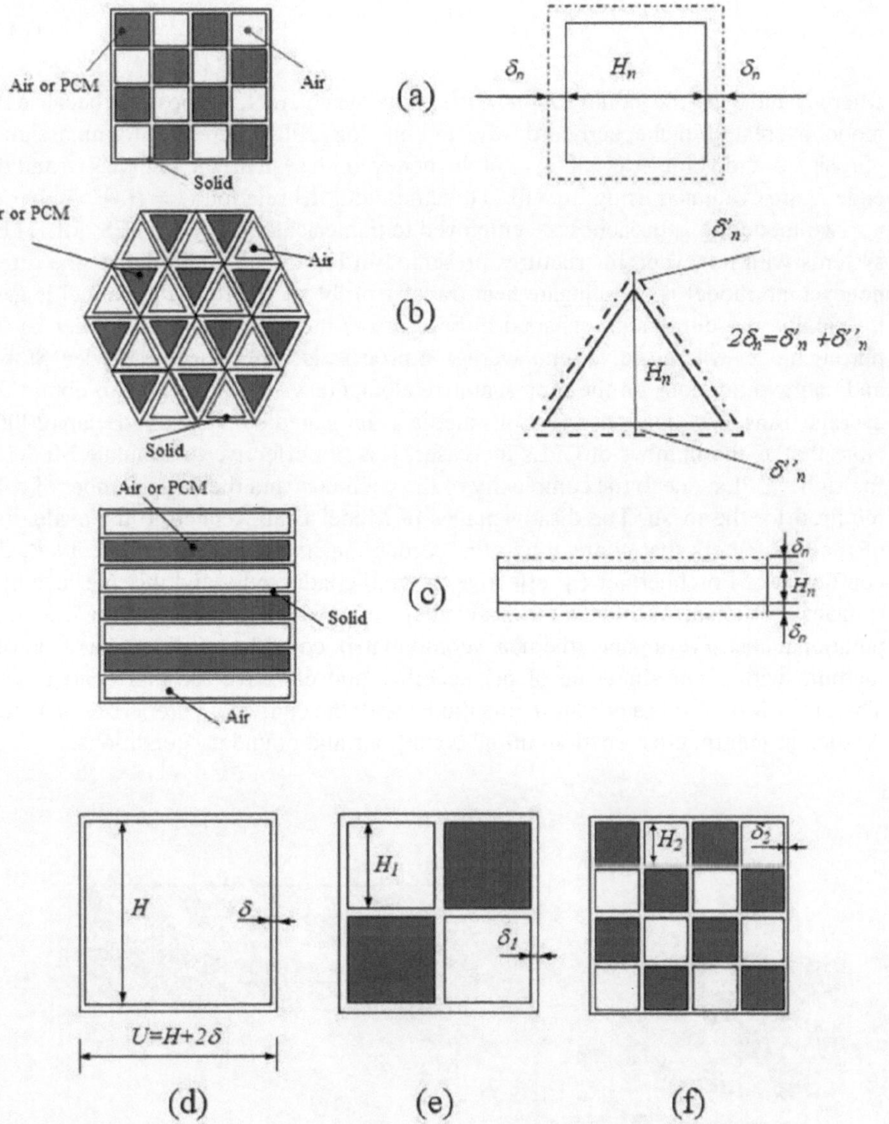

FIGURE 13.2 Geometrical configurations of honeycomb structures: (a) honeycomb with transversal square section, (b) honeycomb with transversal triangular section and (c) honeycomb with parallel plates, and cross section and arrangement of the PCM inside of honeycomb structure for various values of the number channels per unit of length (CPLs): (d) CPL = 1, (e) CPL = 4 and (f) CPL = 16.

The factor η is equal to 1 for the case of sensible heat storage and is equal to ½ for the case of latent heat storage because half of channels are closed. The honeycomb system with different CPLs is reported in Figures 13.2d–g in a unit length U.

The relation between the channel height H_n and the thickness δ_n for different CPLs is:

$$H_n = \frac{H}{2^n};\ \delta_n = \frac{\delta}{2^n} \tag{13.2}$$

where H and δ are the values for one CPL (only one channel, not possible because the model is created in checkerboard way) and $n = \log_2$CPL. Therefore, for an assigned porosity, ε, and volume, $U \times U \times L$, of the honeycomb system, the thickness δ and the edge H are evaluated using Eq. (13.2) together with the relation $U = H + 2\delta$.

Two modeling approaches are employed to numerically study SHTES or LHTES systems with honeycomb structures presented in Figure 13.3: (Model A) the direct honeycomb model is a conjugate heat transfer problem where each channel is geometrically and numerically shaped in base of the number of CPLs; (Model B) the porous model where the volume averaging process is applied to the Navier–Stokes and energy equations on the representative elementary volume (REV) to obtain the average transport equations in porous media as indicated by Nield and Bejan (2006). Note that as the number of CPLs increases, it is not effective to simulate Model A for higher CPLs, due to the complexity of the geometry and the large number of cells required for the mesh. The disadvantages of Model B are related to the evaluation of the parameters that characterize the porous matrix such as permeability K, the coefficient of Forchheimer C_F, effective thermal conductivity and the interface heat transfer coefficient. While the main advantage of Model B is related to the low computational cost. Fortunately, the honeycomb matrix could be considered as a porous medium with a certain value of permeability and effective thermal conductivity. Therefore, Model B is a porous media model with the equivalent properties of Model A, such as length, cross section, initial conditions and boundary conditions.

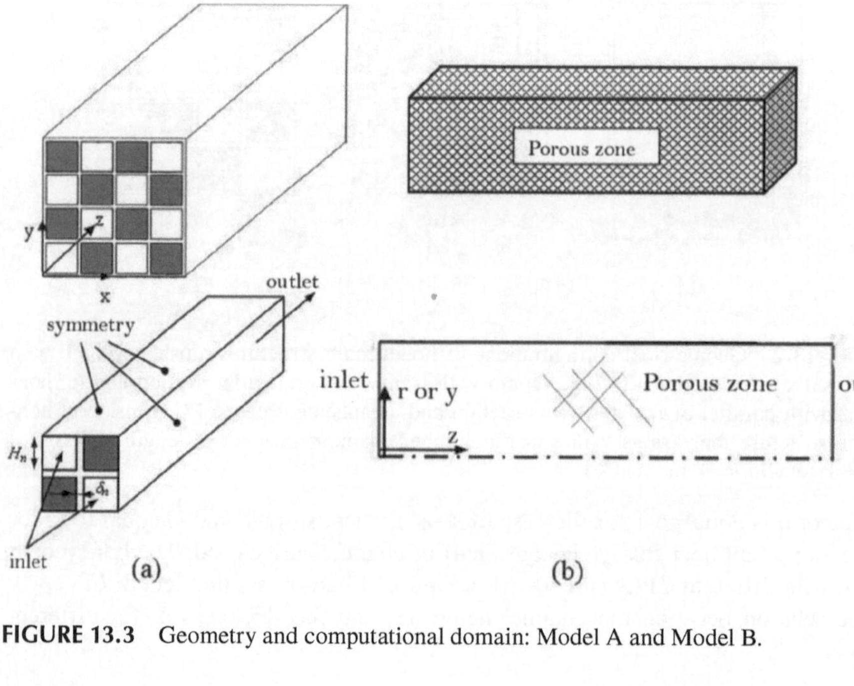

FIGURE 13.3 Geometry and computational domain: Model A and Model B.

The assumptions made for both models (A and B) are the following: (1) the fluid flow is transient, laminar and incompressible, (2) the thermal proprieties of the air are considered temperature dependent, while those of the solid and the PCM are assumed constant, (3) the viscous dissipation and the work done by the pressure forces are neglected, (4) the buoyancy effect in the PCM enclosed within the channels of the honeycomb structure is considered negligible and the velocity of the liquid phase is assumed to be zero, (5) the temperature difference of the encapsulated PCM in a channel and that of the solid shell of the channel are considered negligible. Moreover, for Model B the following assumptions are added: (6) the 2D planar flow is considered for the rectangular storage tank and a 2D axial symmetric flow for the cylindrical storage tank, (7) the extended Darcy–Brinkman law is used to model mass transport in a porous medium, (8) the honeycomb matrix is anisotropic with an effective thermal conductivity and permeability along z direction different respect to x and y directions, (9) both the local thermal non-equilibrium (LTNE) and the local thermal equilibrium model (LTE) are adopted to simulate the heat exchange between the fluid zone and the solid zone in the porous model.

The enthalpy porosity method developed by Voller and Prakash (1987) is used for the mathematical description of melting behavior of PCM in both models. In this method, a "pseudo" porous zone called a mushy region is considered to model the solid–liquid interface and there is no explicit separation surface between the solid–liquid phases. This model is suitable for PCM where melting occurs over a range of temperatures. The mushy region is described using a new parameter – liquid fraction β – varying from 0 to 1. In the fully solid zone the parameter is 1, in the fully liquid zone its value is null and mixed region its value is between 0 and 1:

$$\beta = \begin{cases} 0 & \text{for} \quad T < T_{\text{Solidus}} \\ \dfrac{T - T_{\text{Solidus}}}{T_{\text{Liquidus}} - T_{\text{Solidus}}} & \text{for} \quad T_{\text{Solidus}} < T < T_{\text{Liquidus}} \\ 1 & \text{for} \quad T > T_{\text{Liquidus}} \end{cases} \tag{13.3}$$

where T is the local temperature, $T_{solidus}$ and $T_{liquidus}$ are, respectively, the temperature below which the PCM is liquid and upper which the PCM is solid. The region of melting lies in a range of temperature between $T_{liquidus}$ and $T_{solidus}$.

Under these assumptions, the equations of continuity, momentum and energy are:

- **Direct model (Model A)**

  Continuity equation:

$$\nabla \cdot \vec{V} = 0 \tag{13.4}$$

  Momentum equation for fluid:

$$\frac{\partial(\rho\vec{V})}{\partial t} + \nabla \cdot (\rho\vec{V}\vec{V}) = -\nabla \bar{p} + \nabla \cdot \left[\rho\left(\nabla\vec{V} + \nabla\vec{V}^{T}\right)\right] + \vec{S} \tag{13.5}$$

Energy equation for fluid:

$$\rho_f c_{p,f}\left(\frac{\partial T_f}{\partial t}+u\frac{\partial T_f}{\partial x}+v\frac{\partial T_f}{\partial y}+w\frac{\partial T_f}{\partial z}\right)$$
$$=\frac{\partial}{\partial x}\left(k_f\frac{\partial T_f}{\partial x}\right)+\frac{\partial}{\partial y}\left(k_f\frac{\partial T_f}{\partial y}\right)+\frac{\partial}{\partial z}\left(k_f\frac{\partial T_f}{\partial z}\right) \quad (13.6)$$

Energy equation for the PCM:

$$(\rho c)_{\text{PCM}}\frac{\partial T_{\text{PCM}}}{\partial t}=k_{\text{PCM}}\left(\frac{\partial^2 T_{\text{PCM}}}{\partial x^2}+\frac{\partial^2 T_{\text{PCM}}}{\partial y^2}+\frac{\partial^2 T_{\text{PCM}}}{\partial z^2}\right)-\rho_{\text{PCM}}H_L\frac{\partial \beta}{\partial t} \quad (13.7)$$

Energy equation for the solid:

$$\rho_s c_s\frac{\partial T_s}{\partial t}=k_s\left(\frac{\partial^2 T_s}{\partial x^2}+\frac{\partial^2 T_s}{\partial y^2}+\frac{\partial^2 T_s}{\partial z^2}\right) \quad (13.8)$$

where x, y, and z are the Cartesian coordinate, u, v, and w are the velocity components along x-, y- and z-axes, respectively. T is the temperature, p is the pressure and ρ, c, k and μ are the density, the specific heat, thermal conductivity and the dynamic viscosity. The subscripts "f", "s" and "PCM" denote the HTF, the solid phase and phase change material, respectively. Finally, H_L and β are the latent heat transfer and the liquid fraction, defined in Eq. (13.3), of PCM.

The boundary conditions associated with the governing equations and referred to the computational domain shown in Figure 13.3a are as follows: at inlet air section, the mass flow and temperature is constant and uniform; an hydrodynamic and thermal fully developed flow is assumed at the outlet air sections; the non-slip velocity condition is employed on the interface solid–fluid, and the temperature and heat flux profiles are assumed be continuous; the symmetry condition is used continuous; the symmetry condition is used on lateral planes.

- **Porous model (Model B)**

  Continuity equation:

$$\frac{1}{r^n}\frac{\partial\left(r^n v_p\right)}{\partial r}+\frac{\partial w_p}{\partial z}=0 \quad (13.9)$$

  Momentum equations:

$$\frac{\rho_f}{\varepsilon}\left[\frac{\partial v_p}{\partial t}+\frac{v_p}{\varepsilon}\frac{\partial v_p}{\partial r}+\frac{w_p}{\varepsilon}\frac{\partial v_p}{\partial z}\right]=-\frac{\partial p}{\partial r}+\frac{1}{r^n}\frac{\partial}{\partial r}\left[2r^n\mu\left(\frac{\partial v_p}{\partial r}\right)\right]$$
$$+\frac{\partial}{\partial z}\left[\mu\left(\frac{\partial v_p}{\partial z}+\frac{\partial w_p}{\partial r}\right)\right]-2n\mu\frac{v_p}{r^2}-\frac{\mu}{K_{rr}}v_p-\frac{C_F\rho_f}{\sqrt{K_{rr}}}\sqrt{w_p^2+v_p^2}\,v_p \quad (13.10)$$

$$\frac{\rho_f}{\varepsilon}\left[\frac{\partial w_p}{\partial t}+\frac{v_p}{\varepsilon}\frac{\partial w_p}{\partial r}+\frac{w_p}{\varepsilon}\frac{\partial w_p}{\partial z}\right]=-\frac{\partial p}{\partial z}+\frac{\partial}{\partial z}\left[\left(2\mu\frac{\partial w_p}{\partial z}\right)\right]$$

$$+\frac{1}{r^n}\frac{\partial}{\partial r}\left[r^n\mu\left(\frac{\partial w_p}{\partial r}+\frac{\partial v_p}{\partial z}\right)\right]-\frac{\mu}{K_{zz}}w_p-\frac{C_F\rho_f}{\sqrt{K_{zz}}}\sqrt{w_p^2+v_p^2}\,w_p \quad (13.11)$$

where v_p and w_p are the components of Darcy velocity along r and z. In eqs. (13.9), (13.10) and (13.11), for the case of a cylindrical tank z, r, are the cylindrical coordinates and n is equal to 1, while for the case of rectangular tank and $r = y$, z are the Cartesian coordinates and $n = 0$.

It is observed that the perpendicular components velocity in each channel of the honeycomb system are negligible compared to the velocity component z because the walls of the channel are impermeable. Consequently, the honeycomb system considered as a porous medium is anisotropic with the permeability K only along the axial direction, z-axis in Figure 13.3b, while along the x and y directions, $K \sim 0$. The Darcy law is used to evaluate the permeability along z-axis (Bejan, 1995):

$$\frac{\Delta p}{L}=\frac{\varepsilon\mu_f}{K_{zz}}u_{\text{avg}} \quad (13.12)$$

A fully developed laminar flow is assumed to evaluate the relationship between the pressure drop along the channel and average velocity. The average velocity of the fluid in channels of various cross sections is:

$$u_{\text{avg}}=\beta\frac{\Delta p}{\mu_f L}\left(\frac{H_n}{2}\right)^2 \quad (13.13)$$

where u_{avg} is the average velocity in the single channel, Δp is the pressure drop along the channel, L is the channel length and μ_f is the dynamic viscosity of the fluid. The factor, β, is equal to 0.1415 for squared section, 1/60 for equilateral triangular section and 1/12 for parallel plates (Bahrami et al., 2005; Bejan, 1995).

By combining Eqns. (13.12) and (13.13), the permeability K_{zz} is given:

$$K_{zz}=\varepsilon\beta\left(\frac{H_n}{2}\right)^2 \quad (13.14)$$

- Energy equations under the LTNE assumption ($T_f \neq T_s = T_{\text{PCM}}$) are:

$$\varepsilon(\rho c_p)_f\frac{\partial T_f}{\partial t}+(\rho c_p)_f\left(v_p\frac{\partial T_f}{\partial r}+w_p\frac{\partial T_f}{\partial z}\right)$$

$$=\varepsilon\left(\frac{1}{r}\frac{\partial}{\partial r}\frac{\partial(rk_fT_f)}{\partial r}+\frac{\partial}{\partial z}\frac{\partial(k_fT_f)}{\partial z}\right)+h_{sf}a_{sf}\left(T_s-T_f\right) \quad (13.15)$$

$$(\rho c)_{sp}\frac{\partial T_s}{\partial t}=\left(\frac{1}{r^n}\frac{\partial}{\partial r}\frac{\partial\left(r^n k_{rr,sp}T_s\right)}{\partial r}+\frac{\partial}{\partial z}\frac{\partial\left(k_{zz,sp}T_s\right)}{\partial z}\right)$$

$$-h_{sf}a_{sf}\left(T_s-T_f\right)-\varepsilon\rho_{PCM}H_L\frac{\partial\beta}{\partial t} \tag{13.16}$$

- Energy equation under LTE assumption ($T_f = T_s = T_{\text{PCM}}$) is:

$$(\rho c_p)_{\text{eff}}\frac{\partial T}{\partial t}+(\rho c_p)_f\left(v_p\frac{\partial T}{\partial r}+w_p\frac{\partial T}{\partial z}\right)$$
$$=\left(\frac{1}{r^n}\frac{\partial}{\partial r}\frac{\partial\left(r^n k_{rr,\text{eff}}T\right)}{\partial r}+\frac{\partial}{\partial z}\frac{\partial\left(k_{zz,\text{eff}}T\right)}{\partial z}\right)-\varepsilon\rho_{\text{PCM}}H_L\frac{\partial\beta}{\partial t} \tag{13.17}$$

where k_{rr} and k_{zz} are the components of the conductivity tensor of anisotropic honeycomb structure. h_{sf} and a_{sf} are interface convective heat transfer coefficient and the area density of the surface contact between the fluid phase and the solid phase. The subscript "*sp*" and "eff" denote the effective thermal properties of the mixtures solid–PCM and solid–PCM–fluid, respectively. The solid phase of the porous medium with the LTNE model is considered as an equivalent solid where the PCM is fixed, and its effective heat capacity is evaluated as weighted average of the heat capacities of the solid shell and PCM:

$$(\rho c)_{sp}=(1-2\varepsilon)(\rho c)_s+\varepsilon(\rho c)_{\text{PCM}} \tag{13.18}$$

For the LTE model, the effective heat capacity is:

$$(\rho c)_{sp}=\varepsilon(\rho c)_f+\varepsilon(\rho c)_{\text{PCM}}+(1-2\varepsilon)(\rho c)_s \tag{13.19}$$

In Eqs. (13.18) and (13.19) the porosity, ε, of the fluid zone is weight of the heat capacity of PCM because it occupies the same space of the fluid zone inside the honeycomb matrix. The effective thermal conductivity along x and y directions depends on the shape of the cross section of the channels of the honeycomb matrix. It is calculated using the serial mechanism of the heat conduction between fluid, PCM and solid, as (Andreozzi et al., 2015; Andreozzi et al., 2017a, 2017b):

$$\frac{k_\perp}{k_f}=\begin{cases}(1-2\varepsilon)\kappa_2+\varepsilon\kappa_1\kappa_2+\varepsilon\kappa_1 & \text{for parallel plates channel}\\[2ex] \dfrac{\dfrac{\kappa_1}{\sqrt{\varepsilon}}\left(\dfrac{(1+\kappa_2)}{2}+\dfrac{\kappa_1}{\sqrt{\varepsilon}}-\kappa_1\right)}{\left(\dfrac{1}{\sqrt{\varepsilon}}\left(\dfrac{(1+\kappa_2)}{2}-\kappa_1+\dfrac{\kappa_1}{\sqrt{\varepsilon}}\right)+\left(\kappa_1-\dfrac{(1+\kappa_2)}{2}\right)\right)} & \text{for squared channel}\end{cases} \tag{13.20}$$

Along the z direction the effective thermal conductivity does not depend on the shape of the channel but only on the porosity, and it is calculated considering a parallel mechanism of the heat conduction as:

$$\frac{k_{\parallel}}{k_f} = \varepsilon + (1 - 2\varepsilon)\kappa_1 + \varepsilon\kappa_2 \tag{13.21}$$

where $\kappa_1 = k_s/k_f$, $\kappa_2 = k_{PCM}/k_f$ are the conductivity ratios of the solid phase and PCM phase with respect to the fluid phase. For sensible heat storage all channels are filled with air and κ_2 is equal to 1. The second term, $h_{sf}\, a_{sf}\, (T_s - T_f)$, on the right side of Eqs. (13.15) and (13.16) is related with the local convective heat transfer inside the porous medium due to the local temperature imbalance between the fluid and the solid porous matrix surfaces. The value of the surface density area, a_{sf}, depends on shape of the channel and it is evaluated as a function of porosity, ε, and CPL:

$$a_{sf} = \begin{cases} \dfrac{4\sqrt{\varepsilon}CPL^2}{U} & \text{for squared channels} \\ \dfrac{2\varepsilon CPL}{U} & \text{for parallel plates} \end{cases} \tag{13.22}$$

A 3D numerical solution of laminar and steady state forced convection in a single channel of honeycomb structure can be employed to evaluate the convective heat transfer coefficient, h_{sf}. A second method to evaluate the convective heat transfer coefficient is obtained using the laminar fully developed flow in channel for assigned wall temperature or assigned wall heat transfer. The values of Nusselt number for the laminar fully developed flow in channels at constant temperature and constant heat fluxes are proposed by Ozisik (1984) for squared channel, by Shah (1975), and Schmidt and Newell (1967) for triangular channel and by Bejan (1995) for parallel plates.

At inlet and outlet sections, the boundary conditions for the porous model (Model B) are the same as for the direct model (Model A). Thermal energy losses toward external ambient are neglected and consequently the z-axis is an axis of symmetry for temperature and velocity fields. The initial temperature in the tank is uniform.

The potassium carbonate (K_2CO_3) is used as PCM in high-temperature LHTES system, the HTF is air and the honeycomb structure is cordierite. The values of thermophysical properties of K_2CO_3 and cordierite are reported in the papers of Chase (1998), Wang et al. (2013), and Zalba et al. (2003). The ideal gas model is using to evaluate the density of air, and the heat specific and thermal conductivity are calculated by the relations given by Zalba et al. (2003).

13.5 NUMERICAL PROCEDURE

For both models, the Ansys-Fluent commercial code is employed to solve the governing equations along with the boundary and initial conditions associated. This code uses the finite volume method (FVM) for the numerical solution of governing equations. For Model B, a 2D planar or 2D axisymmetric option is enabled to numerically solve the two-dimensional planar or axisymmetric flow. In addition, the porous medium model is activated in the porous region and the LTNE model is employed. Grid and time-step dependence tests of the numerical solution are performed. Details of the grid and time-step analysis are reported by Andreozzi et al. (2014) and Buonomo et al. (2020). The SIMPLE algorithm (Semi-Implicit Method for Pressure Linked Equations) is employed to solve the coupling between pressure and velocity in the momentum equations. Here, the pressure field is obtained using a relationship between velocity and pressure corrections to satisfy the continuity equation. The PRESTO (PREssure STaggering Option) scheme is employed to compute the pressure on the cell faces. For the energy and momentum equations, the second order upwind and central schemes are employed to discretize the convective and diffusive fluxes, respectively. The residuals for the continuity and momentum equations are fixed to 10^{-6} while for the energy equation is set to 10^{-8}.

13.6 RESULTS

Numerical results are reported for the different boundary conditions and other characteristics for both the sensitive and latent thermal storage systems in order to understand the thermal behavior of the various honeycomb structures. Comparison between Model A and Model B for the LTNE assumption are accomplished. The results are presented in terms of temperature as function of time and energy profiles stored by cordierite in an SHTES system and by cordierite and phase change material in an LHTES system.

13.6.1 Sensible Heat Thermal Energy Storage System (SHTES)

A comparison between Models A and B in terms of average solid temperature rise, $\Delta T_{s,\,avg}$, and heat transfer rate between fluid and solid phases as a function of time are given in Figure 13.4a and b, respectively, for the square honeycomb structure. $\Delta T_{s,\,avg}$ represents the internal energy stored in the solid for assigned $\rho_s c_s V_s$. It is observed that the higher the CPL the lower the time to reach the maximum stored thermal energy (see Figure 13.4a). Heat transfer rate is a decreasing function of time and the decrease is higher as the higher the CPL (see Figure 13.4b). In the case that the complete energy storage capacity is reached, to employ a higher CPL for assigned porosity means only a lower charging time, whereas if the charge is partial, at an assigned partial charge time, i.e. a percentage of the complete energy storage capacity is reached, the higher the CPL the higher the stored thermal energy.

Figure 13.4c and d shows the stored thermal energy as a function of time of a honeycomb structure with parallel squared channels for assigned porosity values equal to 0.9. 0.3 and 0.1, and mass flow values equal to 0.1 and 0.2 kg/m^{2}s, without radiation heat

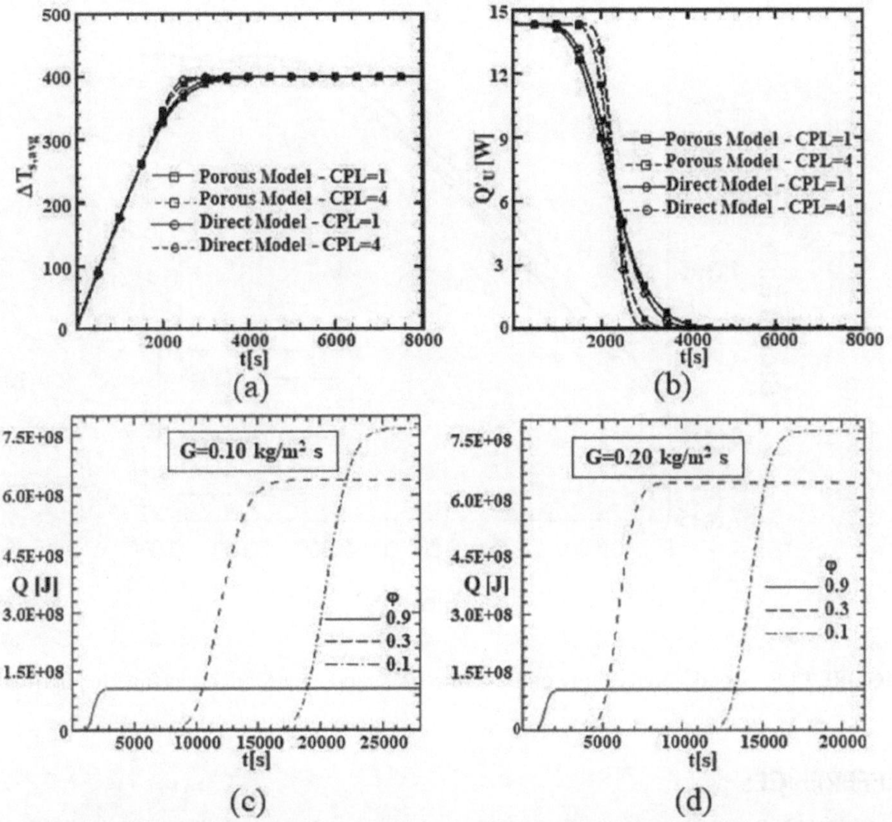

FIGURE 13.4 Comparison between direct and porous medium model for CPL = 1 and 4 in terms of (a) average solid temperature rise; (b) heat transfer rate between air and solid, and stored energy for different porosity values and mass flow: g; equal to (c) 0.1 kg/(m^{2}s) and (d) 0.2 kg/(m^{2}s).

transfer effects. Decreasing the porosity the charging time increases due to the thermal capacity increase. Increasing the porosity value steady state conditions are reached at lower time for the charging phase. However, increasing the mass flow rate any variation in the stored energy values is detected. As expected, for assigned mass flow rate, the porosity value determines also the stored energy level whereas, for assigned porosity, the mass velocity or the mass flow rate affects the charging time of the system.

13.6.2 Latent Heat Thermal Energy Storage System (LHTES)

The total stored energy profiles for different CPLs as function of time is reported in Figure 13.5. It is observed for lower CPLs the stored energy profiles are higher while for CPL = 8 and CPL = 16 the two profiles are overlapped. Furthermore, it is observed that the charging time decreases as the CPL increases, this behavior could be justified by the fact that the exchange contact area between air–solid and solid–PCM is higher.

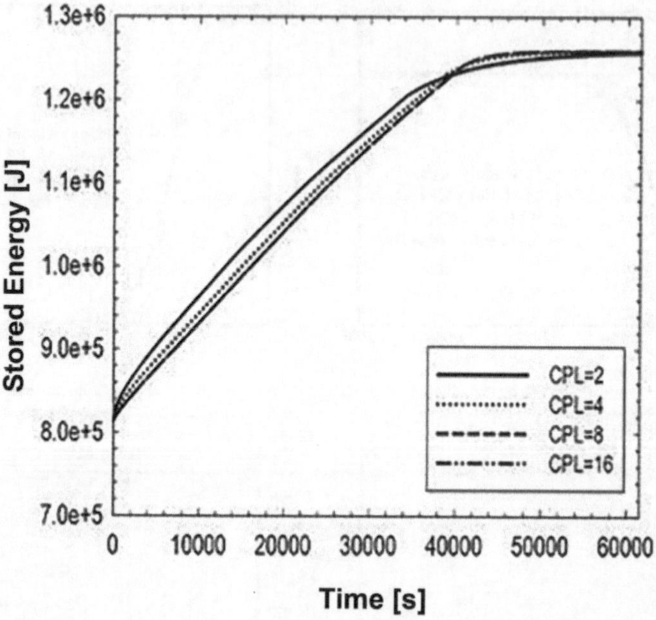

FIGURE 13.5 Total stored energy of the honeycomb model for Model A as function of time.

REFERENCES

S. Afrin, J.D. Ortega, V.K. Umar, D. Bharathan, A computational analysis: a honeycomb flow distributor with porous approximation for a thermocline thermal energy storage system, *Proceedings of ASME 2013 7th International Conference on Energy Sustainability, ES 2013*, Minneapolis, MN, 2013, Article No. V001T03A013.

N.A.M. Amin, M. Belusko, F. Bruno, M. Liu, Optimising PCM thermal storage systems for maximum energy storage effectiveness, *Solar Energy* 86(9) (2012) 2263–2272.

A. Andreozzi, B. Buonomo, A. Di Pasqua, D. Ercole, O. Manca, Heat transfer behaviors of parallel plate systems in sensible thermal energy storage, *Energy Procedia* 126 (2017a) 107–114.

A. Andreozzi, B. Buonomo, D. Ercole, O. Manca, Parallel plate channels for latent heat thermal energy storages, *Proceedings of the 13th International Conference on Heat Transfer, Fluid Mechanics and Thermodynamics (HEFAT2017)*, 17–19 July 2017, Portoroz, Slovenia, 2017b, pp. 446–451.

A. Andreozzi, B. Buonomo, D. Ercole, O. Manca, Solar energy latent thermal storage by phase change materials (PCMs) in a honeycomb system, *Thermal Science and Engineering Progress* 6 (2018a) 410–420.

A. Andreozzi, B. Buonomo, D. Ercole, O. Manca, Parallel triangular channel system for latent heat thermal energy storages, *International Heat Transfer Conference 2018* – August, Beijing, 2018b, pp. 4411–4418.

A. Andreozzi, B. Buonomo, D. Ercole, O. Manca, Parallel triangular channel system for sensible heat thermal energy storages with external heat losses, *Proceedings ASME 2019 Heat Transfer Summer Conference*, Bellevue, WA, 15–18 July 2019, Paper No. HT2019-3607 V001T13A012.

A. Andreozzi, B. Buonomo, O. Manca, S. Nardini, S. Tamburrino, Heat transfer behaviors of thermal energy storages for high temperature solar systems, In Delgado, J. M. P. Q., Ed., *Industrial and Technological Applications of Transport in Porous Materials*, Springer-Verlag, Berlin, 2013, pp. 119–139.

A. Andreozzi, B. Buonomo, O. Manca, S. Tamburrino, Thermal energy storages analysis for high temperature in air solar systems, *Applied Thermal Engineering* 71 (2014) 130–141.

A. Andreozzi, B. Buonomo, O. Manca, S. Tamburrino, Transient analysis of heat transfer in parallel squared channels for high temperature thermal storage, *Computational Thermal Sciences* 7 (2015) 477–489.

T.H. Andriotty, P. Smith Schneider, L. Jenisch Rodrigues, Inverse design methodology to optimize sensible thermal energy storage systems working as rectifiers, *International Journal of Energy Research* 43(12) (2019) 6442–6453.

T.H. Andriotty, P. Smith Schneider, L.J. Rodrigues, Accuracy of lumped element model for cyclic sensible thermal energy storage systems, *Journal of Energy Storage* 28 (2020) 101277.

T.H. Andriotty, L.J. Rodrigues, L.A.O. Rocha, P.S. Schneider, Optimization of a sensible thermal storage system by a lumped approach, *Defect and Diffusion Forum* 366 (2016) 182–191.

J.P.A. Lopez, F. Kuznik, D. Baillis, J. Virgone, Numerical modeling and experimental validation of a PCM to air heat exchanger, *Energy and Buildings* 64 (2013) 415–422.

A. Bejan, *Convection Heat Transfer*, 2nd ed., New York: Wiley, 1995.

M. Bahrami, M.M. Yovanovich, J.R. Culham, Pressure drop of fully-developed, laminar flow in microchannels of arbitrary cross-section, *Proceedings of ICMM 2005 3rd International Conference on Microchannels and Minichannels*, 2005, Toronto, Paper ICMM2005-75109.

M. Belusko, E. Halawa, F. Bruno, Characterising PCM thermal storage systems using the effectiveness-NTU approach, *International Journal of Heat and Mass Transfer* 55(13–14) (2012) 3359–3365.

E. Borri, G. Zsembinszki, L.F. Cabeza, Recent developments of thermal energy storage applications in the built environment: A bibliometric analysis and systematic review. *Applied Thermal Engineering* 189 (2021) 116666.

B. Buonomo, L. Cirillo, A. Diana, A. di Pasqua, D. Ercole, V. Fardella, O. Manca, S. Nardini, Thermal Energy Storage Systems. *TECNICA ITALIANA-Italian Journal of Engineering Science* 64(1) (2020) 39–44.

B. Buonomo, D. Ercole, O. Manca, S. Nardini, Numerical analysis on a latent thermal energy storage system with phase change materials and aluminum foam, *Heat Transfer Engineering* 41(12) (2020) 1075–1084.

L.F. Cabeza, *Advances in Thermal Energy Storage Systems: Methods and Applications*, Woodhead Publishing, Cambridge, 2014.

M.W. Jr. Chase, *NIST-JANAF Thermochemical Tables*, 4th ed., American Institute of Physics, 1998.

A. Crespo, G. Zsembinszki, D. Vérez, E. Borri, C. Fernández, L.F. Cabeza, A. de Gracia, Optimization of design variables of a phase change material storage tank and comparison of a 2D implicit vs. 2D explicit model, *Energies* 14(9) (2021) 2605.

I. Dincer, M.A. Rosen, *Thermal Energy Storage, Systems and Applications*, Wiley, New York, 2002.

C. Ding, Z. Niu, B. Li, D. Hong, Z. Zhang, M. Yu, Analytical modeling and thermal performance analysis of a flat plate latent heat storage unit, *Applied Thermal Engineering* 179 (2020) 115722.

P. Dolado, A. Lazaro, J.M. Marin, B. Zalba, Characterization of melting and solidification in a real scale PCM-air heat exchanger: Numerical model and experimental validation, *Energy Conversion and Management* 52(4) (2011) 1890–1907.

S. Farah, M. Liu, W. Saman, Numerical investigation of phase change material thermal storage for space cooling, *Applied Energy* 239 (2019) 526–535.

M.M. Farid, A.M. Khudhair, S.A.K. Razack, S. Al-Hallaj, A review on phase change energy storage: materials and applications, *Energy Conversion and Management* 45 (2004) 1597–1615.

A. Gil, M. Medrano, I. Martorell, A. Lazaro, P. Dolado, B. Zalba, L.F. Cabeza, State of the art on high temperature thermal energy storage for power generation. Part 1D concepts, materials and modellization, *Renewable Sustainable Energy Review* 14 (2010) 31–55.

E. Halawa, F. Bruno, W. Saman, Numerical analysis of a PCM thermal storage system with varying wall temperature, *Energy Conversion and Management* 46(15–16) (2005) 2592–2604.

E. Halawa, W. Saman, F. Bruno, A phase change processor method for solving a one-dimensional phase change problem with convection boundary, *Renewable Energy* 35(8) (2010) 1688–1695.

M. Hanchen, S. Bruckner, A. Steinfeld, High-temperature thermal storage using a packed bed of rocks – Heat transfer analysis and experimental validation, *Applied Thermal Engineering* 31 (2011) 1798–1806.

G. Hed, R. Bellander, Mathematical modelling of PCM air heat exchanger, *Energy Building* 38 (2006) 82–89.

D. Iaria, X. Zhou, J. Al Zaili, Q. Zhang, G. Xiao, A. Sayma, Development of a model for performance analysis of a honeycomb thermal energy storage for solar power microturbine applications, *Energies* 12(20) (2019) 3968.

K.A.R. Ismail, O.C. Quispe, J.R. Henríquez, A numerical and experimental study on a parallel plate ice bank, *Applied Thermal Engineering* 19 (1999) 163–193.

J. Janchen, T. H. Herzog, K. Gleichmann, B. Unger, A. Brandt, G. Fischer, H. Richter, Performance of an open thermal adsorption storage system with Linde type A zeolites: Beads versus honeycombs, *Microporous Mesoporous Materials* 207 (2015) 179–184.

T.Y. Kim, B.S. Hyun, J.J. Lee, J. Rhee, Numerical study of the spacecraft thermal control hardware combining solid–liquid phase change material and a heat pipe, *Aerospace Science and Technology* 27 (2013) 10–16.

C. Lai, S. Hokoi, Thermal performance of an aluminum honeycomb wallboard incorporating microencapsulated PCM, *Energy Building* 73 (2014) 37–47.

A. Lazaro, P. Dolado, J.M. Marin, B. Zalba, PCM-air heat exchangers for free-cooling applications in buildings: Empirical model and application to design, *Energy Conversion and Management* 50(3) (2009) 444–449.

P.-W. Li, C. Chan, *Thermal Energy Storage Analyses and Designs*, Academic Press, London. 2017.

Q. Li, F. Bai, B. Yang, Z. Wang, B. El Hefni, S. Liu, S. Kubo, H. Kiriki, M. Han, Dynamic simulation and experimental validation of an open-air receiver and a thermal energy storage system for solar thermal power plant, *Applied Energy* 178 (2016) 281–293.

Q. Li, F. Bai, B. Yang, Y. Wang, L. Xu, Z. Chang, Z. Wang, B. El Hefni, Z. Yang, S. Kubo, H. Kiriki, M. Han, Dynamic simulations of a honeycomb ceramic thermal energy storage in a solar thermal power plant using air as the heat transfer fluid, *Applied Thermal Engineering* 129 (2018) 636–645.

Y. Li, B. Guo, G. Huang, S. Kubo, P. Shu, Characterization and thermal performance of nitrate mixture/SiC ceramic honeycomb composite phase change materials for thermal energy storage, *Applied Thermal Engineering* 81 (2015) 193–197.

Z. Liao, C. Xu, Xu C., Ju X., Gao F., Wei G., Cyclic performance analysis of a high temperature flat plate thermal energy storage unit with phase change material, *Applied Thermal Engineering* 144 (2018) 1126–1136.

M. Liu, M. Belusko, N.H. Steven Tay, F. Bruno, Impact of the heat transfer fluid in a flat plate phase change thermal storage unit for concentrated solar tower plants, *Solar Energy* 101 (2014) 220–231.

H. Liu, K. Nagano, Numerical simulation of an open sorption thermal energy storage system using composite sorbents built into a honeycomb structure, *International Journal of Heat and Mass Transfer* 78 (2014) 648–661.

H. Liu, K. Nagano, D. Sugiyama, J. Togawa, M. Nakamura, Honeycomb filters made from mesoporous composite material for an open sorption thermal energy storage system to store low-temperature industrial waste heat, *International Journal of Heat and Mass Transfer* 65 (2013) 471–480.

H. Liu, K. Nagano, J. Togawa, A composite material made of mesoporous siliceous shale impregnated with lithium chloride for an open sorption thermal energy storage system, *Solar Energy* 111 (2015) 186–200.

M. Liu, W. Saman, F. Bruno, Validation of a mathematical model for encapsulated phase change material flat slabs for cooling applications, *Applied Thermal Engineering* 31(14–15) (2011) 2340–2347.

Z. Luo, C. Wang, G. Xiao, M. Ni, K. Cen, Simulation and experimental study on honeycomb-ceramic thermal energy storage for solar thermal systems, *Applied Thermal Engineering* 73 (2014) 620–626.

M.S. Mahdi, H.B. Mahood, A.A. Khadom, A.N. Campbell, Numerical simulations and experimental verification of the thermal performance of phase change materials in a tube-bundle latent heat thermal energy storage system, *Applied Thermal Engineering* 194 (2021) 117079.

M. Medrano, A. Gil, I. Martorell, X. Potau, L.F. Cabeza, State of the art on high temperature thermal energy storage for power generation. Part 2D case studies, *Renewable Sustainable Energy Review* 14 (2010) 56–72.

H. Mehling, L.F. Cabeza, *Heat and Cold Storage with PCM: An Up to Date Introduction into Basics and Applications*, Springer, Verlag, Berlin, Heidelberg, 2008.

S. Nekoonam, R. Roshandel, Modeling and optimization of a multiple (cascading) phase change material solar storage system, *Thermal Science and Engineering Progress* 23 (2021) 100873.

D.A. Nield, A. Bejan, *Convection in Porous Media*, 3rd ed., Springer, New York, 2006.

M.N. Ozisik, *Heat Transfer: A Basic Approach*, McGraw-Hill, Singapore, 1984.

N. Rafidi, W. Blasiak, Thermal performance analysis on two composite material honeycomb heat regenerators used for HiTAC burners, *Applied Thermal Engineering* 25 (2005) 2966–2982.

REN21, Renewables 2020 global status report 2020. 2020.

J. Sacharczuk, D. Taler, Numerical and experimental study on the thermal performance of the concrete accumulator for solar heating systems, *Energy* 170 (2019) 967–977.

W. Saman, F. Bruno, E. Halawa, Thermal performance of PCM thermal storage unit for a roof integrated solar heating system, *Solar Energy* 78(2) (2005) 341–349.

F.W. Schmidt, M.E. Newell, Heat transfer in fully developed laminar flow through rectangular and isosceles triangular ducts, *International Journal of Heat and Mass Transfer* 10 (1967) 1121–1123.

R.K. Shah, Laminar flow friction and forced convection heat transfer in ducts of arbitrary geometry, *International Journal of Heat and Mass Transfer* 18 (1975) 849–862.

O. Srikanth, S.D. Khivsara, R. Aswathi, C.D. Madhusoodana, R.N. Das, V. Srinivasan, P. Dutta, Numerical and experimental evaluation of ceramic honeycombs for thermal energy storage, *Transactions of the Indian Ceramic Society* 76(2) (2017) 102–107.

D. Taler, P. Dzierwa, M. Trojan, J. Sacharczuk, K. Kaczmarski, J. Taler, Numerical modeling of transient heat transfer in heat storage unit with channel structure, *Applied Thermal Engineering* 149 (2019) 841–853.

S.M. Vakilaltojjar, W. Saman, Analysis and modelling of a phase change storage system for air conditioning applications, *Applied Thermal Engineering* 21(3) (2001) 249–263.

V.R. Voller, C. Prakash, A fixed grid numerical modelling methodology for convection-diffusion mushy region phase-change problems, *International Journal of Heat and Mass Transfer* 30 (1987) 1709–1719.

F.Q. Wang, Y. Shuai, H.P. Tan, C.L. Yu, Thermal performance analysis of porous media receiver with concentrated solar irradiation, *International Journal of Heat and Mass Transfer* 62 (2013) 247–254.

B. Zalba, J.M. Marin, L.F. Cabeza, H. Mehling, Review on thermal energy storage with phase change: materials, heat transfer analysis and applications, *Applied Thermal Engineering* 23 (2003) 251–283.

B. Zalba, J.M. Marín, L.F. Cabeza, H. Mehling, Free-cooling of buildings with phase change materials, *International Journal of Refrigeration* 27(8) (2004) 839–849.

Y Zhang, A. Faghri, Heat transfer enhancement in latent heat thermal energy storage system by using the internally finned tube, *International Journal of Heat and Mass Transfer* 39(15) (1996) 3165–3173.

C.Y. Zhao, W. Lu, Y. Tian, Heat transfer enhancement for thermal energy storage using metal foams embedded within phase change materials (PCMs), *Solar Energy* 84 (2010) 1402–1412.

G. Zsembinszki, P. Moreno, C. Solé, A. Castell, L.F. Cabeza, Numerical model evaluation of a PCM cold storage tank and uncertainty analysis of the parameters, *Applied Thermal Engineering* 67(1–2) (2014) 16–23.

M. Zukowski, Mathematical modeling and numerical simulation of a short term thermal energy storage system using phase change material for heating applications, *Energy Conversion and Management* 48(1) (2007) 155–165.

14 Recent Progress of Phase Change Materials and a Novel Application to Cylindrical Lithium-Ion Battery Thermal Management

Yiwei Wang
Chinese Academy of Sciences
University of Chinese Academy of Sciences

Peng Peng, Wenjiong Cao, and Fangming Jiang
Chinese Academy of Sciences

CONTENTS

DOI: 10.1201/9781003213260-14

14.1 INTRODUCTION

Energy is not only the "engine" of global economy, but also the basis for human survival on the earth [1]. With the development of industrial technology and the continuous progress of human society, the traditional fossil fuel is increasingly exhausted and the environmental pollution due to the consumption of fossil fuel is becoming more and more serious [2]. Under the dual pressure of energy shortage and environmental pollution, it is urgent to develop new clean energy and improve energy utilization efficiency [3].

Traditional vehicles are powered by diesel or gasoline internal combustion engines. Both diesel and gasoline are petroleum products. On the one hand, the consumption of petroleum products will produce greenhouse gases such as carbon dioxide (CO_2). On the other hand, the produced sulfur dioxide (SO_2) and other gases are harmful to the environment and human beings, resulting in a series of pollutions including air pollution, soil pollution and so on. Moreover, petroleum is a non-renewable energy, which has to take a long time to form, and the massive use is bound to consume it all sometime in the not very far future [4–5].

Andersen et al. [6] indicated that 95% of motor vehicles depend on oil, accounting for more than 50% of the world's carbon emissions. Since late twentieth century, the transportation sector, as a large consumer of energy and a source of environmental pollution, is imperative to implement energy conservation and emission reduction for traditional internal combustion engine vehicles. The electric vehicles have at least the two advantages: low energy consumption and zero emissions on road, and thus the use of electric vehicles can significantly reduce carbon emissions. The key to the development of electric vehicles is to improve the working performance of power batteries [7–8], mostly lithium-ion batteries.

Much research and practice indicated the failure rate of lithium-ion battery directly relates to its working temperature. Too low temperature will lead to rapid capacity fading, and ease of overcharge or even formation of metallic lithium dendrite, causing internal short circuit. When the working temperature is too high, great vulnerability there exists to thermal runaway or even fire and explosion [9]. In addition, when the lithium-ion battery encounters extrusion, over-discharge, nail penetration and other abuse conditions, its temperature can also rise quickly, causing thermal runaway if the temperature rises beyond the safety threshold.

For the power battery pack that contains hundreds or thousands of batteries within a compact configuration, a timely heat removal is commonly very difficult to attain and the heat dissipation problem becomes more serious. Local non-uniform heat accumulation within the pack or even within a single battery causes temperature difference, i.e. thermal non-uniformity, resulting in non-uniform utilization, premature degradation, and even thermal runaway and safety accidents [10]. Therefore, an effective and reliable thermal management system is crucial to lithium-ion battery modules and packs.

A battery thermal management system (BTMS) aims to control the battery pack working temperature within a suitable temperature range and to minimize the temperature difference across the pack [11]. There are at least three categories of battery thermal management strategies commonly in use: air-cooling, liquid-cooling, phase change material (PCM) cooling [12].

The air-cooling means that the external air enters the inside of the battery pack, and removes heat convectively from the battery surface to the outside. According to whether the air flow is spontaneous or not, air-cooling can be divided into natural cooling and forced cooling [13]. Generally, the natural cooling of air has a convective heat transfer coefficient around 5 W/m/K. The forced air-cooling is commonly driven by fan, and the convective heat transfer coefficient is dependent on the air speed, generally greater than 10 W/m/K [14]. The shape and size of the air flow conduits and the air flow rate are the main factors affecting the cooling effects such as the uniformity of battery temperature [15].

The liquid-cooling uses liquid as the heat transfer medium. The heat transfer coefficient of liquid-cooling is usually much higher than that of air-cooling. Liquid-cooling can be classified into the direct cooling and indirect cooling. In the direct liquid-cooling system, the battery pack is immersed in the liquid directly [16] and the usable liquid must be dielectric and of sufficiently high thermal conductivity, e.g. mineral oil or silicone oil. Different from the direct liquid-cooling system, the indirect liquid-cooling system constrains the liquid in a closed loop and the liquid has thus no direct contact with the battery. The indirect liquid-cooling system is not that picky at the fluid medium, but requires the liquid pathway to be strictly liquid-proof [17–18]. Additionally, a pump is used to drive the liquid flow, which causes extra energy consumption.

PCM absorbs heat when phase change occurs, and can effectively lower the temperature of the battery module due to the high latent heat it has. The PCM cooling may achieve a more uniform temperature distribution in batteries than the air- or liquid-cooling. Moreover, as the heat is stored in the PCM itself, at lowered temperatures the battery can be heated by the PCM [19]. The PCM BTMS was first proposed by Al-Hallaj and Selman [20], and it is attracting more and more attention in recent years.

14.2 PHASE CHANGE MATERIALS (PCMs)

PCMs can absorb or release latent heat during phase change. When the temperature rises up to the phase transition temperature, PCM absorbs and stores the heat in the form of latent heat [21]. When the temperature decreases to be lower than the phase transition temperature, the PCM will return to the initial phase state and the heat will be released [22]. There are a large variety of classification standards for PCMs. According to the phase change process of physical state, the PCMs can be divided into solid–liquid PCMs, liquid–gas PCMs, solid–gas PCMs, and solid–solid PCMs [23]. In terms of the phase transition temperature, the PCMs are classified as low-temperature (≤100°C), medium-temperature (100°C–250°C), and high-temperature (>250°C) PCMs [24]. Based on the chemical constituents, there are three classes of PCMs: the organic, inorganic and eutectic mixture PCMS [25]. The solid–liquid PCMs have relatively small volume change (compared to liquid–gas and solid–gas PCMs) and relatively large phase change heat (compared to solid–solid PCMs) during phase transition, so they have attracted extensive attention [26].

The organic PCMs include paraffin, fatty acids, alcohols, lipids, and diols [27]. Paraffin is the most widely used PCM currently. It is a mixture of long-chain alkanes. Its melting point depends on the chain length. The shorter the chain length is, the higher is the melting point. Therefore, paraffin with different melting points can be obtained to meet the needs of BTMS. However, paraffin wax usually has low thermal conductivity, which limits its application [28]. Inorganic PCMs mainly refer to molten salt, hydrated salt, metal or metal alloy [29–30]. Hydrated salt phase change process has high supercooling and is prone to phase separation [31], resulting in thermal instability and uneven internal temperature distribution. All these greatly reduce the heat storage capacity of hydrated salt PCM. To eliminate the supercooling problem of hydrated salt, nucleating agent (such as nano-Al_2O_3) is usually added. To restrain the phase separation, addition of thickener (e.g. sodium betonies) can significantly improve the performance of hydrated salt [32]. The metal-based PCMs not only have high thermal conductivity, but also have low thermal expansion and large phase change latent heat. However, the corrosion and unstable chemical properties impair their long-term performance [33]. The eutectic PCM is a mixture of two or more kinds of substances, so as to obtain the required phase change temperature [34]. The eutectic PCMs have ideal physicochemical properties, such as high latent heat and demanded melting point, whereas due mainly to the unstable thermal properties and high cost, this class of PCMs are rarely used in practice.

For BTMS applications, the latent heat and thermal conductivity of PCM should be large enough, and its phase change temperature should be within the optimum working temperature range of battery. Due to the frequent charge–discharge, the thermal properties of the PCM need to be very stable. Moreover, the PCM should be dielectric, non-corrosive, and environmentally friendly [35].

14.2.1 PCM Heat Transfer Enhancement Methods

Except for the metal-based PCMs, almost all the PCMs may have the same disadvantages, e.g. low thermal conductivity, poor cycle stability, and high degree of supercooling [36]. Therefore, how to improve the heat transfer performance of PCMs is the main problem to be solved. The enhancement methods can be: encapsulation, nanoparticle enhancement, porous material enhancement, fins-based enhancement, or some combinations of the former four.

The encapsulated PCM consists of two parts: the PCM as the core plus the capsule shell. Compared with the ordinary PCM, the encapsulated PCMs have many advantages, such as the increased specific surface area and the decreased volume dilatation/contraction during phase change [37]. Depending on the capsule size, the encapsulated PCMs can be classified into macro- (>1 mm), micro- (1–1000 μm), and nano-sized capsules (1–1000 nm). Frazzica et al. [38] studied a small-scale hybrid sensible/latent storage system (nominal volume 48.6 dm^3), consisting of water added by two different macro-encapsulated PCMs (a commercial paraffin and a hydrated salts mixture). Both of them guarantee a remarkable enhancement of the volumetric heat storage capacity, about 10% increase with respect to water. Chai et al. [39] designed and synthesized a bifunctional microencapsulation, which encapsulates eicosane into a crystalline TiO_2 shell. Through the Scanning Electronic Microscope (SEM) and Transmission Electron Microscope (TEM), all the microcapsules present a perfect spherical shape with a uniform particle size of 1.5–2 μm. Moreover, the microcapsules achieve a high encapsulation efficiency and a high thermal-storage capability. The supercooling of these microcapsule samples is also suppressed due to the encapsulation by the inorganic wall material with a high thermal conductivity (1.324 W/m/K). The thermal conductivity of pure eicosane is 0.16 W/m/K, which increases to 0.75–0.86 W/m/K depending on the weight ratio of eicosane/tetrabutyl titanate. Due to the small size of microchannel and the relatively large size of microcapsules, pipeline blockage happens easily. Nano-capsule PCMs were proposed and developed rapidly. Peng et al. [40] summarized the structure, principle and function of nano-encapsulated PCMs, discussed the thermophysical properties and chemical properties of nano-capsules. Compared with the macro- or microcapsule, nano-capsules have better thermal and mechanical properties and are more valuable in nanoscale systems.

The concept of nanofluid was first proposed by Choi and Eastman in 1995. Nanoscale solid, usually carbon nanotubes, graphite, graphene, metal (e.g. Cu, Ag), or metal oxide (e.g. TiO_2, CuO, ZnO, MgO, and Al_2O_3) particles [41], can enhance the heat transfer performance of fluid owing to their high surface area to volume ratio [42]. Nano-enhanced PCM was firstly proposed by Khodadadi and Hosseinizadeh [43] in 2007. With a small portion of nanoparticles being added to the PCM, not only the thermal conductivity characteristic of PCM will be improved, but also other characteristics including latent heat capacity, sub-cooling, phase change temperature and viscosity, will be improved. Guo et al. [44] prepared composites of erythritol-expanded graphite (EG) and erythritol-carbon nanotubes (CNTs) by using melting dispersion method. The melting point, latent heat and thermal conductivity of the composites were evaluated. Bahiraei et al. [45] explored the heat transfer enhancement of the phase change nanocomposites using three different types of carbon-based nanostructures (carbon nanofiber, graphene nanoplatelets, and graphite nano-powder) with mass fractions from 2.5% to 10%. The thermal conductivity of all the solid phase nano-enhanced PCM sample is increased by adding some mass fraction of nanoparticles and the graphite-based nano-enhanced PCM exhibits the best thermal conductivity. Lin and Al-Kayiem [46] dispersed 20 nm copper nanoparticles into paraffin to synthesis Cu-PCM nanocomposites. The thermal conductivity of the paraffin wax was improved by 14%, 23.9%, 42.5%, and 46.3% when the mass fraction of copper particles is 0.5%, 1.0%, 1.5%, and 2.0%, respectively. Rufuss et al. [47] added 0.3 wt.% nanoparticles (TiO_2, CuO) to the paraffin to form nanoparticle-enhanced PCM. Compared with the paraffin wax, there is an increase in thermal conductivity (25% increase for TiO_2, 28.8% for CuO). Despite the thermal conductivity of PCM can be increased by adding nanoparticles, its viscosity may significantly rise in liquid state, resulting in weakened natural convection during phase change heat transfer. The weakened natural convection may offset or even exceed the improvement of thermal conductivity, worsening the thermal performance of PCM [48].

The metal-based (e.g. aluminum foam, copper foam, and nickel foam) and carbon-based (e.g. carbon foam, EG, and graphite foams) porous materials possess good mechanical strength and thermophysical properties, high porosity, stable solid skeleton structure, and especially high thermal conductivity. Therefore, these porous materials can be used to enhance the performance of PCMs. Porous material-enhanced PCMs have been used for solar collectors [49], thermal energy storage systems [50], cooling/heating sinks [51], and so on.

Due to low cost and ease of manufacturing, using fins is one common and effective way to improve heat transfer. Inserting some fins directly into the PCM and placing some fins on the PCM surface are the two main methods to increase the heat transfer area of the PCM [52]. Costa et al. [53] developed a theoretical model to predict the thermal performance of the latent heat energy storage system with or without fins. With fins added, a balance between weight and performance may be required. Therefore, it is very important to study the performance optimization related to number of fins, fins' geometry (thickness, height, length and width) and arrangement, etc. To obtain the best fin-based PCM heat sink, some novel fins, such as non-uniform fins [54], uneven tree-like fins [55], angle-inclined fins [56] and the tee fin [57], were proposed.

Some hybrid reinforcements can be used to enhance the heat transfer performance. Accounting for the PCM poor cycle stability and high degree of supercooling, Lin et al. [58] modified the microencapsulated PCMs with nanoparticles, i.e. nanoparticles were placed on the outer surface of microencapsulated PCMs to effectively alleviate the external impact or mechanical defects in the materials. Min et al. [59] put stearyl alcohol into carbon nanotubes and mixed them with microencapsulated PCMs by in situ polymerization method. After 100 heating and cooling cycles, compared with the un-modified microencapsulated PCMs, the composite microencapsulated PCMs possess excellent durability and thermal stability (the decomposition temperature only decreases by 4.7%). Zhu et al. [60] prepared composite nano-encapsulated PCMs with SiO_2/graphene hybrid shell. The thermal conductivity of the composite nano-encapsulated PCMs with SiO_2/graphene hybrid shell increases by 132.9% in comparison to the nano-encapsulated PCMs with SiO_2 shell. The supercooling phenomenon is completely eliminated in the produced nano-encapsulated PCMs, while the latent heat slightly changes from 3.914 to 3.727 J/g/K within 10°C–60°C temperatures. Zhu et al. [61] fabricated a carbon nanotubes and Cu foam hybrid composite through a high-temperature tube furnace process. The thermal conductivity of the composite increases from 0.105 W/m/K (paraffin) to 3.49 W/m/K. Moreover, the supercooling degree is reduced from 9.44°C (paraffin) to 4.84°C (Cu: 41.3 wt.%, carbon nanotubes: 0.09 wt.%).

14.2.2 Application of PCMs to BTMS

The various PCMs presented above have been applied to BTMS. Ng et al. [62] selected aluminum nitride and aluminum oxide nanoparticles as additives to the copolymer formed by methyl methacrylate and triethoxyvinylsilane. Microcapsule PCMs of high thermal conductivity and good electrical insulation were synthesized by mini-suspension polymerization method. When applied to a BTMS, the central temperature of the battery module was decreased up to 7.3°C. Ling et al. [63] proposed a BTMS using SAT($CH_3COONa{\cdot}3H_2O$)-Urea/EG composite PCM. The SAT-Urea/EG composite was made by the multiscale encapsulation method. The thermal conductivity of the composite can be up to 4.96 W/m/K.

Li et al. [64] designed a silicone sealant (SS) immersed BTMS for an 18650-type lithium-ion power battery module. Compared with other BTMSs (air-cooling and pure SS cooling), the SS/BN (boron nitride) composite with 10 wt.% BN added shows more effective heat dissipation and better long-term corrosion protection. In addition, the water-immersed SS/BN BTMS can effectively improve the temperature uniformity. Lv et al. [65] developed a nanosilica (NS)-enhanced composite PCM for BTMS. The

NS (30–100 nm nanoscale pores) was added to paraffin to form the composite. The composite PCM with 5.5 wt.% mass fraction NS shows better heat dissipation performance than the PCM without NS. Xiao et al. [66] explored a polymer PCM (PoPCM) being applied to the BTMS. The battery module shows an excellent and stable heat dissipation performance.

Ling et al. [67] compared the thermal performance of two kinds of composite PCMs (one is the highly thermally conductive 60 wt.% RT44HC/EG composite and the other 60 wt.% RT44HC/fumed silica composite with a lower thermal conductivity) applied to the BTMS of a 20-cell battery pack at low temperatures (5 and −10°C). The results indicate that the low thermal conductivity composite PCM shows better heat conservation performance, but causes a higher thermal non-uniformity in the battery module. The highly thermally conductive composite PCM improves the temperature uniformity in the battery pack. To enhance the thermal conductivity and heat dissipation capability of the PCM in BTMS, He et al. [68] developed a composite PCM of a binary thermal conductive skeleton of EG/copper foam. Jiang et al. [69] designed a tube-shell battery pack with a passive thermal management system using composite PCM, which was made of the EG and paraffin and can significantly reduce the temperature rise during 3C and 5C discharge process. Situ et al. [70] used a double copper mesh, paraffin, EG and a low-density polyethylene formed plate to enhance the thermal performance of the rectangular battery module. The battery module under forced air-cooling shows an excellent thermal performance. The performance of PCM-based BTMS with graphene and multi-walled carbon nanotubes additives was experimentally studied by Zou et al. [71] The composite PCM effectively decreases the temperature rise rate of the battery module.

Sun et al. [72] and Fan et al. [73] proposed and optimized a composite PCM-fin structures for thermal management of cylindrical battery module. A ternary composite material with four aluminum fins was developed for BTMS by Lv et al. [74]. The composite PCM consists of EG, paraffin and low-density polyethylene, and presents better heat dissipation capability. Wang et al. [75] investigated experimentally the thermal behavior of cylindrical battery with BTMS of vertical straight fins submerged composite paraffin. The numerical study by Heyhat et al. [76] unraveled that porous-material-enhanced PCM BTMS could be more effective than the fins-based and the nano-enhanced PCM BTMSs.

The PCM cooling is intrinsically a passive cooling strategy. Releasing the heat generation of batteries to ambient is more effective if some active cooling strategies are introduced, such as forced air convection assisted cooling [77–78], or indirect liquid assisted cooling [79–80], or heat pipe assisted cooling [81]. Especially for batteries in harsh working conditions (e.g. high C-rate charge/discharge or high ambient temperature), combining passive cooling strategies with active strategies is crucial to effective thermal management.

14.3 A NOVEL APPLICATION OF PCM TO TMS OF CYLINDRICAL BATTERY MODULE

This section presents a thermal management system (TMS) for 18650-type lithium-ion battery module. The cooling system is a combination of PCM tubes and heat pipes with expanded-fin structure (i.e. air-cooling) at the condensation section.

Lightweight and compactness are the major points considered during the design and manufacturing of the hybrid cooling system. The shape and structure of the cooling system can fully utilize the empty space in-between the tightly assembled cylindrical batteries, and the assembling of the cooling system will not add extra volume to the battery module. Both the PCM tube and the heat pipe (including the fin structure) are made of aluminum, which is light weight and relatively not costly. Experiments and numerical simulations are specially designed to study the effects of PCM tube and heat pipe on the thermal performance of the battery module during charge/discharge processes of various C-rates and at different ambient temperatures.

14.3.1 Materials Preparation and Characterization

The BAK-produced commercial 18650 Li-ion batteries of 2750 mAh nominal capacity are used in this work. The battery chemistry is graphite anode and nickel–cobalt–aluminum (NCA) cathode. The battery discharge/charge operations are controlled by an Arbin BT-2000 equipment. The cut-off voltage of discharge process is 2.8 V. The charge operations follow the conventional constant current–constant voltage (CC–CV) protocol. A direct current (2.75 A, 1C) is applied to charge the battery until the voltage reaches 4.2 V, and then keeping the voltage at 4.2 V to charge the battery until the charging current drops to 1/30C (~90 mA).

A batch of fresh batteries was cycled a few times to determine the related properties. With the Arbin battery test station, the open-circuit voltage (OCV), capacity and internal resistance for each of these batteries were obtained from hybrid pulse power characterization (HPPC) tests. The batteries that were chosen to assemble the experimental battery module must suffice the following conditions: (1) the capacity is within 2750 ± 50 mAh; (2) the internal resistance is about 30–40 mΩ (100% SOC @25°C); (3) the OCV is in the range of 4.18–4.185 V (100% SOC @25°C).

The heat pipe is made of aluminum and weighs about 12 ± 1 g. Figure 14.1 details the shape and structure including the geometric dimensions of the heat pipe. It mainly consists of two sections: an evaporation section and a condensation section. The heat pipe internal is a columned cavity of 6.5 mm diameter. In the evaporation section, the outer shape of the heat pipe is specially designed to have four circular arc surfaces, which enable the tight assembling with maximum four 18650-type batteries. In the condensation section, four annular flake fins are manufactured to extend the heat transfer area exposed to the ambient. The fins and the pipe are integrally molded and the contact thermal resistance is thus eliminated.

The present work uses acetone as the working fluid of the heat pipe. Pretests show that there exists an optimal charging ratio, 40%, which gives the minimum heat transfer temperature difference between the evaporation and condensation section of the heat pipe. The overall heat transfer performance of the heat pipe at 40% charging ratio is quantified to be about 0.19–0.97 W/K. The PCM tubes are also made of aluminum. As shown in Figure 14.2, the PCM tubes have the same shape and structure as the heat pipes except without the fin-structured condensation section. The PCM tubes are filled with the industrial grade paraffin. A differential scanning calorimeter, DSC 823E, manufactured by METTLER TOLEDO (Zurich, Switzerland), was used to evaluate the properties of the PCM under a heating rate of 5°C/min with a 60 mL/min argon stream. The differential scanning calorimeter results indicate the phase change of the PCM

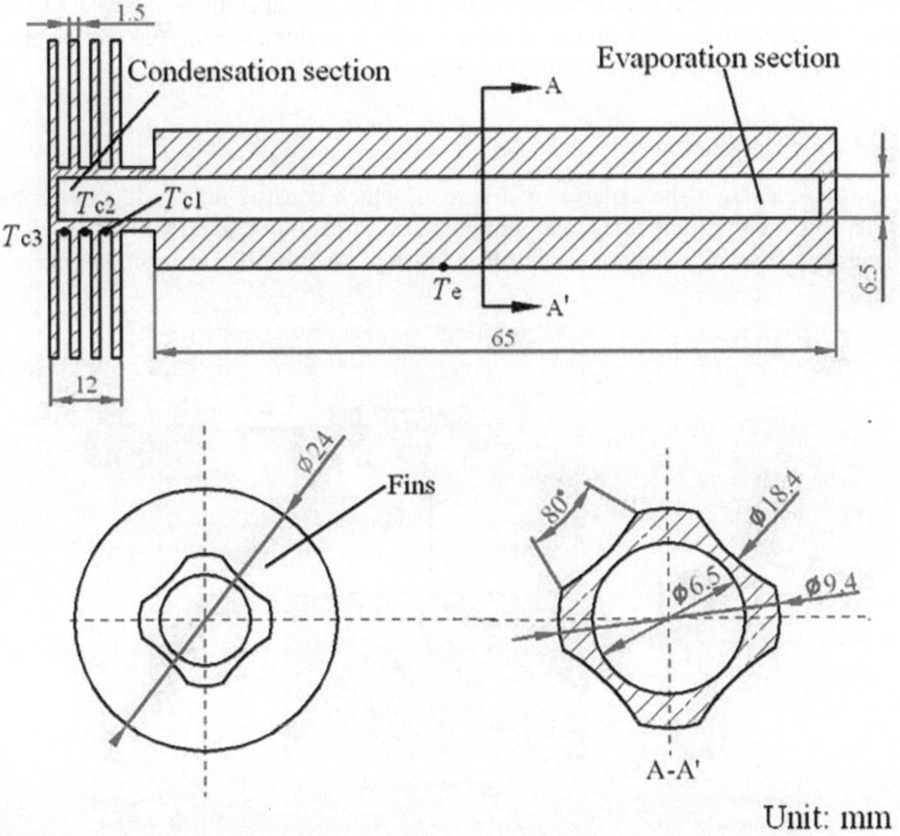

FIGURE 14.1 The shape and structure of the heat pipes.

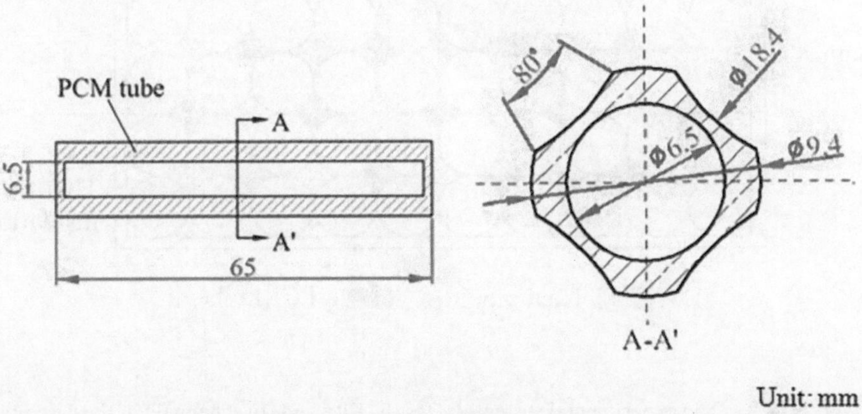

FIGURE 14.2 The shape and structure of the PCM tubes.

occurs within a temperature range, the onset and ending temperatures are 29.89°C and 40.02°C, respectively. The peak temperature of PCM melting is 36.59°C. The weights of the PCM tube and the filled paraffin are about 5.5 and 1.9 g, respectively.

14.3.2 Experimental

Figure 14.3a shows the experimental system, which consists mainly of three subsystems: the battery module (containing batteries, heat pipes, and PCM tubes, etc.), the charge–discharge operation controlling system, and the data acquisition system.

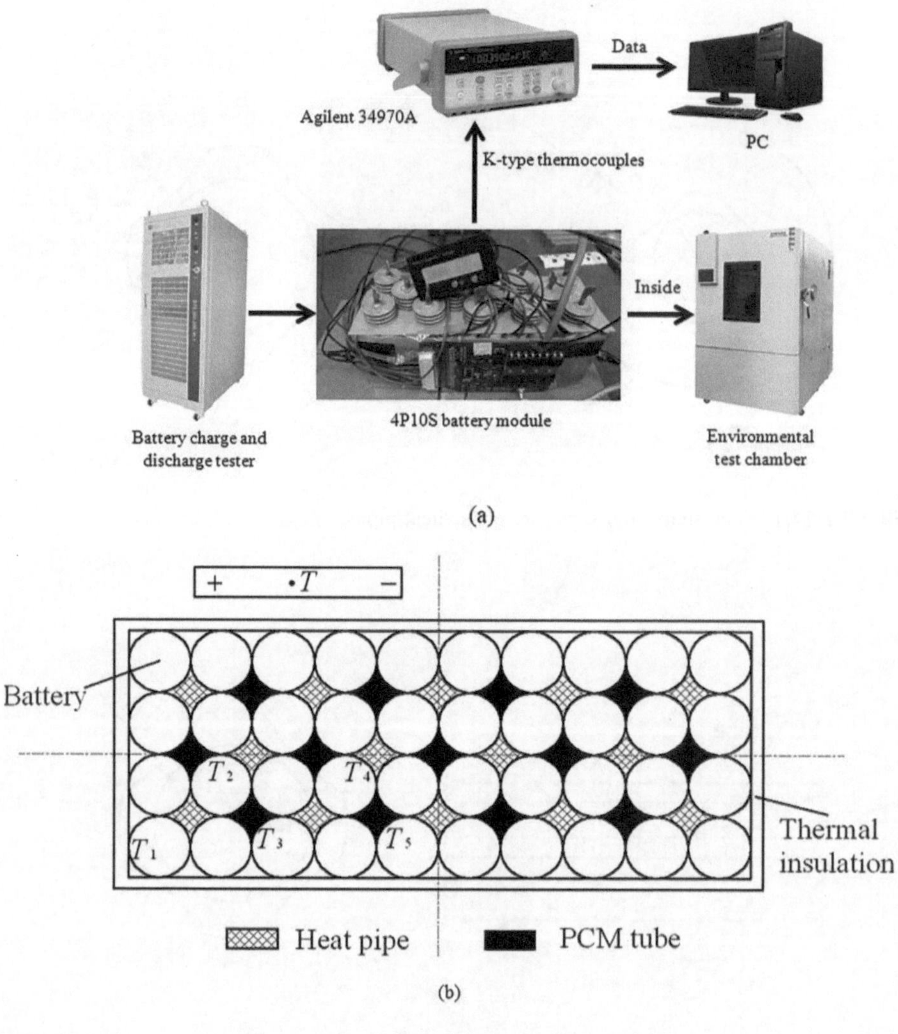

FIGURE 14.3 The experimental system: (a) test facility and (b) schematic of the assembled battery module and temperature measuring points.

The battery module contains 40 18650-type batteries, 4 batteries in parallel connection to form a single group and ten groups in serial connection to form the module (i.e. 4P10S). The 40 cylindrical batteries are tightly assembled and the empty spaces in-between batteries are embedded by either the heat pipes or the PCM tube, as displayed in Figure 14.3b. Totally, there are 14 heat pipes and 13 PCM tubes are inserted into the battery module. The condensation section of the heat pipe is exposed to the ambient. The battery module is wrapped by an epoxy board (65 mm in height and 0.5 mm in thickness; its resistivity is about 10^{13} Ω·cm) for electric insulation and also by some thermal insulation material (5 mm thick, with 0.034 W/m/K thermal conductivity) to reduce heat loss directly to ambient. The PCM tube and the wrapped thermal insulation material can, to some extent, maintain the battery temperature when the battery module is in idle state or operated in low-temperature environment. Some nickel bars of 0.12 mm thickness are used to connect the batteries to realize the desired serial or parallel connection. To improve thermal contact between the heat pipes and batteries, some heat-conducting silicone grease is coated on the outer surface of the batteries. The assembling of the heat pipes and PCM tubes with the batteries will not increase the volume of the battery module, ensuring a compact configuration.

The charge–discharge operations are controlled by the YKY-BTS 600V-200A-2 equipment and a battery management system (BMS). The BMS monitors the electric state of each battery and prevents over-discharge/charge; it can also actively balance the electric amount stored among the batteries to ensure even utilization. All the controlling parameters are preset in the personal computer (PC), which guides the YKY-BTS equipment to execute the charge/discharge operations.

The data acquisition system contains six K-type thermocouples (±0.3°C inaccuracy), an Agilent 34970A data collector, and the PC. Temperatures measured by the six thermocouples are collected by the Agilent data collector, and then delivered to the PC. Figure 14.3b also illustrates the locations of the six thermocouples. T_1–T_5 are used to measure the temperature of batteries in the module and T is to monitor the ambient temperature. Due to the geometric symmetry, temperature monitoring points are arranged only to one-fourth region of the battery module. The temperature sensors T_1–T_5 are positioned at the middle of the outer side surface of the corresponding battery.

The battery module is placed at the thermostat environmental chamber to regulate its ambient temperature. Before the test, the module was conditioned by cycling three times at standard charging rate namely ~0.5C (5.5 A) with the cut-off voltage of 27.5 and 42 V during discharging and charging, respectively. These conditioning cycles allow the full formation of solid electrolyte interface and eliminate the impact from irreversible capacity fade of fresh Li-ion batteries [33]. From the tests, the capacity of the module is obtained, which is about 10.5 Ah.

The experimental study includes various tests with respect to five different combinations of the 40 18650-type batteries and the 14 heat pipes and the 13 PCM tubes. The first (i.e. BM-0) is the naked battery module, which contains only the 40 18650-type batteries without any heat pipes or PCM tubes embedded. The second is named the BM-1 module, in which all the 14 heat pipes and 13 PCM tubes have been assembled with the 40 18650-type batteries, but the heat pipes and PCM tubes all

TABLE 14.1
Configurational Setups of the Five Experimental Modules

Module #	Configurational Setups	Mass Fraction of Batteries (%)
BM-0	40 Batteries only	100
BM-1	40 Batteries + 14 heat pipes and 13 PCM tubes (all empty)	87.5
BM-2	40 Batteries + 14 heat pipes and 13 empty PCM tubes	88
BM-3	40 Batteries + 13 PCM tubes and 14 empty heat pipes	87.4
BM-4	40 Batteries + 14 heat pipes and 13 PCM tubes	87

are empty, without working fluid or PCM injected. The third module, BM-2 module, is similar in structure to BM-1, whereas all the 14 heat pipes are filled with acetone at 40% charging ratio and the PCM tubes remain still empty. The fourth module, namely BM-3, has all the 14 heat pipes and all the 13 PCM tubes positioned inside like BM-1 and BM-2, while it is different from BM-1 and BM-2 as all the 13 PCM tubes are filled with PCM and all the 14 heat pipes are empty without filling any working fluid. The last, BM-4 module, contains the 40 batteries and all the well-equipped 14 heat pipes and 13 PCM tubes. The 40 batteries used in the five module configurations: BM-0, BM-1, BM-2, BM-3, and BM-4 are the same. The configurational setups of the five experimental modules are summarized in Table 14.1.

Table 14.1 also lists the calculated mass fraction of batteries for all the five module configurations. BM-0 only has all the 40 batteries included in and therefore the mass fraction of batteries is 100%; for the other four module configurations, this mass fraction values are different, but the difference is small, they are ranging within 87%–88%. The liquid (or PCM) filled into the heat pipes (or PCM tubes) is of negligible mass in comparison with the mass of the heat pipe (or PCM tube) itself.

The electrical-thermal performance of BM0–BM4 is tested under different ambient temperatures and various discharge C-rates. The charge operations follow the conventional CC–CV protocol. During the CC charge phase, the charge C-rate is 0.75C unchanged for all cases, and the CV charge phase has a fixed voltage of 42 V and it terminates when the charging current drops to 1/30C (~365 mA).

14.3.3 Experimental Results and Discussion

14.3.3.1 The Electrical-Thermal Performance of BM-0

The battery module BM-0 relies only on natural air convection to cool the battery module during charging/discharging. Figure 14.4 presents the temperature and voltage evolution of the battery module during the CC–CV charge process. The battery temperature continuously increases until the charge process turns to the CV phase. The much-lowered charge current during CV charge phase slows down the heat generation in batteries. The time instant of temperature rise-to-decrease turning point is a little lagging behind the transition of CC charge to CV charge, indicating the thermal inertia of the battery module.

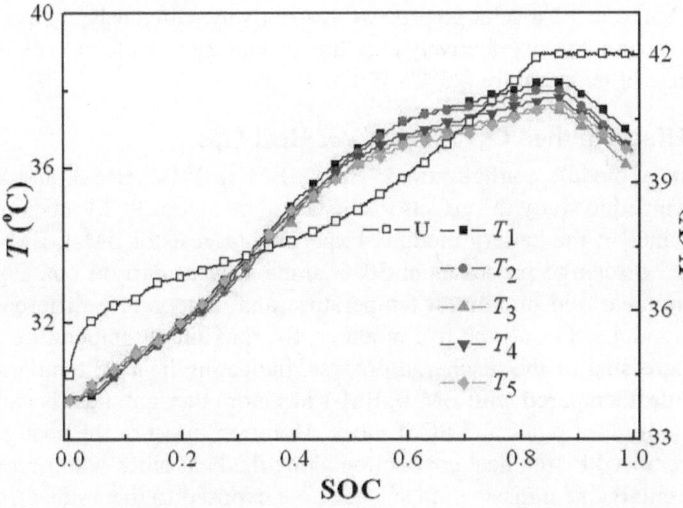

FIGURE 14.4 The temperature and voltage evolution of the BM-0 during charging at 30°C ambient.

Experiments of discharging the fully charged BM-0 at 0.5C and 1C were also conducted. Figure 14.5 shows the measured temperature and voltage evolution curves. Upon discharging, the output voltage suddenly drops at the very beginning, then gradually diminishes for a long time, and lastly quickly decreases in the ending period. The 1C discharge process shows generally lower voltage values and slightly less electricity amount discharged (the ending DOD does not reach 1) than the 0.5C discharge process. The battery temperature continuously increases during the discharge processes. The 1C discharge process appears to have evidently larger temperature rises. At the end of the 1C discharge process (DOD 1), the maximum battery temperature is about 50.9°C. Some differences are seen between the temperatures measured at different batteries and the difference is enlarged with the progressing of the discharge process. The maximum temperature differences (ΔT_{max}) are 2.4°C and

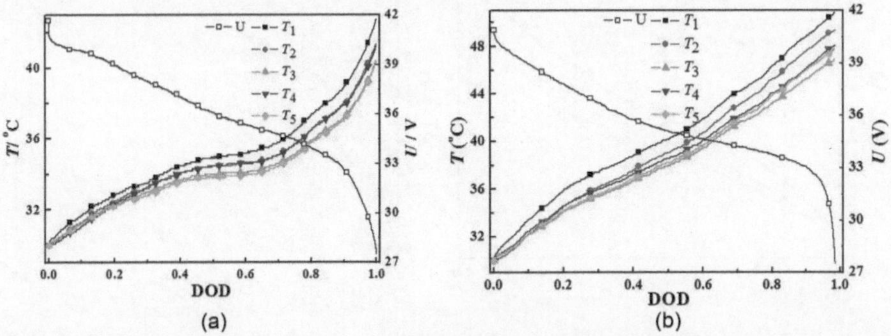

FIGURE 14.5 The voltage and temperature characteristic of BM-0 during 0.5C (a) and 1C (b) discharge process at 30°C ambient.

4°C for the 0.5C and 1C discharge process, respectively. Obviously, the pure natural air convection cannot very effectively regulate the battery module to work within the optimum range of temperature (25°C–45°C).

14.3.3.2 Effects of the PCM Tube and/or Heat Pipe

The five battery module configurations: BM-0, BM-1, BM-2, BM-3, and BM-4 are specially designed to study the effects of the heat pipe and/or PCM tube on the thermal performance of the battery module. Experimental tests of BM-1, BM-2, BM-3, and BM-4 1C discharge processes at 30°C ambient were carried out. Figure 14.6 compares the measured maximum temperature and temperature difference in the five battery modules. For all the five modules, the maximum temperature increases with the progressing of the discharge process, indicating heat accumulation in the battery module. Compared with BM-0, BM-1 has more thermal mass because of the embedded empty heat pipes and PCM tubes. Moreover, most of the heat generation in BM-1 is removed by the heat conduction along the heat pipes' aluminum shell to the fins that enlarge heat dissipation surface area exposed to the ambient. All these factors result in large decrease of the maximum temperature in BM-1, compared with that in BM-0. At the end of the 1C discharge process (DOD 1), the maximum temperatures in BM-1 and BM-0 are 50.9°C and 47.3°C, respectively.

Figure 14.6a also indicates that the operation of heat pipes and the melting of PCM clearly improve the cooling effects, BM-4 has the lowest maximum temperatures and BM-2 and BM-3 both have maximum temperatures lower than BM-1. At the end of the discharge process (DOD 1), the maximum temperatures in BM-2, BM-3, and BM-4 are 44.2°C, 45.5°C, and 42.8°C, respectively.

The temperature difference (ΔT) across the battery module signifies the temperature non-uniformity. Minimizing ΔT to achieve better temperature uniformity is one major target of battery TMSs. It is seen from Figure 14.6b that ΔT increases with the increase of DOD and the BM-0 has always the largest ΔT, BM-1 the second largest, BM-2 the third largest, BM-3 the fourth largest, and BM-4 the smallest. At the end of discharge process (DOD 1), ΔT-s for BM-0, BM-1, BM-2, BM-3, and BM-4 are

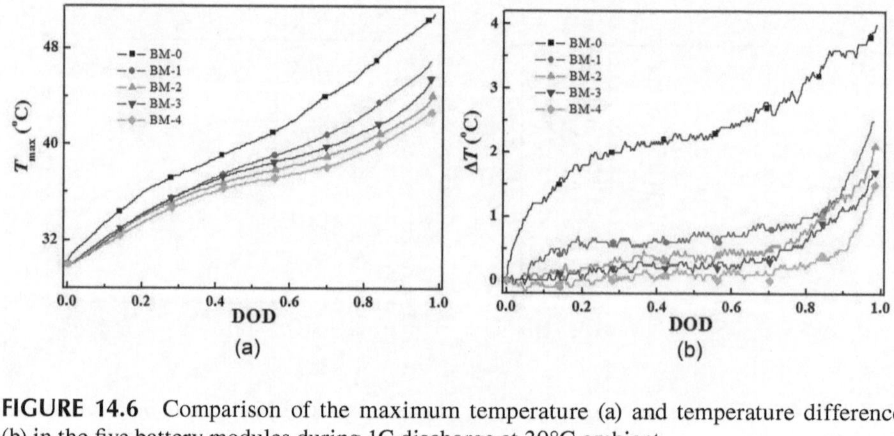

FIGURE 14.6 Comparison of the maximum temperature (a) and temperature difference (b) in the five battery modules during 1C discharge at 30°C ambient.

TABLE 14.2
The T_{max} and ΔT_{max} of Different BM Configurations During 1C Discharge

	BM-0	BM-1	BM-2	BM-3	BM-4
T_{max} (°C)	50.9	47.3	44.2	45.5	42.8
ΔT_{max} (°C)	4.0	2.5	2.1	1.7	1.5

4°C, 2.5°C, 2.1°C, 1.7°C, and 1.5°C, respectively. Table 14.2 summarizes the T_{max} and ΔT_{max} values for all the five modules during 1C discharge.

For BM-2 and BM-3, either the heat pipes or the PCM tubes take effects. Inspection of the measured temperature curves shown in Figure 14.6 unravels that the heat pipe seems to be more effective at lowering the battery temperature rise while the PCM tube is more effective at reducing the temperature non-uniformity in the battery module. For BM-4, the PCM tubes and the heat pipes both take effects. The thermal performance of the module during 1C discharge is greatly enhanced. As the melting of the PCM does not occur at a fixed temperature point, but within a temperature range (29.89°C–40.02°C), there is no clear clue about the progress of PCM melting process seen from the temperature curves of BM-3 and BM-4 in Figure 14.6a.

In practical applications, the heat pipes used in BTMSs can be designed in different shapes and the inside of the heat pipes may be arranged with some capillary wicks or some mini-channels (or mini-grooves), to facilitate the installation and further improve the performance. For example, Wang et al. [82] reported an L-shaped heat pipe used for prismatic lithium-ion batteries thermal management; the experimental results showed that the heat pipe TMS can very effectively manage the battery thermal state even under abusive or sub-zero temperature (battery heating is needed) conditions. The paraffin is of low thermal conductivity. Wu et al. [83] used a copper mesh-enhanced paraffin/EG composite to develop a PCM BTMS. As a result, the copper mesh-enhanced PCM of paraffin/EG plate (PCMP) presents much better heat dissipation performance and temperature uniformity compared to PCMP without copper mesh, especially in harsh working conditions.

14.3.3.3 Effects of the Ambient Temperature

The ambient temperature is actually the initial temperature of battery module during discharge process. For a given battery module, if the ambient temperature is low, it may not experience too high temperature during a low or mild high C-rate discharge process. As the PCM melts within a temperature range of 29.89°C–40.02°C, the ambient temperature may greatly influence the effects of PCM tubes on the thermal performance of battery module during discharge. We conducted the experimental tests of BM-4 1C discharge processes at various ambient temperatures: 10°C, 20°C, 30°C, and 40°C. Figure 14.7 presents the maximum temperature rise (T_{rise}) and temperature difference (ΔT) in the module after the 1C discharge process. The thermal management system shows the best performance in the 30°C ambient temperature case.

The temperature ranges that the battery module has experienced in the discharge processes are 10°C–26.8°C, 20°C–33.9°C, 30°C–42.8°C, and 40°C–55.3°C for the four cases, respectively. It is easy to figure out that the PCM does not melt at all in the

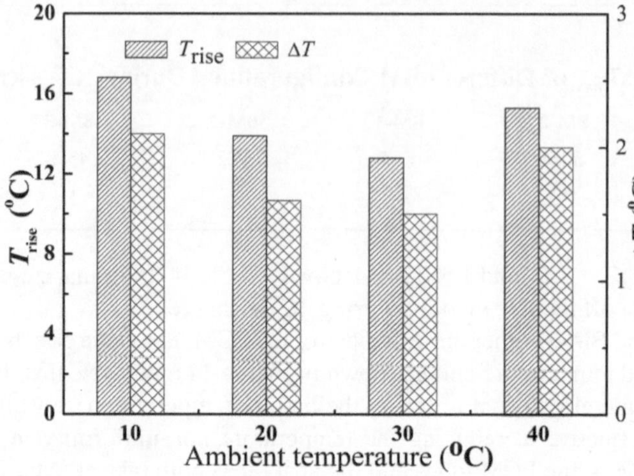

FIGURE 14.7 The temperature characteristics of BM-4 after 1C discharge processes at different ambient temperatures.

10°C case, only partially melts in the 20°C case, and completely melts in the 30°C case, while in the 40°C case the PCM is already in liquid state from the beginning. The maximum temperature differences in the battery module are 2.1°C, 1.6°C, 1.5°C, and 2°C for 10°C, 20°C, 30°C, and 40°C ambient temperature case, respectively. Besides the different working temperatures cause different heat generation rates (e.g. the 10°C ambient case should have slightly more heat generation due to the more inefficiency of electrochemical reactions), it is mainly because the PCM tubes take different effects in the four cases that makes the battery module emerge different thermal behaviors.

14.3.3.4 Effects of the Discharge C-Rate

The discharge C-rate dictates the heat generation rate (amount) as the heat generation rate is approximately proportional to the squared discharge current. Tests of the 0.5C, 1.0C, and 2C discharge processes were carried out for BM-4 and the measured output voltage and temperature as function of DOD are presented in Figure 14.8. The ambient temperature is 25°C for all the three discharge processes. Figure 14.8a presents the voltage curves for the three cases of different discharge C-rates. Higher C-rate causes lower output voltage and the case of higher C-rate discharges less electricity, i.e. the ending DOD is further away to 1. The temperature curves in Figure 14.8b–d show that the discharge process of larger C-rate gives higher battery temperature and larger temperature rise.

It is indicated by the temperature curves presented in Figure 14.8b–d that the hybrid system of heat pipes and PCM tubes has remarkable thermal management capacity. Even during the 2C discharge process, the highest battery temperature is controlled below 50°C, and meanwhile, the maximum temperature difference is seen to be about 2.5°C, indicating prominent effectiveness of thermal management. The temperature uniformity in the battery module during the 1C and 0.5C is good as expected, and the maximum temperature differences are about 1.5°C and 1.0°C, respectively.

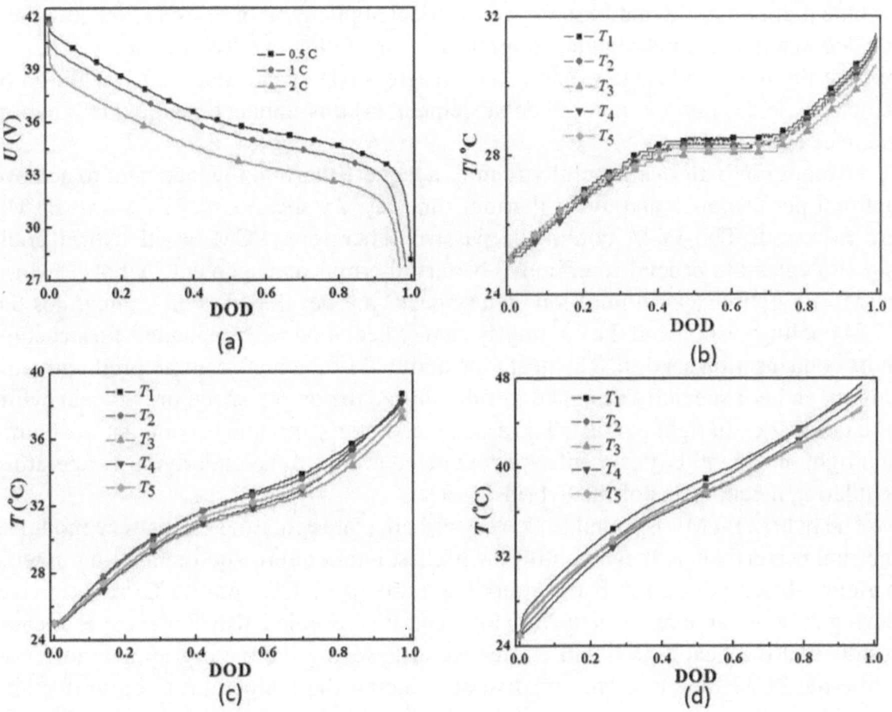

FIGURE 14.8 The output voltage (a) and the measured temperatures (b–d) as function of DOD for BM-4 during (b) 0.5C, (c) 1C, and (d) 2C discharge process at 25°C ambient.

Checking the measured temperatures, one may find that the highest temperature in the battery module is 31.6°C, 38.9°C, and 47.7°C during the 0.5C, 1C, and 2C discharge process, respectively. The melting of PCM begins at 29.89°C and ends at 40.02°C, and the peak melting point is 36.59°C. During the 0.5C and 1C discharge processes, the PCM only partially melts and during the 2C discharge process the PCM completes melting. Partially due to the different effects that the PCM tubes take, the temperature curves presented in Figure 14.8b–d show qualitatively different trends, though they all continually rise through the whole discharge process.

In reality, the lithium-ion battery modules/packs equipped with the present heat pipe and PCM hybrid BTMS can be used to power vehicles such as motorbikes or scooters. When the vehicles in operation, the heat release from the fins arranged at the heat pipe condensation section is in forced air convection, which can greatly enhance the thermal management effectiveness, enabling the battery module/pack can work under harsh environmental and operation conditions.

14.4 CONCLUSIONS

PCM absorbs or releases a lot heat in the process of phase transition due to its high latent heat. This chapter presents a general review and summarizes the current development status of PCMs, and exemplifies a novel application of PCM to a cylindrical lithium-ion battery thermal management.

One main obstacle that hinders the practical application of PCMs is the low thermal conductivity. This chapter summarizes the following five methods/routes to resolve the issue and reviews the recent progress: (1) encapsulation, (2) addition of nanoparticle, (3) porous-material enhancement, (4) fins-enhancement, and (5) hybrid enhancement.

Lithium-ion batteries generally require adequate thermal management to achieve optimal performance and avoid thermal runaway. Applications of PCMs in BTMS are reviewed. The PCM cooling is passive. Therefore, PCM-based hybrid cooling strategies are crucial to effective battery thermal management. A novel hybrid BTMS for cylindrical lithium-ion battery packs are developed, which integrates the PCM cooling system and the cooling system of heat pipe with expanded-fin structure at its condensation section. The heat pipe and PCM tube are manufactured with aluminum and are specially designed to fully utilize the empty space in-between cylindrical batteries in tight contact. The assembled battery module is compact in volume and light in weight. Experiments are conducted to reveal the underlying temperature regulating mechanisms of the hybrid TMS.

The hybrid BTMS is found to be very effective at regulating the battery module's thermal performance. It can control the highest temperature and the maximum temperature difference in the battery module at about 47.7°C and 2.5°C, respectively, during 2C discharge process in 25°C ambient. It is revealed also by the experimental results that the heat pipe is more effective at lowering the battery temperature rise while the PCM tube is more effective at reducing the temperature non-uniformity in the battery module. In addition, the effects of PCM tubes are found to be very sensitive to the ambient temperature, due mainly to the fixed temperature (range), at (within) which the solid–to–liquid phase change of the PCM occurs. To fully utilize the PCM cooling effect, the working temperature range of battery module during discharging/charging, which is related with the ambient temperature and the operating conditions, etc., is required to rightly ensure the complete melting of PCM. In other words, the choice/preparation of appropriate PCM is of significant importance.

ACKNOWLEDGMENTS

Financial support received from the China National Key R&D Project (2018YFB0905300, 2018YFB0905303), Guangzhou Science and Technology project (202102080433), and Guangdong Key Laboratory of New and Renewable Energy Research and Development Foundation (E1390304) is gratefully acknowledged.

REFERENCES

1. X. Hu, S. Chang, J. Li, et al., *Energy* 35 (2010) 4289–4301.
2. N. Kannan, D. Vakeesan, *Renew. Sust. Energ. Rev.* 62 (2016) 1092–1105.
3. T. Ma, H. Yang, Y. Zhang, et al., *Renew. Sust. Energ. Rev.* 43 (2015) 1273–1284.
4. C. Lan, J. Xu, Y. Qiao, et al., *Appl. Therm. Eng.* 101 (2016) 284–292.
5. G. Li, X.F. Zheng, *Renew. Sust. Energ. Rev.* 62 (2016) 736–757.
6. P.H. Andersen, J.A. Mathews, M. Rask, *Energy Policy* 37 (2009) 2481–2486.
7. T. Zhang, C. Gao, Q. Gao, et al., *Appl. Therm. Eng.* 88 (2015) 398–409.
8. J. Chen, S. Kang, E. Jiaqiang, et al., *J. Power Sources* 442 (2019) 227228.

9. S. Wang, L. Lu, D. Ren, et al., *J. Electrochem. Energy Convers. Storage* 16 (2019) 031006.
10. F.M. Nasir, M.Z. Abdullah, M.A. Ismail, *Arabian J. Sci. Eng.* 44 (2019) 7541–7552.
11. H. Zhou, F. Zhou, L. Xu, et al., *Int. J. Heat Mass Transfer* 131 (2019) 984–998.
12. J. Kim, J. Oh, H. Lee, *Appl. Therm. Eng.* 149 (2019) 192–212.
13. H. Liu, Z. Wei, W. He, et al., *Energy Convers. Manage.* 150 (2017) 304–330.
14. J.P. Holman, *Heat Transfer*, 4th Edition, New York, McGraw-Hill (1976).
15. K. Chen, M. Song, W. Wei, et al., *Energy* 145 (2018) 603–613
16. H. Hirano, T. Tajima, T. Hasegawa, et al., *ITEC Asia-Pacific* (2014) 1–4.
17. G. Karimi, A.R. Dehghan, *Int. J. Energy Res.* 38 (2014) 1793–1811.
18. K. Li, J. Yan, H. Chen, et al., *Appl. Therm. Eng.* 132 (2018) 575–585.
19. T.U. Rehman, H.M. Ali, M.M. Janjua, et al., *Int. J. Heat Mass Transfer* 135 (2019) 649–673.
20. S. Al-Hallaj, J.R. Selman, *J. Electrochem. Soc.* 147 (2000) 3231–3236.
21. S.S. Chandel, T. Agarwal, *Renew. Sust. Energ. Rev.* 67 (2017) 581–596.
22. B. Zalba, J.M. Marín, L.F. Cabeza, et al., *Appl. Therm. Eng.* 23 (2003) 251–283.
23. K.Y. Leong, M. Rahman, B.A. Gurunathan, *J. Energy Storage* 21 (2019) 18–31.
24. X. Chen, H. Gao, Z. Tang, et al., *Cell Rep. Phys. Sci.* 1 (2020) 100218.
25. L. Liu, D. Su, Y. Tang, et al., *Renew. Sust. Energ. Rev.* 62 (2016) 305–317.
26. J. Yan, K. Li, H. Chen, et al., *Energy Convers. Manage.* 128 (2016) 12–19.
27. M.M. Farid, A.M. Khudhair, S. Razack, et al., *Energy Convers. Manage.* 45 (2004) 1597–1615.
28. M. Li, M. Chen, A. Wu, *Appl. Energy* 127 (2014) 166–171.
29. M. Liu, W. Saman, F. Bruno, *Renew. Sust. Energ. Rev.* 16(2012) 2118–2132.
30. J. Jaguemont, N. Omar, P. Van den Bossche, et al., *Appl. Therm. Eng.* 132 (2018) 308–320.
31. Y. Dutil, D.R. Rousse, N.B. Salah, et al., *Renew. Sust. Energ. Rev.* 15 (2011) 112–130.
32. W. Su, J. Darkwa, G. Kokogiannakis, *Renew. Sust. Energ. Rev.* 48 (2015) 373–391.
33. F. Souayfane, F. Fardoun, P.H. Biwole, *Energy Build.* 129 (2016) 396–431.
34. L. Ianniciello, P.H. Biwole, P. Achard, *J. Power Sources* 378 (2018) 383–403.
35. Z. Rao, S. Wang, *Renew. Sust. Energ. Rev.* 15 (2011) 4554–4571.
36. K. Pielichowska, K. Pielichowski, *Prog. Mater. Sci.* 65 (2014) 67–123.
37. C.Y. Zhao, G.H. Zhang, *Renew. Sust. Energ. Rev.* 15 (2011) 3813–3832.
38. A. Frazzica, M. Manzan, A. Sapienza, et al., *Appl. Energy* 183 (2016) 1157–1167.
39. L. Chai, X. Wang, D. Wu, *Appl. Energy* 138 (2015) 661–674.
40. H. Peng, J. Wang, X. Zhang, et al., *Appl. Therm. Eng.* 185 (2020) 116326.
41. K.W. Shah, *Energy Build.* 175 (2018) 57–68.
42. S.U.S. Choi, J.A. Eastman, *Proc. IMECE San Francisco USA* 1995 99–105.
43. J.M. Khodadadi, S.F. Hosseinizadeh. *Int. Commun. Heat Mass Transfer* 34 (2007) 534–543.
44. S. Guo, Q. Liu, J. Zhao, et al., *Appl. Energy* 205 (2017) 703–709.
45. F. Bahiraei, A. Fartaj, G. Nazri, *Energy Convers. Manage.* 153 (2017) 115–128.
46. S.C. Lin, H.H. Al-Kayiem, *Sol. Energy* 132 (2016) 267–278.
47. D. Rufuss, L. Suganthi, S. Iniyan, et al., *J. Cleaner Prod.* 192 (2018) 9–29.
48. L.W. Fan, Z.Q. Zhu, Y. Zeng, et al., *Appl. Therm. Eng.* 75 (2015) 532–540.
49. P. Zhang, X. Xiao, Z.N. Meng, et al., *Appl. Energy* 137 (2015) 758–772.
50. X. Meng, L. Yan, F. He, *J. Energy Storage* 32 (2020) 101867.
51. S.T. Hong, D.R. Herling, *Scr. Mater.* 55 (2006) 887–890.
52. B. Amaa, A. Sm, B. Jaa, et al., *Renew. Sust. Energ. Rev.* 82 (2018) 1620–1635.
53. M. Costa, D. Buddhi, A. Oliva, *Energy Convers. Manage.* 39 (1998) 319–330.
54. S.Z Tang, H.Q. Tian, J.J. Zhou, et al., *J. Energy Storage* 33 (2021) 102124.
55. X. Liu, Y. Huang, X. Zhang, et al., *Appl. Therm. Eng.* 179 (2020) 115749.
56. C. Ji, Q. Zhen, Z. Low, et al., *Appl. Therm. Eng.* 129 (2018) 269–279.

57. A.H.N. Al-Mudhafar, A.F. Nowakowski, F.C.G. Nicolleau, *Energy Rep.* 7 (2021) 120–126.
58. X. Lin, X. Zhang, J. Ji, et al., *Int. J. Energy Res.* 45 (2021) 9831–9857.
59. L. Min, M. Chen, Z. Wu, *Appl. Energy* 127 (2014) 166–171.
60. Y. Zhu, Y. Qin, S. Liang, et al., *Appl. Energy* 250 (2019) 98–108.
61. W. Zhu, N. Hu, Q. Wei, et al., *Mater. Des.* 172 (2019) 107709.
62. D.Q. Ng, Y.L. Tseng, Y.F. Shih, et al., *Polymer* 133 (2017) 250–262.
63. Z. Ling, S. Li, C. Cai, et al., *Appl. Therm. Eng.* 193 (2021) 117002.
64. X. Li, Q. Huang, J. Deng, et al., *J. Power Sources* 451 (2020) 227820.
65. Y. Lv, W. Situ, X. Yang, et al., *Energy Convers. Manage.* 163 (2018) 250–259.
66. C. Xiao, G. Zhang, Z. Li, et al., *J. Mater. Chem. A* 8 (2020) 14624–14633.
67. Z. Ling, X. Wen, Z. Zhang, et al., *Energy* 144 (2018) 977–983.
68. J.S. He, X.Q. Yang, G.Q. Zhang, *Appl. Therm. Eng.* 148 (2019) 984–991.
69. G. Jiang, J. Huang, M. Liu, et al., *Appl. Therm. Eng.* 120(2017) 1–9.
70. W. Situ, G. Zhang, X. Li, et al., *Energy* 141 (2017) 613–623.
71. D. Zou, X. Ma, X. Liu, et al., *Int. J. Heat Mass Transfer* 120 (2018) 33–41.
72. Z. Sun, R. Fan, F. Yan, et al., *Int. J. Heat Mass Transfer* 145 (2019) 118739.
73. R. Fan, N. Zheng, Z. Sun, *Int. J. Heat Mass Transfer* 166 (2021) 120753.
74. Y. Lv, X. Yang, X. Li, et al., *Appl. Energy* 178 (2016) 376–382.
75. Z.W. Wang, H.Y. Zhang, X. Xia, *Int. J. Heat Mass Transfer* 109 (2017) 958–970.
76. M.M. Heyhat, S. Mousavi, M. Siavashi, *J. Energy Storage* 28 (2020) 101235.
77. R. Sabbah, R. Kizilel, J.R. Selman, et al., *J. Power Sources* 182 (2008) 630–638.
78. F. Chen, R. Huang, C. Wang, et al., *Appl. Therm. Eng.* 173 (2020) 115154.
79. J. Cao, Y. He, J. Feng, et al., *Appl. Energy* 279 (2020) 115808.
80. J. Cao, M. Luo, X. Fang, et al., *Energy* 191 (2020) 116565
81. W. Zhang, J. Qiu, X. Yin, et al., *Appl. Therm. Eng.* 165 (2019) 114571.
82. Q. Wang, B. Jiang, Q.F. Xue, et al., *Appl. Therm. Eng.* 88 (2015) 54–60.
83. W. Wu, X. Yang, G. Zhang, et al., *Energy* 113 (2016) 909–916.

15 Phase Change Material-Based Thermal Energy Storage for Cold Chain Applications – From Materials to Systems

Yelaman Maksum, Lin Cong, Boyang Zou, Binjian Nie, Siyuan Dai, and Yongliang Li
University of Birmingham

Yanqi Zhao
University of Birmingham
Jiangsu University
Changzhou University

Bakytzhan Akhmetov
Nanyang Technological University

Lige Tong and Li Wang
University of Science & Technology Beijing

Yulong Ding
University of Birmingham
University of Science & Technology Beijing

CONTENTS

DOI: 10.1201/9781003213260-15

15.1 INTRODUCTION

A cold chain is a supply chain with controlled temperature across production, transportation, storage, distribution, and delivery processes [1]. It is a term used to symbolize the continuity of means that successively provide the refrigerated preservation of perishable foodstuffs from production to consumption [2]. The main objective of the cold chain is to ensure that the thermal performance of every component in the chain is operated in an uninterrupted manner [3]. Clearly, these operations depend on many factors, e.g. product logistic circuits, and temperature and humidity levels at each step of the circuit [4]. Unfortunately, cold chain is not always easily accessible and does not operate effectively due to economic and political reasons. Globally, more than 200 million tons of perishable foods per year could be saved and consumed in developing countries if refrigeration and cold chain were accessible as in developed countries [5]. Even in developed countries, due to unoptimized operations, food spoilage and losses are considerable, which amount to ~13% of global food production [6]. In addition, pharmaceutical products are also highly sensitive to temperature and humidity. Therefore, it requires strict thermal control during storage and transportation. The cold chain provides safe products of high organoleptic quality, which in turn influences the environment, water, and land [7]. Moreover, the cold chain also has made substantial contribution to the economic boom of rural areas as an extremely important sub-branch of the logistics industry [8].

Refrigerated foods can be classified into four broad categories: frozen (−18°C or below), cold-chilled (0°C to +1°C), medium-chilled (+5°C), and exotic-chilled (+10°C to +15°C) [4,9]. On the other hand, pharma products are mostly kept within +2°C to +8°C temperature range [10]. Such diversity of temperature requirements results in the application of different types of technologies along cold chain logistics depending on the process at a given point of the chain (see Figure 15.1). One can see that a cold chain has many steps that requires thermal management [3], with each of these steps consisting mainly of transportation and storage.

The transportation can be land-, air- or sea-based. Cold energy is usually produced by an on-board refrigeration system of a transport unit, where the energy source is

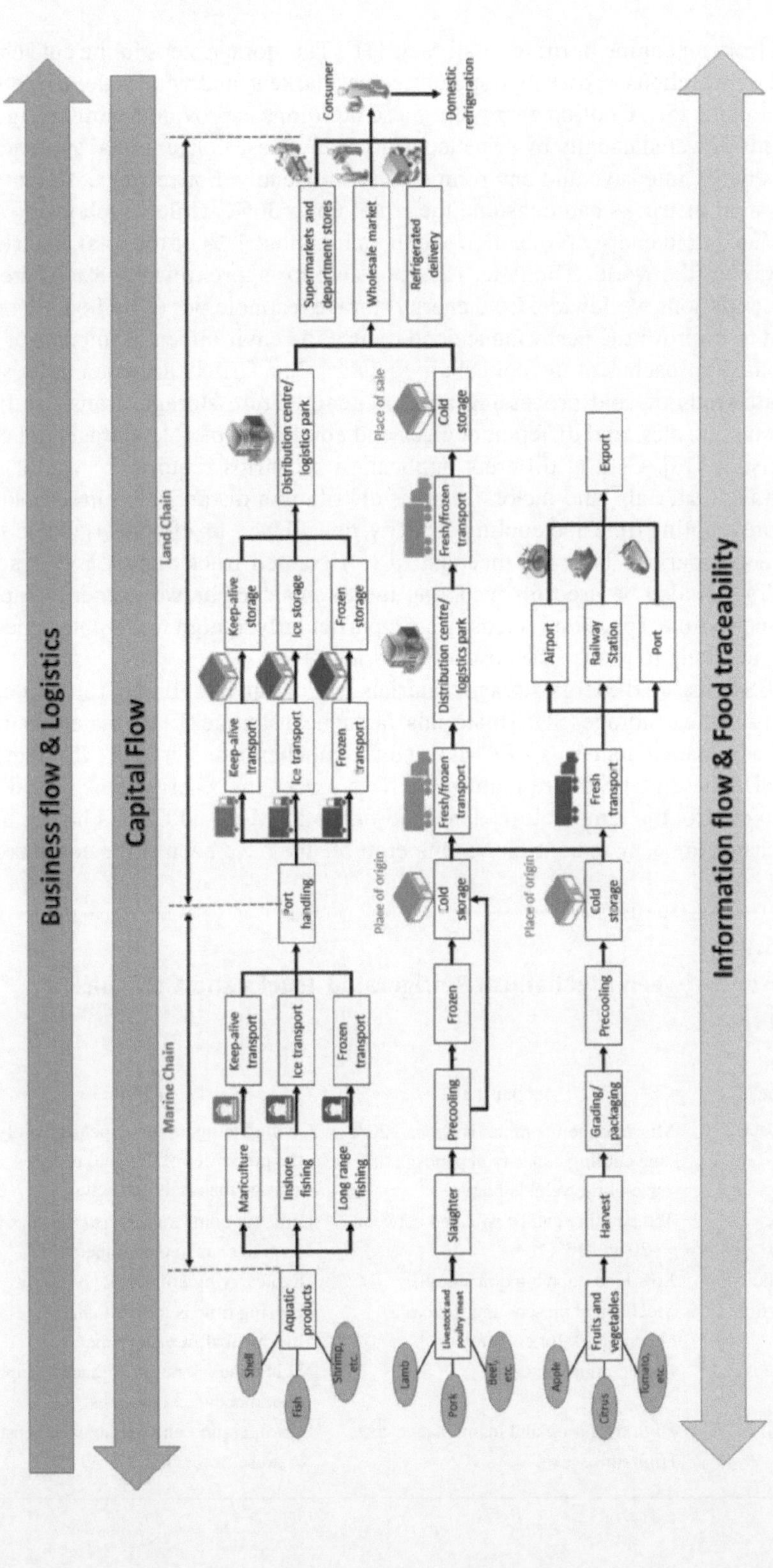

FIGURE 15.1 Cold chain of fresh products [3].

delivered from an engine burning fossil fuels [11]. The storage steps in the cold chain are based on warehouses, distribution hubs, supermarkets, and wholesale markets as shown in Figure 15.1. Cooling energy for these buildings is provided using refrigeration systems powered usually by grid electricity [12]. These conventional approaches are often energy intensive and environmentally unfriendly. For instance, the refrigeration system of trucks can consume their fuel up to 30%, while warehousing and storage related cold energy production accounts for about 17% of the total electricity consumption in the world. The latter is responsible for approximately 8% of greenhouse gas emissions worldwide. Cold energy storage technologies (CESTs) have been developed to improve the performance and reduce the environmental impacts of the conventional approaches of the cold chain logistics. The CESTs are potentially suitable for all kinds of cold processing and low-temperature storage, transportation, distribution and sales, and different devices and equipment of cold storage and cold chain logistics [13]. Clearly different application scenarios require the use of different storage materials, and factors, such as distribution distance, required holding temperature, cooling time and cooling capacity, can all have an effect [14]. Table 15.1 provides a comparison between mechanical refrigerated trucks and CEST trucks. The CESTs can also be used for resolving the mismatches between energy supply and demand in time, space and intensity [15], particularly at high renewable penetrations [16], and help to reduce the costs of electricity [17].

The CESTs use cold energy storage materials, which can be split into three categories of sensible heat storage (SHS) materials, latent heat storage (LHS) materials (also termed phase change materials – PCMs), and thermochemical storage (TCS) materials. The SHS-based CESTs are mature but have a very low energy density and are difficult to control the temperature. The LHS or the PCM-based CESTs have a high energy density, are easy to control the temperature, and have been at the commercial

TABLE 15.1
Comparison between Mechanical Refrigerated Trucks and CEST-Based Trucks [14]

Refrigerated Vehicle Type	Mechanical	CEST
Cooling capacity	Minimum temperature is about −20°C; the cooling capacity depends on the size of the vehicle body	Cooling temperature depends mainly on the properties of the cold energy storage materials and setup
Temperature adjustment	Temperature can be adjusted between −20 and 15°C	Different cold storage materials can be used to change the temperature
Carrying efficiency	Low load-carrying efficiency	High carrying efficiency
Pre-cooling time	Mechanical pre-cooling time as a benchmark for comparison	Freezing time is longer than the mechanical freezing time
Equipment cost	High equipment costs	Night-time operation is much cheaper than mechanical systems
Operation costs	High operation and maintenance costs	Low operation and maintenance costs
Other	High noise level	No noise or low noise level

demonstration stage. The TCS-based CESTs have a very high energy density but are still at an early stage of technology development. This chapter focuses on the use of PCM-based CESTs and there have been a considerable amount of research on the area [18]. As mentioned above, PCMs have a large energy density, ~5–14 times that of the SHS materials [19]. Another salient feature of PCMs is the function of self-regulating the temperature close to their phase transition point, and, as a result, they have been actively investigated for use in cold chain and refrigeration systems [20].

For the LHS-based CESTs, PCMs are the key, which should have a high specific heat capacity, a high latent heat, a good chemical stability, a low supercooling degree, and a high thermal conductivity, and are non-toxic and harmless [21]. For cold chain transportation applications, PCMs are encapsulated in packs which are placed/installed within the thermally insulated vehicle container to ensure the products transported at a desired temperature. The amount of the PCM, the shape of the PCM containers/packs, the arrangement of the packs within the container and the weather conditions define the duration that the temperature can be maintained. The slower the heat ingress is, the longer the duration is [18]. This reduces the extent of the dependence on an on-board refrigeration unit, leading to a reduced greenhouse gas emission. Indeed, the integration of PCM into truck refrigeration units has been shown to give a much enhanced efficiency [22]. For stationary applications, such as cold chain warehouses and supermarket cabinets, the use of PCMs has been demonstrated to achieve load-shifting, leading to energy saving and cost reduction [23–25].

15.2 PCM-BASED COLD ENERGY STORAGE MATERIALS

The past two decades have seen rapid developments in PCMs for thermal energy storage (TES) applications. This section focuses on PCMs for cold energy storage applications, which can be subdivided into solid–liquid, solid–solid, solid–gas, and liquid–gas categories. PCMs with solid–liquid transition is mainly used in cold chain applications due to small volume changes and hence reduced system complexity, particularly for cold chain transportation applications.

15.2.1 Desirable Properties and Classification of PCMs for Cold Energy Storage

A PCM for cold energy storage should ideally meet some desirable thermophysical, kinetics, chemical and environmental requirements as illustrated in Figure 15.2. There are many such PCMs and they are often classified according to their chemical composition as shown in Figure 15.3.

Inorganic PCMs include salt hydrates and salt solutions with the latter often used for sub-zero energy storage applications. This category of PCMs has been extensively researched and developed due to good thermal conductivity, high heat of fusion, low costs, and abundance. However, corrosivity, supercooling and volume change due to phase change are the main concerns.

Organic PCMs consist of paraffin- and non-paraffin-based categories. The paraffin-based PCMs are commercially available at a large scale with a high heat of fusion and phase transition temperatures depending on the length of the hydrocarbon

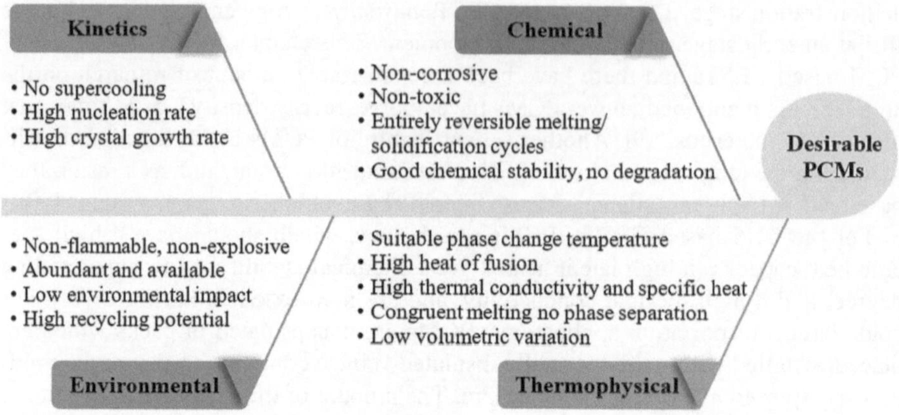

FIGURE 15.2 Desirable properties of PCMs for cold energy storage [66,67].

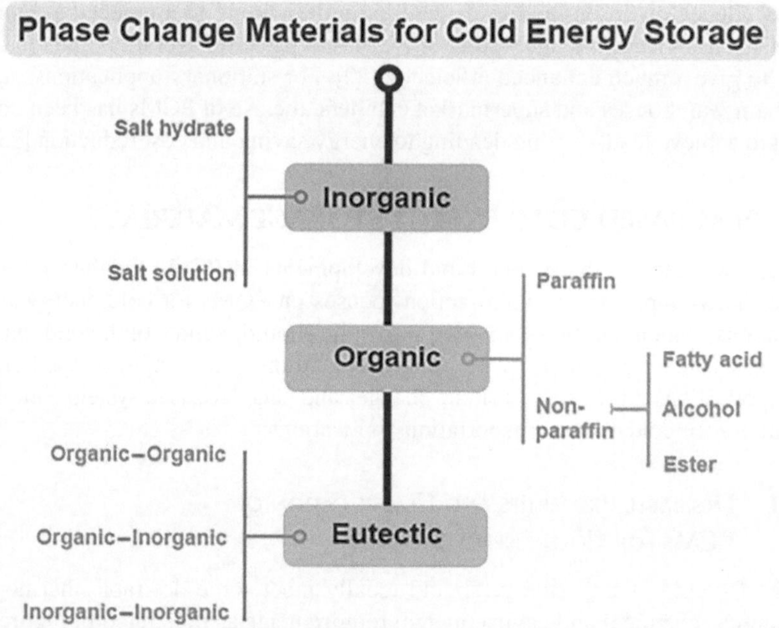

FIGURE 15.3 Classification of PCMs for cold thermal energy storage.

chain [26]. They are chemically stable and non-corrosive to most metallic materials, but their thermal conductivity is very low, which requires a well-designed heat exchanger to enhance the efficacy of heat transfer. Non-paraffin PCMs are often costly compared to paraffin-based PCMs and have been less investigated [27].

Eutectic PCMs are a mixture of two or more components with a minimum melting point lower than any of the constituents. This minimum melting point is called the eutectic point, at which all constituents melt or solidify congruently with no phase separation. Eutectic PCMs are often formulated to meet specific phase temperature or heat of fusion requirements.

Table 15.2 provides a list of cold energy storage PCMs under different categories discussed above. Some challenges for the commercialization of these PCMs are also listed.

15.2.2 Performance Enhancement of PCMs

As discussed above, several drawbacks of PCMs have been hindering the widespread applications of CESTs. These are discussed together with ways to address the issues as follows.

Thermal conductivity Thermal conductivity is one of the most important thermophysical properties which affects the charging/discharging rates. Various attempts have been made to enhance this property. Examples include the incorporation of highly thermally conductive materials into PCM, e.g. graphite, metal foams and nanomaterials, and the extension of the contact surfaces between PCMs and heat transfer fluids [28]. He et al. [29] studied the thermal conductivity of $BaCl_2/H_2O$. They showed that the addition of 16.74 wt.% of TiO_2 nanoparticles led to an increase in the thermal conductivity by 12.76% and the enhancement was proportional to the concentration of nanoparticles.

Supercooling Poor nucleation characteristics and insufficient crystal growth rate have been proposed to be the main reasons for the supercooling of low-temperature PCMs [30]. It can be addressed by various methods such as the addition of nucleating agents, the use of electrical field, and the application of a mechanical action [31]. Wu et al. [32] dispersed multi-walled carbon nanotubes (MWCNTs) into a eutectic $MgCl_2$-H_2O solution, and found that with 1–3 wt.% MWCNTs, the supercooling of the PCM can be lowered 18%–37%. However, the amount of nucleating agent addition should be controlled in a reasonable range in order not to affect much of the energy density.

Stability The phase separation and incongruent phase change are the main reasons that affect the stability of some low-temperature PCMs particularly the salt hydrate-based and composite PCMs with multi-density components. Therefore, various approaches have been investigated to increase the stability of these cold storage materials. Johansen et al. [33] employed both carboxymethyl cellulose (CMC) and xanthan gum to decrease the phase separation of sodium acetate trihydrate (SAT)-based PCM with a thermal conductivity enhancer. Their results showed that the thickened composites stay in a homogeneous state. With the same concentration (5 wt.%), xanthan appeared to have a higher viscosity than CMC.

Corrosion Although the corrosion impact of the cold energy storage is less severe than that at medium and high temperatures, it is an essential factor to consider for long-term applications. It is reported that the use of encapsulation [34], the addition of thickening/gelling agents [35] and the application of coating [36] are effective to reduce the corrosion effects, although very limited work has been done in the cold storage field. Oró et al. [35] studied the compatibility of nine low-temperature PCMs with four metals. The results showed that copper and carbon steel had the highest corrosion rate; bubbles and pitting occurred on the surface of the aluminium; and little corrosion was observed on stainless steel and all the four polymers.

TABLE 15.2
Potential Pure and Formulated PCMs for Cold Thermal Energy Storage [70,71]

	Material/Formulation	(°C)	(kJ/kg)	Material/Formulation	(°C)	(kJ/kg)
Inorganic	K_2CO_3 (39.6%)/H_2O	−36.5	165	NH_4Cl (19.5%)/H_2O	−16	248
	$Mg(NO_3)_2$ (34.6%)/H_2O	−29	187	KCl (19.5%)/H_2O	−10.7	253
	NaCl (22.4%)/H_2O	−21.2	235	$KHCO_3$ (17%)/H_2O	−5.4	268.54
	$MgCl_2$ (25 wt.%)/H_2O	−19.4	223	Na_2CO_3 (5.9%)/H_2O	−2.1	281
	Advantages: • High thermal conductivity • Low cost • Abundant in nature			• Non-corrosive to metals • Less supercooling • Low volumetric changes		
	Challenges: • Physically instability			• Corrosive to metals		
Organic	n-Decane	−29.85	194	n-Tetradecane	5–6	231
	2-Octanone	−21.55	190	n-Pentadecane	10	212
	Diethylene glycol	−10.4	247	Glycerine	17.9	199
	n-Tridecane	−5.15	157	Erythritol tetrapalmitate	21.9	201
	Triethylene glycol	−7	247	D-Lactic acid	26	184
	Advantages: Paraffins: • High fusion heat • Good stability • Non-corrosive to metals • Less supercooling • Recyclable			Non-paraffins: • High fusion heat • Physically stable • Less supercooling		
	Challenges: Paraffins: • Low thermal conductivity • Infiltrate plastic • Relatively expensive • Low volumetric storage density			Non-paraffins: • Low thermal conductivity • Hazard • Mild corrosive • Flammable • Expensive		
Eutectic	Ethylene glycol/sodium formate/H_2O (1:1:8)	−25	173.1	Tridecane/tetradecane (50:50)	−4.5	156
	NaCl:KCl:H_2O (15:6.5:78.5)	−23	264	Capric acid:myristic acid (74:26)	23	155
	Glycerol/sodium lactate/H_2O (1:1:8)	−10	159	Methyl palmitate:lauric acid (60:40)	25.6	205.4
	Advantages: • Sharp melting behaviour					
	Challenges: • Incongruently melting • Phase separation • Limited available database • Others corresponding to the challenges in organic and inorganic PCMs					

15.3 PCM-BASED COLD ENERGY STORAGE DEVICES

The next step after selecting the PCM material is to find a convenient and applicable device design that will be later integrated into the system. Device design can affect charging and discharging processes and containment of the material. Thus, good device design requires knowledge of heat transfer and material properties. TES device often consists of two main elements: a PCM suitable for the desired temperature range and the containment of the PCM [37]. There are different containment methods for the PCM, and one of the frequently used ones are packed-bed-based, shell-and-tube-based and plate-type-based configurations. They are discussed as follows.

Packed-bed configurations Figure 15.4 shows a packed-bed configuration with the PCM encapsulated within PCM components. The PCM components can be in different shapes. However, mostly spherical shaped components are used. The charging/discharging processes occur through a heat transfer fluid (HTF) passing through the gaps and voids among the components in the packed bed [38]. During charging, the HTF (at a higher temperature than the PCM) flows through the voids of the bed and exchanges heat with PCM components (at a lower temperature than the HTF temperature), resulting in the melting of the PCM. During discharging, the PCM components heat the HTF passing through the bed, leading to the solidification of the PCM. Apart from properties of the PCM and HTF, other parameters including packing ratio, component size and shape, and operating conditions could also have important effects and need to be carefully considered for a specific application. For example, the right packing can give relatively high heat transfer contact surface for a comparatively small volume of the bed. Here, heat transfer is a very important aspect that affects charging/discharging kinetics.

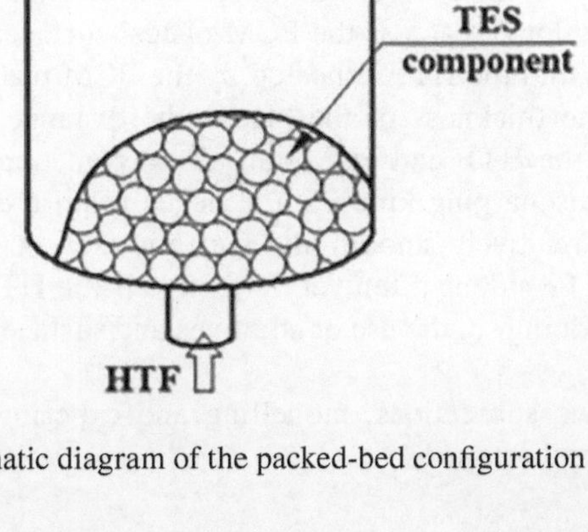

FIGURE 15.4 Schematic diagram of the packed-bed configuration of the TES device [37].

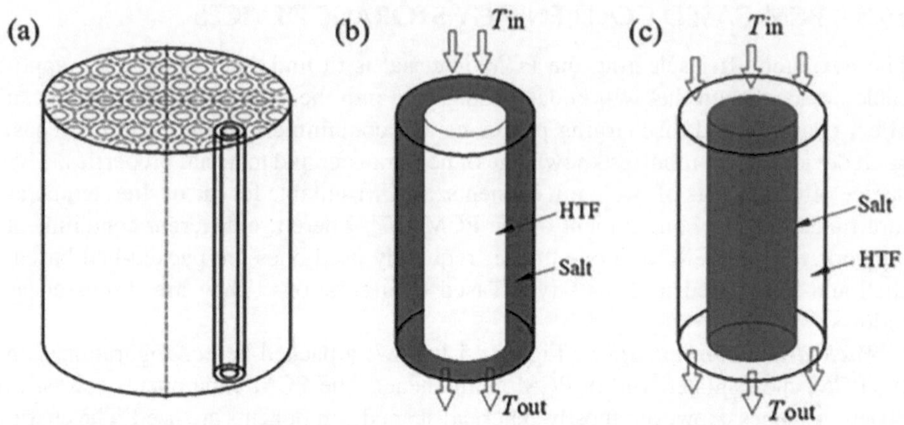

FIGURE 15.5 Schematic diagram of (a) shell and tube configuration of the TES device and (b and c) shell and tube components [68].

Shell-and-tube configurations This configuration consists of a shell that houses tubes as illustrated in Figure 15.5a. The PCM can be placed in the space between the shell and tubes, and HTF flows through the tubes as shown in Figure 15.5b. The PCM can also be encapsulated inside the tube, with the HTF passing through the space between the shell and tubes (see Figure 15.5c). During charging, the HTF at a high temperature melts the PCM by transferring heat to the PCM; whereas during discharging the PCM solidifies, releasing heat to the HTF. In the shell and tube configuration, parameters including aspect ratio, number of tubes, and flow rate, etc. can all affect the charging/discharging behaviour. Similar to the packed-bed configuration, low thermal conductivity of the PCM is one of the main barriers that limit the kinetics of shell-and-tube-based PCM devices. Many studies have been conducted to address this issue. These include the use of fins, the cascaded use of PCMs, the integration of porous matrixes, and the addition of the nanoparticles to the HTF and PCMs.

Plate-type configurations Figure 15.6 illustrates this type of configurations with PCM filled interior of the plates and HTF flowing between the plates transferring the cold energy. Similar to the two configurations outlined above, the charging process involves the HTF flowing across the PCM plates' surface, transferring heat to and melting the PCM, and the HTF is heated by the PCM plates during discharge. Parameters such as the thickness of the plates, the distance between the plates, and the properties of the HTF and PCM can all play an important role in determining the charging/discharging kinetics and hence require careful consideration in the design. Due to relatively smooth and inerratic flow between the plates, the enhancement of the HTF side is often very effective if the HTF has a low thermal conductivity [37]. For doing so, the use of additives and surface patterned structures have been explored.

In the following two subsections, modelling and experimental studies on the PCM-based devices are discussed.

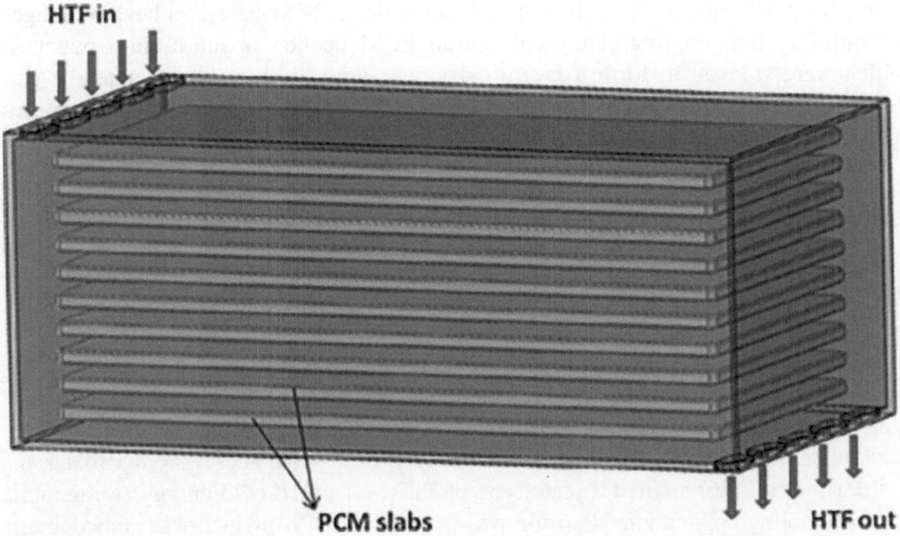

FIGURE 15.6 A schematic diagram of a plate-type configuration of a PCM-based device [69].

15.3.1 Modelling of PCM-Based Cold Energy Storage Devices

Numerical simulation has been a popular method for the design and optimization of energy storage devices. In the current sub-section, a selection of the literature on the modelling of the three types of PCM-based cold energy storage devices is reviewed.

Bourne and Novoselac [39] modelled a TES device for shifting cooling loads from peak to off-peak hours. Their proposed device design was shell-and-tube type with the containment tank packed with cylindrical tubes filled with PCM. They found that the PCM device had a competitive edge over a chilled water tank in terms of thermal capacity. Mosaffa et al. [40] numerically investigated the freezing behaviour of a PCM in a shell-and-tube device with radial fins for air-conditioning applications. The results indicated the PCM solidifies faster in a cylindrical shell store than in a rectangular store. Sciacovelli et al. [41] studied a shell-and-tube-based PCM TES device, particularly the fin shape optimization for increasing energy transfer rate. Proposed optimization of fins allowed to increase the system efficiency up to 24%. The use of fins for PCM-based storage device was also studied by Tao and He [42] who found that the use of a right type of fins can reduce the charging time by 71.2% compared with no-fin cases.

Tay et al. [43] used a method based on the effectiveness – number of transfer units (NTU) – to optimize a tube-in-tank TES unit for low energy cooling of building, where the average NTU represents the average thermal resistance to heat transfer between the HTF and PCM at the phase change front. In their study, the PCM was placed inside the cylindrical tank, with copper tubes submerged in the PCM and water flowing inside the tubes for heat transfer. They did the optimization for different tube diameters and their data suggest a volumetric energy density some 18 times higher than an SHS device.

Hed and Bellander [44] mathematically modelled a PCM-based air heat exchanger for building free cooling. They considered PCM packed in aluminium pouches, which were stacked to form a layered structure with air flowing through the gap between the plates. The modelling results showed an effect of specific heat capacity on the cooling power of the heat exchanger.

15.3.2 Experimental Studies of PCM-Based Cold Energy Storage Devices

There have been many experimental studies on PCM-based cold energy storage devices. The current section gives a brief review of a selection of experimental studies on PCM-based cold energy storage devices.

Allouche et al. [45] experimentally investigated the performance of a 100 L PCM-based TES tank containing two tube bundles (each containing seven tubes) inside the tank for heat transfer between a HTF and the PCM. Their data showed a better discharging kinetics than the tank containing water only as of the cold energy storage material.

Marín et al. [46] studied a plate-type of PCM-air-based cold energy storage unit for free cooling applications. Paraffin was used as the PCM, which was embedded in a graphite matrix with a PCM-matrix ratio of 80%–20% to enhance the conductivity of the PCM. The external shell consisted of parallel plates with an air chamber in between. Their data suggested a reduction of the power consumption of the fan by 50%.

Castell et al. [47] studied a PCM-based cold energy storage device, consisting of a tank filled with PCM and containing coils. Two configurations with different packings were explored. Their results indicated that the performance of storage device decreased with increased water cycling rate, but increased with an increase in heat transfer area.

Garg et al. [48] proposed a radiant heat exchanger design for building thermal management. They used high-density polyethylene containers for the encapsulation of PCM and copper pipes for the provision of channels for HTF (see Figure 15.7). The PCM-based device was installed on the ceiling of a testing room. It was observed that the heat gain of the testing room decreased by approximately 50% and the mean air temperature was decreased by more than 6%.

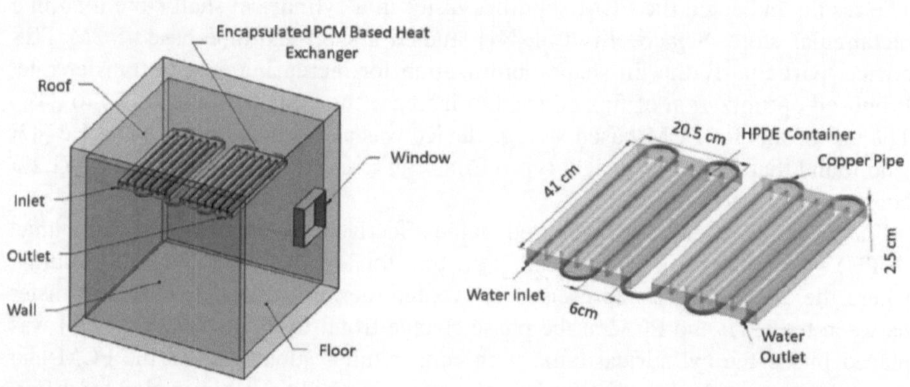

FIGURE 15.7 A schematic diagram of the encapsulated PCM-based cooling device (right) installed in a test chamber (left) [48].

Stritih [49] conducted an experimental study of rectangular PCM-based cold energy storage device with finned surface. The researcher evaluated the effectiveness of the fins, defined as the ratio of the heat flux with fins and that without the fins. The results showed that the addition of the fins enhanced the total heat transfer, leading to the reduction of cold energy release time by 40%.

15.4 APPLICATIONS OF THE PCM-BASED COLD ENERGY STORAGE DEVICES THROUGH INTEGRATION

With years of research and development, PCM-based cold energy storage devices have been deployed through integration to refrigerated warehouses, refrigerated transportation, vaccine storage and transport, and cooling systems of buildings to name but a few. PCM-based cold energy storage has also been studied for use for ice core storage and transport and cooling systems for data centres and telecommunication base stations. The following subsections will discuss some of these applications.

15.4.1 PCM-Based Cold Storage for Warehouse Applications

As an important infrastructure in the cold chain, refrigerated warehouses can greatly reduce the risk of food quality degradation and slow down the rate of microbial growth by providing a suitable temperature and humidity environment [50,51]. However, high energy consumption and high operation cost of warehouse present a considerable challenge, mainly due to the refrigeration system. The use of PCM-based storage enables both the economic benefits due to the peak-to-valley price difference and intermittent renewable energy utilization.

There are two types of integration of a PCM store with a warehouse. One is to integrate the PCM device into the refrigeration system, and the other is to use PCM as part of building structure materials. Wang et al. [52] integrated a set of PCM units into a refrigeration system of a warehouse. The PCM units were installed onto the inner wall surface of the cold storage room for peak load shift. It was shown that the system-level coefficient of performance (COP) increased by 6%–8% and the payback period of the PCM units was 4.1 years. Fioretti et al. [53] added a PCM layer to the sandwich panel of a cold storage room to improve its envelope thermal performance. The PCM layer contained paraffin RT35HC from Rubitherm (Germany) (see Figure 15.8). It was found that the peak cooling load was reduced by 20% compared with the traditional

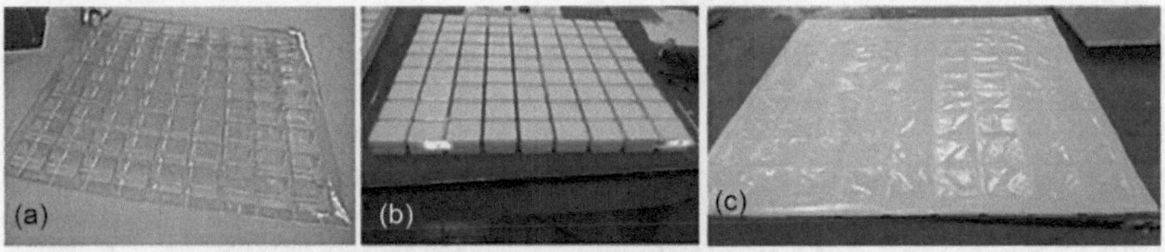

FIGURE 15.8 PCM layer for a cold room: (a) polyethylene panel, (b) RT35HC-based PCM encapsulation, and (c) polyvinyl chloride closing layer [53].

cold room, the total energy consumption was reduced by 4.7%, and the cooling peak demand was delayed for 4 hours on average.

Yang et al. [54] examined the installation of PCM panels on the surface of insulation walls and ceilings of a warehouse. A tube-in-plate heat exchanger was incorporated inside PCM panels, which implies that the recirculating refrigerant during the charging process stayed inside the tubes. It was shown that the payback period for such a system was about 2.6 years.

In parallel to the research, commercialization of the PCM-based cold energy storage technology has been taken place for cold warehouses. An example is the cold energy storage-based refrigeration system developed by Viking Cold Solutions Inc. The PCM plates are installed at the top of frozen food racks inside the warehouses, which was reported to be able to prevent some 85% of heat infiltration. They also claimed an energy cost reduction of up to 50% and an increase in the refrigeration system efficiency by ~26%.

15.4.2 PCM-Based Cold Energy Storage for Cold Chain Transportation Applications

As the main route for cold chain transportation, the highway has the advantages of flexibility and convenience. Commercial refrigerated vehicles rely on diesel engine powered mechanical refrigeration technology to provide temperature regulation to the compartment. This is clearly a high carbon and polluting technology and has high operating costs. The use of PCM-based cold energy storage technology offers a cost-effective, zero-carbon and zero-emission alternative to the provision of cold energy, which is charged to the vehicle and produced using off-peak and/or renewable electricity from the grid. Moreover, a PCM layer can be integrated into the vehicle compartment to enhance thermal insulation and prolong the cooling duration. Some recent progresses in the area are summarized below.

Ahmed et al. [55] proposed a method to reduce heat transfer through refrigerated truck trailers by using a paraffin-based PCM. The PCM was packaged in a copper tube which was evenly distributed in polyurethane foam for use in 1.22 m^3 refrigerated simulation cars. Compared with a standard refrigerated compartment, the peak heat transfer of the simulated compartment was reduced by 29.1% and the total heat transfer was reduced by 16.3%. Copertaro et al. [56] sandwiched an RT35HC PCM layer between the outer skin and its ordinary thermal insulation layer of a refrigerated vehicle and showed a reduction of peak heat load by ~20% in summer and energy saving by ~5%.

Liu et al. [22,57] proposed a PCM-based cooling system as illustrated in Figure 15.9 for refrigerated vehicles. A PCM with a melting temperature of 26.7°C was packed into 19 parallel flat slabs forming a phase change thermal storage unit (PCTSU). An off-vehicle refrigeration unit was used to charge the PCTSU during off-peak hours with a low electricity tariff. Their results demonstrated that the PCTSU could keep the refrigerated goods at −18°C for 10 hours and an energy cost-saving ratio was up to 86.4% compared to a conventional system.

Through collaboration with the University of Birmingham, CRRC Shijiazhuang has developed a passively cooled container technology for integrated road and rail

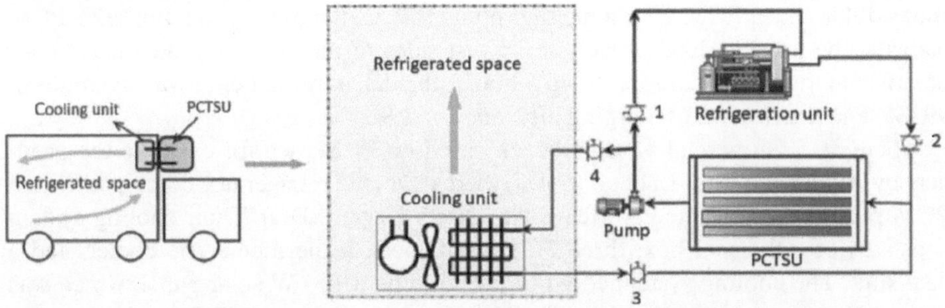

FIGURE 15.9 The configuration of a refrigeration system integrated with a PCM unit [22].

applications [58]. PCM plates were used in the container, which had a size of 1.8 m×1 m×0.1 m containing finned tubes inside plates for connecting to a charging loop. Ten plates were installed on the ceiling of the container. Their results showed that technology maintained the internal temperature needed for fresh produce for over 94 hours, even though the external ambient temperature varied with the highest temperature up to 35°C. Their data also showed a system-level COP of 1.84, much better than diesel-powered cooling systems. The result suggested that the energy consumption, operating cost, and emission could be reduced, respectively, by 86.7%, 91.6%, and 78.5%. As a result of the collaborative effort, the technology has now been in the commercial trial stage. The details can be found in [59].

15.4.3 PCM-Based Cold Energy Storage Technology for Vaccine Storage and Transport

Vaccine, as a special biological product, has the characteristics of high sensitivity to temperature. Therefore, the entire process of vaccine production, storage, transportation and distribution may lead to irrevocable potency losses due to temperature-induced morphological changes in the antigens and/or adjuvants [60]. PCMs can be integrated with the whole cold chain associated with vaccines, particularly vaccine storage devices to help ensure the temperature of the vaccine within the required temperature range during storage, distribution and utilization.

Devrani et al. [61] studied numerically and experimentally the performance of traditional cold chain boxes for optimizing the boxes for vaccine use with PCMs. Their results showed the contributions to heat leakage due to different heat transfer mechanisms as 51.43% from conduction, 27.51% from convection and 23.7% from radiation. A cylindrical shaped box would minimize the surface to volume ratio for a reduced heat transfer rate. Zhao et al. [62] studied a vaccine cold storage device, which combined a cold energy storage PCM with a cold storage box. With a loaded box, the longest time for the vaccine to maintain 2–8°C was found to be 52.36 hours.

Du et al. [63] numerically analysed the influences of PCM location and PCM melting point on the insulation performance of a portable box made of vacuum insulation board. The portable cold box had an internal size of 355 mm×215 mm×265 mm and was filled with five PCM plates. Their simulation results, validated by experiments, showed that, for a PCM with a melting point of 2°C, the arrangement of 20% PCM plates at the top and 20% at each of the rest sides of the box gave the longest temperature-maintaining duration of 46.5 hours; the discharging depth was the highest, 99.4%, and maximum discharging efficiency was 90.7%.

There are commercial vaccine boxes based on PCM; see for example the products by Aucma Co. Ltd. Li et al. [64] studied a vaccine refrigerator combined with a PCM pack freezer, termed Arktek Solar Direct Drive (SDD). Their cooling system consisted of a thermal box, three PCM containers, a thermoelectric cooler, and a heat sink. The cooling system could be powered by a 100 W solar panel, which was reported to have an ample margin. Their results showed that the SDD could maintain a 10 litre vaccine storage temperature between 3°C and 5°C for 8 days.

15.4.4 PCM-Based Cold Energy Storage for Ice Core Storage and Transport

Ice core refers to a core sample taken from the ice inside a glacier, which contains the information of natural changes of climate and environment, as well as the influence of human activities on the climate in the past. Ice core plays an important role in the research of global climate change. Ice cores are cylindrical shaped, usually drilled out of the Antarctic and the Arctic [65], a major challenge has been to transport them in thermally insulated boxes in refrigerated containers powered by diesel engines; see Figure 15.10 for an example of transporting ice cores from Greenland and Antarctica to the UK. It is well known that diesel-powered refrigerators are highly polluting. The University of Birmingham Centre for Energy Storage has been collaborating with British Antarctic Survey to perform a feasibility study on the use of passively cooled PCM technology for ice core transportation. The study used cold boxes integrated with PCMs, which are charged with cold energy from the locations where ice cores are drilled. The stored cold energy is discharged during transportation. It was observed that an ordinary ice core storage box, if combined with vacuum insulation and CPCM, could maintain the inner temperature below −45°C for over 20 hours.

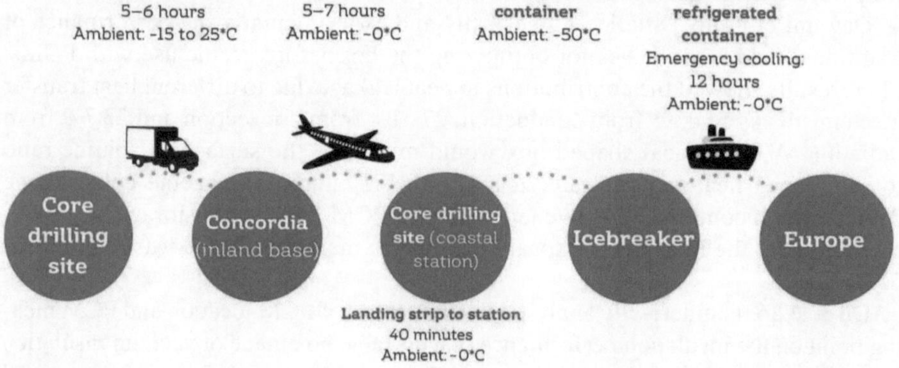

FIGURE 15.10 Transporting ice core from Antarctica to Europe.

15.5 CONCLUDING REMARKS

This chapter summarizes recent progress in PCMs for cold chain applications. It covers materials, devices, and applications through system integration.

- First, a brief review is given on PCM-based cold energy storage materials. These materials are classified in terms of material types, the need for additives and temperature range. Challenges are highlighted in the material formulation and characterization, thermal performance, and property enhancement.
- Second, PCM-based devices are reviewed, including both modelling and experimental aspects. The device design and optimization are also discussed.
- Third, application examples are given about PCM-based cold energy storage devices through integration with different systems including warehouses and transportation.

Although there are numerous studies on both fundamental and applied aspects of PCM-based TES, most of them are on thermal (heat) storage. It is the hope of the authors of this chapter to motivate more researchers from both academic and industrial communities to pay more attention to PCM for cold energy storage applications – an area that matters for food and medicine safety and is important for energy system decarbonization.

ACKNOWLEDGEMENTS

The authors would like to thank the partial funding of this work by the UK Engineering and Physical Sciences Research Council (EPSRC) under EP/P003605/1, EP/V012053/1, EP/T022701/1, and EP/N021142/1, and the British Council under 2019-RLWK11-10724 and 2020-RLWK12-10478.

REFERENCES

1. S. M. Hosseini Bamakan, S. Ghasemzadeh Moghaddam, and S. Dehghan Manshadi, Blockchain-enabled pharmaceutical cold chain: Applications, key challenges, and future trends, *Journal of Cleaner Production*, 302, 2021.
2. IIF-IIR, International dictionary of refrigeration. English-French: terms and definitions, 2007. https://iifiir.org/en/fridoc/international-dictionary-of-refrigeration-english-french-terms-and-4102
3. J. W. Han, M. Zuo, W. Y. Zhu, J. H. Zuo, E. L. Lü, and X. T. Yang, A comprehensive review of cold chain logistics for fresh agricultural products: Current status, challenges, and future trends, *Trends in Food Science and Technology*, 109, 2021, pp. 536–551.
4. E. Derens-Bertheau, V. Osswald, O. Laguerre, and G. Alvarez, Cold chain of chilled food in France, *International Journal of Refrigeration*, 52, 2015, pp. 161–167.
5. O. Laguerre, H. M. Hoang, and D. Flick, Experimental investigation and modelling in the food cold chain: Thermal and quality evolution, *Trends in Food Science and Technology*, 29 (2), 2013, pp. 87–97.
6. IIR, The deployment of an efficient cold chain is essential for global food security, 2020. https://iifiir.org/en/news/the-iir-publishes-a-new-informatory-note-on-the-role-of-refrigeration-in-worldwide-nutrition (accessed February 2022)

7. D. Coulomb, The cold chain: A key component in the development process, *International Journal of Refrigeration*, 67, 2016, pp. v–vi.
8. H. Zhao, S. Liu, C. Tian, G. Yan, and D. Wang, An overview of current status of cold chain in China, *International Journal of Refrigeration*, 88, 2018, pp. 483–495.
9. N. Ndraha, H. I. Hsiao, J. Vlajic, M. F. Yang, and H. T. V. Lin, Time-temperature abuse in the food cold chain: Review of issues, challenges, and recommendations, *Food Control*, 89, 2018, pp. 12–21.
10. A. Ashok, M. Brison, and Y. LeTallec, Improving cold chain systems: Challenges and solutions, *Vaccine*, 35 (17), 2017, pp. 2217–2223.
11. T. L. De Micheaux, M. Ducoulombier, J. Moureh, V. Sartre, and J. Bonjour, Experimental and numerical investigation of the infiltration heat load during the opening of a refrigerated truck body, *International Journal of Refrigeration*, 54, 2015, pp. 170–189.
12. Y. Salehy et al., Energy performances assessment for sustainable design recommendations: Case study of a supermarket's refrigeration system, *Procedia CIRP*, 90, 2020, pp. 328–333.
13. Y. Huang and X. Zhang, Research progress of the application of cold storage technology in food cold chain logistics, *Packaging Engineering*, (15), 2015, pp. 23–29.
14. Y. Zhao, X. Zhang, and X. Xu, Application and research progress of cold storage technology in cold chain transportation and distribution, *Journal of Thermal Analysis and Calorimetry*, 139 (2), 2020, pp. 1419–1434.
15. L. Liu et al., Development of low-temperature eutectic phase change material with expanded graphite for vaccine cold chain logistics, *Renewable Energy*, 179, 2021, pp. 2348–2358.
16. X. Ma, M. Sheikholeslami, M. Jafaryar, A. Shafee, T. Nguyen-Thoi, and Z. Li, Solidification inside a clean energy storage unit utilizing phase change material with copper oxide nanoparticles, *Journal of Cleaner Production*, 245, 2020, p. 118888.
17. X. Li, B. Ma, and Y. Li, Review of development of phase change cold storage technology for food cold chain application, in *Proceedings of the 2015 International Forum on Energy, Environment Science and Materials*, Shenzhen, China, 2015.
18. Y. Zhao, X. Zhang, X. Xu, and S. Zhang, Research progress of phase change cold storage materials used in cold chain transportation and their different cold storage packaging structures, *Journal of Molecular Liquids*, 319, 2020, p. 114360.
19. C. Veerakumar and A. Sreekumar, Phase change material based cold thermal energy storage: Materials, techniques and applications – A review, *International Journal of Refrigeration*, 67, 2016, pp. 271–289.
20. H. Selvnes, Y. Allouche, R. I. Manescu, and A. Hafner, Review on cold thermal energy storage applied to refrigeration systems using phase change materials, *Thermal Science and Engineering Progress*, 22, 2021, p. 100807.
21. W. Lin, Z. Ma, S. Wang, M. I. Sohel, and E. Lo Cascio, Experimental investigation and two-level model-based optimisation of a solar photovoltaic thermal collector coupled with phase change material thermal energy storage, *Applied Thermal Engineering*, 182, 2021, p. 116098.
22. M. Liu, W. Saman, and F. Bruno, Development of a novel refrigeration system for refrigerated trucks incorporating phase change material, *Applied Energy*, 92, 2012, pp. 336–342.
23. F. Alzuwaid, Y. T. Ge, S. A. Tassou, A. Raeisi, and L. Gowreesunker, The novel use of phase change materials in a refrigerated display cabinet: An experimental investigation, *Applied Thermal Engineering*, 75, 2015, pp. 770–778.
24. A. Abhishek et al., Comparison of the performance of ice-on-coil LTES tanks with horizontal and vertical tubes, *Energy and Buildings*, 183, 2019, pp. 45–53.
25. A. López-Navarro et al., Performance characterization of a PCM storage tank, *Applied Energy*, 119, 2014, pp. 151–162.

26. S. Himran, A. Suwono, and G. A. Mansoori, Characterization of alkanes and paraffin waxes for application as phase change energy storage medium, *Energy Sources*, 16 (1), 1994, pp. 117–128.
27. M. Casini, *Phase-Change Materials*, Woodhead Publishing, Cambridge, 2016.
28. J. Gasia, L. Miró, and L. F. Cabeza, Materials and system requirements of high temperature thermal energy storage systems: A review. Part 2: Thermal conductivity enhancement techniques, *Renewable and Sustainable Energy Reviews*, 60, 2016, pp. 1584–1601.
29. Q. He, S. Wang, M. Tong, and Y. Liu, Experimental study on thermophysical properties of nanofluids as phase-change material (PCM) in low temperature cool storage, *Energy Conversion and Management*, 2012, 64, pp. 199–205.
30. P. G. Grodzka, Space thermal control by freezing and melting, Interim report. No. HREC-1123-2, 1969.
31. N. Beaupere, U. Soupremanien, and L. Zalewski, Nucleation triggering methods in supercooled phase change materials (PCM), a review, *Thermochimica Acta*, 670, 2018, pp. 184–201.
32. T. Wu et al., Preparation of a low-temperature nanofluid phase change material: $MgCl_2$–H_2O eutectic salt solution system with multi-walled carbon nanotubes (MWCNTs), *International Journal of Refrigeration*, 113, 2020, pp. 136–144.
33. J. B. Johansen, M. Dannemand, W. Kong, J. Fan, J. Dragsted, and S. Furbo, Thermal conductivity enhancement of sodium acetate trihydrate by adding graphite powder and the effect on stability of supercooling, *Energy Procedia*, 70, 2015, pp. 249–256.
34. Y. E. Milián, A. Gutiérrez, M. Grágeda, and S. Ushak, A review on encapsulation techniques for inorganic phase change materials and the influence on their thermophysical properties, *Renewable and Sustainable Energy Reviews*, 73, 2017, pp. 983–999.
35. E. Oró, L. Miró, C. Barreneche, I. Martorell, M. M. Farid, and L. F. Cabeza, Corrosion of metal and polymer containers for use in PCM cold storage, *Applied Energy*, 109, 2013, pp. 449–453.
36. S. M. Hassani-Gangaraj, A. Moridi, and M. Guagliano, Critical review of corrosion protection by cold spray coatings, *Surface Engineering*, 31 (11), 2015, pp. 803–815.
37. C. Li, P. Zhang, Q. Li, L. Tong, L. Wang, and Y. Ding, Chapter 10. Latent heat storage devices, in *Thermal Energy Storage: Materials, Devices, Systems and Applications*, edited by Yulong Ding, The Royal Society of Chemistry, 2021, pp. 265–328.
38. C. Li, Q. Li, and Y. Ding, Investigation on the thermal performance of a high temperature packed bed thermal energy storage system containing carbonate salt based composite phase change materials, *Applied Energy*, 247 (5), 2019, pp. 374–388.
39. S. Bourne and A. Novoselac, Compact PCM-based thermal stores for shifting peak cooling loads, *Building Simulation*, 8 (6), 2015, pp. 673–688.
40. A. H. Mosaffa, F. Talati, H. Basirat Tabrizi, and M. A. Rosen, Analytical modeling of PCM solidification in a shell and tube finned thermal storage for air conditioning systems, *Energy and Buildings*, 49, 2012, pp. 356–361.
41. A. Sciacovelli, F. Gagliardi, and V. Verda, Maximization of performance of a PCM latent heat storage system with innovative fins, *Applied Energy*, 137, 2015, pp. 707–715.
42. Y. B. Tao and Y. L. He, Numerical study on performance enhancement of shell-and-tube latent heat storage unit, *International Communications in Heat and Mass Transfer*, 67, 2015, pp. 147–152.
43. N. H. S. Tay, M. Belusko, and F. Bruno, Designing a PCM storage system using the effectiveness-number of transfer units method in low energy cooling of buildings, *Energy and Buildings*, 50, 2012, pp. 234–242.
44. G. Hed and R. Bellander, Mathematical modelling of PCM air heat exchanger, *Energy and Buildings*, 38 (2), 2006, pp. 82–89.

45. Y. Allouche, S. Varga, C. Bouden, and A. C. Oliveira, Experimental determination of the heat transfer and cold storage characteristics of a microencapsulated phase change material in a horizontal tank, *Energy Conversion and Management*, 94, 2015, pp. 275–285.
46. J. M. Marín, B. Zalba, L. F. Cabeza, and H. Mehling, Improvement of a thermal energy storage using plates with paraffin-graphite composite, *International Journal of Heat and Mass Transfer*, 48 (12), 2005, pp. 2561–2570.
47. A. Castell, M. Belusko, F. Bruno, and L. F. Cabeza, Maximisation of heat transfer in a coil in tank PCM cold storage system, *Applied Energy*, 88 (11), 2011, pp. 4120–4127.
48. H. Garg, B. Pandey, S. K. Saha, S. Singh, and R. Banerjee, Design and analysis of PCM based radiant heat exchanger for thermal management of buildings, *Energy and Buildings*, 169, 2018, pp. 84–96.
49. U. Stritih, An experimental study of enhanced heat transfer in rectangular PCM thermal storage, *International Journal of Heat and Mass Transfer*, 47 (12–13), 2004, pp. 2841–2847.
50. M. Rokka, S. Eerola, M. Smolander, H. L. Alakomi, and R. Ahvenainen, Monitoring of the quality of modified atmosphere packaged broiler chicken cuts stored in different temperature conditions B. Biogenic amines as quality-indicating metabolites, *Food Control*, 15 (8), 2004, pp. 601–607.
51. S. Mercier, S. Villeneuve, M. Mondor, and I. Uysal, Time–temperature management along the food cold chain: A review of recent developments, *Comprehensive Reviews in Food Science and Food Safety*, 16 (4), 2017, pp. 647–667.
52. C. Wang, Z. He, H. Li, R. Wennerstern, and Q. Sun, Evaluation on performance of a phase change material based cold storage house, *Energy Procedia*, 105, 2017, pp. 3947–3952.
53. R. Fioretti, P. Principi, and B. Copertaro, A refrigerated container envelope with a PCM (phase change material) layer: Experimental and theoretical investigation in a representative town in Central Italy, *Energy Conversion and Management*, 122, 2016, pp. 131–141.
54. T. Yang, C. Wang, Q. Sun, and R. Wennersten, Study on the application of latent heat cold storage in a refrigerated warehouse, *Energy Procedia*, 142, 2017, pp. 3546–3552.
55. M. Ahmed, O. Meade, and M. A. Medina, Reducing heat transfer across the insulated walls of refrigerated truck trailers by the application of phase change materials, *Energy Conversion and Management*, 51 (3), 2010, pp. 383–392.
56. B. Copertaro, P. Principi, and R. Fioretti, Thermal performance analysis of PCM in refrigerated container envelopes in the Italian context – Numerical modeling and validation, *Applied Thermal Engineering*, 102, 2016, pp. 873–881.
57. M. Liu, W. Saman, and F. Bruno, Computer simulation with TRNSYS for a mobile refrigeration system incorporating a phase change thermal storage unit, *Applied Energy*, 132, 2014, pp. 226–235.
58. S. Tong et al., A phase change material (PCM) based passively cooled container for integrated road-rail cold chain transportation – An experimental study, *Applied Thermal Engineering*, 195, 2021, p. 117204.
59. University of Birmingham, Passively cooled containers being delivered for integrated rail and road cold chain transportation following world's first commercial demonstration, University of Birmingham, 2019. https://www.birmingham.ac.uk/research/energy/news/2019/passively-cooled-containers-being-delivered-for-integrated-rail-and-road-cold-chain-transportation.aspx.
60. O. S. Kumru, S. B. Joshi, D. E. Smith, C. R. Middaugh, T. Prusik, and D. B. Volkin, Vaccine instability in the cold chain: Mechanisms, analysis and formulation strategies, *Biologicals*, 42 (5), 2014, pp. 237–259.

61. S. Devrani, S. Pandey, S. Chaturvedi, K. Sankar, S. Patil, and K. Sridhar, Design and analysis of an efficient vaccine cold chain box, in *ASME International Mechanical Engineering Congress and Exposition, Proceedings (IMECE)*, Phoenix, AZ, 2016, 3, pp. 1–7.
62. Y. Zhao, X. Zhang, X. Xu, and S. Zhang, Development of composite phase change cold storage material and its application in vaccine cold storage equipment, *Journal of Energy Storage*, 30, 2020, p. 101455.
63. J. Du, B. Nie, Y. Zhang, Z. Du, li Wang, and Y. Ding, Cooling performance of a thermal energy storage-based portable box for cold chain applications, *Journal of Energy Storage*, 28, 2020, p. 101238.
64. J. M. Li, M. Friend, A. Miller, and S. Stone, A SDD and PCM solution for vaccine storage and outreach, in *GHTC 2016 – IEEE Global Humanitarian Technology Conference: Technology for the Benefit of Humanity, Conference Proceedings*, Seattle, WA, 2016, pp. 555–562.
65. USGS, About ice cores, 2021. https://icecores.org/about-ice-cores.
66. E. Onder and N. Sarier, Thermal regulation finishes for textiles, in *Functional Finishes for Textiles: Improving Comfort, Performance and Protection*, edited by Roshan Paul, Woodhead Publishing, 2015, pp. 17–98.
67. Z. Ge, Z. Jiang, L. Cong, B. Zou, and Y. Ding, Chapter 4. Latent heat storage materials, in *Thermal Energy Storage: Materials, Devices, Systems and Applications*, edited by Yulong Ding, The Royal Society of Chemistry, 2021, pp. 55–90.
68. Q. Li, C. Li, Z. Du, F. Jiang, and Y. Ding, A review of performance investigation and enhancement of shell and tube thermal energy storage device containing molten salt based phase change materials for medium and high temperature applications, *Applied Energy*, 255, 2019.
69. M. Liu, M. Belusko, N. H. Steven Tay, and F. Bruno, Impact of the heat transfer fluid in a flat plate phase change thermal storage unit for concentrated solar tower plants, *Solar Energy*, 101, 2014, pp. 220–231.
70. L. Yang et al., A comprehensive review on sub-zero temperature cold thermal energy storage materials, technologies, and applications: State of the art and recent developments, *Applied Energy*, 288, 2021, p. 116555.
71. L. Cong, X. She, G. Leng, G. Qiao, C. Li, and Y. Ding, Formulation and characterisation of ternary salt based solutions as phase change materials for cold chain applications, *Energy Procedia*, 158, 2019, pp. 5103–5108.

Index